D0889740

DATE DUE

Handbook of
MODERN SOLID-STATE
AMPLIFIERS

PRENTICE-HALL SERIES IN ELECTRONIC TECHNOLOGY

Dr. Irving L. Kosow, *editor*

Charles M. Thomson, Joseph J. Gershon, and Joseph A. Labok, *consulting editors*

PRENTICE-HALL INTERNATIONAL, INC., *London*
PRENTICE-HALL OF AUSTRALIA, PTY. LTD., *Sydney*
PRENTICE-HALL OF CANADA, LTD., *Toronto*
PRENTICE-HALL OF INDIA PRIVATE LIMITED, *New Delhi*
PRENTICE-HALL OF JAPAN, INC., *Tokyo*

Handbook of
MODERN SOLID-STATE
AMPLIFIERS

JOHN D. LENK

Prentice-Hall, Inc., Englewood Cliffs, New Jersey

Library of Congress Cataloging in Publication Data

LENK, JOHN D
   Handbook of modern solid-state amplifiers.

   1. Transistor amplifiers.   I. Title.
TK7871.2.L385   621.3815'35   73-21834
ISBN   0-13-380394-5

TK
787l.2
.L385

© 1974 by
PRENTICE-HALL, INC.
Englewood Cliffs, New Jersey

Printed in the United States of America

10 9 8 7 6 5 4 3

Dedicated to
Irene, Karen, Mark, Lambie, and Kris

CONTENTS

3. RADIO-FREQUENCY AMPLIFIERS 130

4. DIRECT-COUPLED AND
COMPOUND AMPLIFIERS   216

5. DIFFERENTIAL AMPLIFIERS   270

6. OPERATIONAL AMPLIFIERS   299

7. AMPLIFIER TEST AND
TROUBLESHOOTING   356

INDEX   409

PREFACE

This handbook is devoted to the subject of modern electronic amplifiers. Virtually all fields of electronics require the use of amplifiers in one form or another. The handbook is written at the "middle level," combining both theory and practice.

Most existing books on amplifiers fall into one of three classes: the oversimplified beginner's introduction, the technician's "how" book limited to a specific type of amplifier, or the engineer's highly theoretical, math-oriented, design book. Although all this information can be useful, it still leaves many questions unanswered, and creates a gap for those who must work with amplifiers, at all levels.

The *Handbook of Modern Solid-State Amplifiers* fills the gap, since it covers the four areas most needed: theory (including circuit analysis of both theoretical and commercial amplifiers), proved design practices, comprehensive test procedures, and practical troubleshooting techniques. Thus, the handbook is suitable for the student who wants basic theory, the designer who needs simplified design approaches, the technician who must test and troubleshoot amplifiers, or anyone who wants an all-around source book for solid-state amplifiers.

The handbook describes all types of amplifiers in current use: audio, radio frequency, direct coupled, differential, compounds, and operational amplifiers (op-amps). It covers both discrete amplifier circuits (two-junction transistor and field-effect transistor, or FET), as well as selected integrated-circuits (ICs). Of course no book can cover all amplifier circuits in existence. For this reason, this handbook describes the time-tested, yet up-to-date, amplifier circuits in common use.

Chapter 1 is devoted to basic amplifier theory. Chapters 2 through 6 describe theory and simplified design for audio, radio-frequency, direct-coupled, differential, and op-amps, respectively. Chapter 7 summarizes test procedures and troubleshooting techniques for all types of solid-state amplifiers.

The author has received much help from various organizations in writing this book. He wishes to give special thanks to the following: General Electric, Hewlett-Packard, Motorola, Texas Instruments, and Radio Corporation of America.

JOHN D. LENK

Handbook of
MODERN SOLID-STATE
AMPLIFIERS

1. BASIC AMPLIFIER THEORY

In this chapter we shall review and summarize basic amplifier theory. The first question that may arise is "What is an amplifier?" For our discussion, the purpose of an amplifier is to increase the amplitude of a voltage, current, or power. Secondary functions are to isolate signal sources from other circuits, and provide impedance matching.

In both simple and complex amplifiers, an input signal (consisting of a voltage or current having a definite waveform) produces a corresponding output signal with the same waveform, but in amplified form. Thus, an amplifier is a circuit that develops a voltage or current (or power) having an amplitude greater than the control factor (or input).

1-1. CLASSIFYING AMPLIFIERS

There are many ways in which amplifiers are classified. One method is to classify amplifiers as to *basic circuit connections*. Depending upon the circuit configuration, there are three basic classifications: *common emitter*, *common base*, and *common collector*. Note that the word *grounded* can be substituted for *common* in describing the three amplifier circuit configurations.

Amplifiers can also be classified in terms of *operating point* (or bias that establishes the operating point). There are four such operating-point classifications: class A, class B, class AB, and class C.

Another method of amplifier classification is based on the *function* they are to serve; under this system we have *voltage amplifiers* and *power amplifiers*.

Sometimes amplifiers are grouped according to frequency of operation: AF (audio frequency), RF (radio frequency), and IF (intermediate frequency). The frequency system is further broken down into *narrowband* (where amplification is limited to a specific frequency or narrow range of frequen-

1

cies), *wideband* (where a wide range of frequencies is amplified at about the same level, *tuned* (where amplification is limited to a specific frequency, or to a narrow range, by tuning controls), and *untuned* (where the frequency range is set by nonadjustable component values).

Amplifiers can also be classified by the *type of signals* they amplify. For example, there are ac (alternating current), dc (direct current), and pulse amplifiers. From a practical standpoint, a dc amplifier will also amplify ac signals. (However, the reverse is not true.) Likewise, many ac and dc amplifiers will pass pulse or squarewave signals without difficulty.

In addition to the various classifications, amplifiers can be assigned some specific title that is descriptive of the function or circuit configuration. The most important of these are DC (direct-coupled) amplifier, differential amplifier, and operational amplifier (op-amp). Note that the term "DC amplifier" can mean either direct-coupled amplifier or direct-current amplifier. This confusion arises from the fact that all direct-current amplifiers must be direct coupled. (That is, there must be no capacitors between amplifier stages.) However, note that direct-coupled amplifiers are not limited to direct-current amplification. In general, direct-coupled amplifiers will operate with alternating current and pulse signals, as well as direct current.

1-2. THE BASIC AMPLIFIER

Before going into the various amplifier classifications and types, let us review how transistors amplify signals. Figure 1-1 shows a *PNP* transistor connected in a basic common-emitter circuit. The base–emitter circuit is forward biased, whereas the base–collector circuit is reverse biased. An input signal is applied across resistor R_1; the output is taken from across resistor R_2.

Under no-signal (quiescent) conditions, current flows in the input circuit, causing a steady value of current to flow in the output circuit. When one alternation of an input signal is applied, as shown in Fig. 1-1a, a voltage is developed across input resistor R_1. This voltage, positive at the base end of R_1, subtracts from the bias voltage provided by B_1, causing the base-to-emitter voltage, V_{BE}, to become less negative.

Assume that the B_1 battery voltage is 3 V and the signal voltage at its peak is 0.5 V. Since the signal voltage drop opposes the B_1 battery voltage, the net voltage between the base and emitter will decrease to 2.5 V from the original 3 V. The input current will also decrease, and there will be less current carriers (holes, in this case) available to the collector. Less current will flow through the output resistor R_2, and the voltage drop across R_2 will decrease.

The voltage from the collector to emitter will increase, the collector

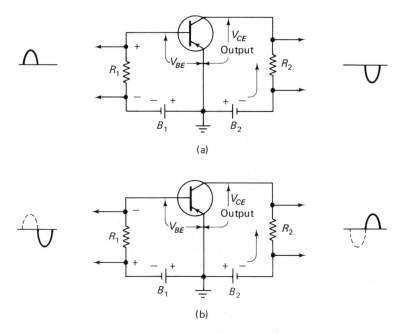

Fig. 1-1   Operation of transistors in amplifier circuits

becoming more negative than it was during the no-signal condition. As the signal voltage increases in a positive direction, the output voltage (taken from collector to ground) increases in a negative direction. This action illustrates signal-phase reversal, a characteristic of the common-emitter stage. Amplification occurs because the collector current is many times greater than the base–emitter current.

When the next alternation of the input signal is applied (Fig. 1-1b), the voltage is again developed across R_1. However, this voltage is negative at the base. The voltage drop adds to the B_1 battery voltage, so that the net voltage between base and emitter will increase to 3.5 V from the original 3 V. The input current will increase, and there will be more current carriers (holes) available to the collector. More current flows through R_2, and the voltage drop across R_2 increases. This increase will still be out of phase with the input, and is amplified by the same amount as the previous alternation.

The total voltage V_{BE} is a combination of the input signal and B_1; the total voltage V_{CE} is a combination of output signal and B_2. In practical circuits, the input is measured from base to ground, with the output measured from collector to ground.

If an *NPN* transistor is placed in the same circuit, operation will be identical, except that polarities of the voltages are reversed.

1-3. RULES FOR LABELING, BIASING, AND POLARITIES IN TRANSISTORS

Before starting the discussion of two-junction or bipolar transistors, it may be well to review some of the basic principles of transistors. The following general rules can be helpful in a practical analysis of how transistor amplifiers operate. The rules apply primarily to a class A amplifier, but also remain true for many other transistor amplifier circuits.

In *PNP* transistors, holes flow from the emitter to the collector. Thus, the collector must be negative relative to the emitter. In the external circuit, electrons flow from emitter to collector.

In the *NPN* transistor, electrons flow from the emitter to the collector. Hence, the collector must be positive relative to the emitter. In the external circuit, aided by the bias voltage, electrons flow from the collector to emitter.

The middle letter in *PNP* or *NPN* always applies to the base.

The first two letters in *PNP* or *NPN* refer to the *relative bias* polarities of the *emitter* with respect to either the base or collector.

For example, the letters *PN* (in *PNP*) indicate that the emitter is positive with respect to both the base and collector. The letters *NP* (in *NPN*) indicate that the emitter is negative with respect to both the base and collector.

The dc *electron current flow* is always against the direction of the arrow in the emitter.

If electron flow is into the emitter, electron flow will be out from the collector.

If electron flow is out from the emitter, electron flow will be into the collector.

The collector–base junction is always reverse biased.

The emitter–base junction is generally forward biased.

A *base-input* voltage that opposes or decreases the forward bias also decreases the emitter and collector currents. For example, a negative input to the base of an *NPN*, or a positive input to a *PNP* base, decreases both currents.

A *base-input* voltage that aids or increases the forward bias also increases the emitter and collector currents (positive to *NPN*, negative to *PNP*).

1-4. COMMON-EMITTER AMPLIFIER

The common-emitter circuit shown in Fig. 1-2 is the most widely used amplifier configuration. Figure 1-2 shows the power-supply connections for both *NPN* and *PNP* transistors. In this configuration the emitter is common to both the input and output circuits. Although the emitter is frequently called the grounded element, the emitter is not neces-

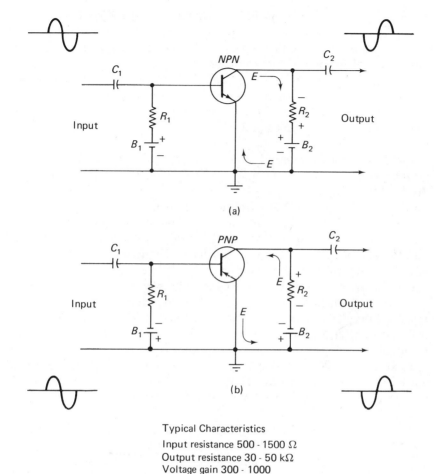

Typical Characteristics

Input resistance 500 - 1500 Ω
Output resistance 30 - 50 kΩ
Voltage gain 300 - 1000
Current gain 20 - 50
Power gain 25 - 40 dB

Fig. 1-2   Basic common-emitter amplifier stage

sarily connected to ground. Battery B_1 forward biases the emitter–base junction. Battery B_2 reverse biases the collector–base junction. In practice, the voltage of B_2 is generally larger than that of B_1.

The input signal is applied between the base and emitter, and the output signal appears between emitter and collector. This provides a very low input impedance, and a very high output impedance. However, the output signal is out of phase with the input.

Common-emitter circuits are the most often used since there is *current gain* and *voltage gain*, thus resulting in a large power gain. For example, if the output impedance is 10 times higher than the input impedance, there will be a power gain of 10, even if the base–emitter (input) and emitter–collector (output) currents are equal (no current gain). Since it is possible to control a large output current with a small input current in a practical common-emitter circuit, there is also current gain.

Capacitor C_1 is used to block out any dc components in the input signal. The input signal voltage is developed across resistor R_1, which also serves to provide a closed circuit for current flow in the emitter–base circuit. This is essential in any transistor amplifier since a transistor is a *current-operated* device.

The flow of collector current produces a voltage drop, with the indicated polarity, across R_2. The ac component of this voltage is the output signal of the amplifier. Capacitor C_2 is the coupling capacitor, which blocks out the steady dc component and passes the ac component.

The phase relationship between the input and output signals is shown by the sinewaves appearing in Fig. 1-2. If the input voltage increases the forward bias of the base–emitter junction, the total emitter current is increased, producing an increase in the collector current in like degree. This increase in collector current, in turn, produces a larger drop across R_2.

In the *NPN* amplifier (Fig. 1-2a) the input voltage must be positive if it is to aid the forward bias of the emitter–base junction. The larger voltage drop makes the top of R_2 more negative with respect to its bottom. Thus, as the input voltage goes up, the output voltage goes down.

To decrease the forward bias of the emitter–base junction, the input voltage must become negative. Thus, the total emitter current is decreased, producing a corresponding decrease in the collector current and voltage drop across R_2. As a result, the top of R_2 becomes less negative with respect to the bottom, and the output voltage becomes less negative. Thus, as the input voltage goes down, the output voltage goes up.

In the *PNP* amplifier (Fig. 1-2b) a similar action takes place, but with reversed polarities. A negative input voltage that increases the forward bias of the emitter–base junction causes the output voltage to become more positive. As a positive input voltage decreases the forward bias of the emitter–base junction, the output voltage becomes less positive.

Thus, in all cases where the common-emitter amplifier configuration is used, as the input voltage rises, the output voltage falls, and vice versa. Hence, there is a *phase reversal* between input and output signals (that is, input and output signals are 180° out of phase).

In a practical amplifier, both junctions are biased from a single power supply (or battery). There are several ways of supplying the bias, as is discussed in Sec. 1-7. A typical method of supplying bias to a common-emitter amplifier stage from a single supply is shown in Fig. 1-3. The circuit is a common-emitter amplifier using an *NPN* transistor. Battery V_{CC} furnishes the bias for both junctions.

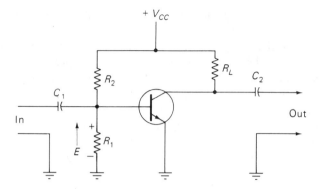

Fig. 1-3   Practical common-emitter amplifier operated from a single
power supply

Since the emitter–base junction must be forward biased and the collector–base junction reverse biased, in the *NPN* transistor this means that the base must be negative relative to the collector, and positive relative to the emitter. Thus, the collector is at the most positive potential of the circuit and the emitter is at the most negative potential.

Note that when we say that the base must be less positive than the collector, and more positive than the emitter, we are talking in relative terms. For example, if the base voltage is $+2$ V and the emitter voltage is $+1.5$ V, the emitter is more negative than the base, even though both voltages are positive. Similarly, the emitter is more negative than the base if the emitter voltage is -2 V and the base voltage is -1.5 V.

Since the collector is connected to the positive terminal of battery V_{CC}, it is at the most positive potential. Similarly, since the emitter is connected to the negative terminal of V_{CC}, it is at the most negative potential. Resistors R_1 and R_2 form a voltage divider across V_{CC}. The flow of current through this divider produces a voltage drop across R_1 with the indicated polarity. Note that this voltage opposes that of V_{CC}. Hence, the voltage applied to the base is less positive than the collector voltage, although more positive than the emitter voltage.

If a *PNP* transistor is used, the action is the same except that the polarities of all voltages and the direction of current flow are reversed.

Because the emitter–base junction in the common-emitter configuration is forward biased, the input resistance of this circuit is low. Because the collector–base junction is reverse biased, the output resistance is high. Typically, the input resistance of the common-emitter amplifier stage is between 500 and 1500 Ω, and the output resistance is between 30 and 50 kΩ. However, as is discussed in later sections, it is possible to set input and output resistances of an amplifier stage by proper selection of resistors (typically, R_1 and R_L of Fig. 1-3).

Typically, the current gain, voltage gain, and power gain of common-emitter amplifiers are 25 to 50, 300 to 1000, and 25 to 40 dB, respectively.

To summarize the common-emitter amplifier, the input signal is applied between the base and emitter, and the output signal appears between emitter and collector. This provides a moderately low input impedance and a very high output impedance. However, the output signal is out of phase with the input. The common-emitter amplifier produces the highest power gain of all three configurations. Its voltage and current gains are fairly high. Common-emitter circuits are the most often used since there is current gain, voltage gain, resistance gain, and power gain.

1-4.1. Gain in Amplifiers

Many terms are used to express the gain of common-emitter amplifiers, as well as other amplifier configurations. For example, there are the terms *alpha* and *beta*, in addition to *current gain, voltage gain, resistance (or impedance) gain, power gain, voltage amplifier,* and *power amplifier.* The terms are interrelated, and are often interchanged (properly and improperly) by technicians and engineers. To minimize this confusion, the following is a summary of how these terms are used throughout this handbook.

Alpha and *beta* are applied to transistors connected in the common-base and common-emitter configurations, respectively. Both terms are a measure of the *current gain of the transistor.* Alpha is always less than 1 (typically 0.9 to 0.99). Beta is more than 1 and can be as high as several hundred (or more). The relationships between alpha and beta are

$$\text{alpha} = \frac{\text{beta}}{\text{beta} + 1}, \qquad \text{beta} = \frac{\text{alpha}}{\text{alpha} - 1}$$

The terms alpha and beta do not necessarily represent the current gain of the stage in which the transistors are used. Instead, the current gain of the stage *cannot* be greater than the alpha or beta of the transistor.

Current gain can be applied to the transistor or stage, and is a measure of change in current at the output for a given change in current at the input. Current gain is equal to the change in output current divided by change in input current. For example, when a 1-mA change in input current produces a 10-mA change in output current, the current gain is 10. Since alpha (common base) is always less than 1, there is no current gain in common-base amplifier circuits. (Instead, there is a slight loss.)

Resistance gain is simply the ratio of output resistance (or impedance) divided by input resistance (or impedance). For example, if the input resistance is 1 kΩ and the output resistance is 15 kΩ, the resistance gain is 15. The input and output resistances (or impedances) are dependent upon circuit values, as well as transistor characteristics. As discussed throughout this handbook, many amplifier characteristics are directly dependent upon the relationship between transistor characteristics and values of circuit components.

Resistance gain, by itself, produces no usable gain for the amplifier stage. However, the resistance gain has a direct effect on the stage voltage and power gain. For example, using the previous values (current gain of 10, resistance gain of 15), it is possible to have a voltage gain of 150. Assume that the input resistance is 1 kΩ, and a 1-mV signal is applied at the input. This results in an input current change of 1 μA. With a current gain of 10, the output current is 10 μA. This 10-μA current passes through a 15-kΩ output resistance to produce a voltage change of 150 mV. Thus, the 1-mV input signal produces a 150-mV output signal (a voltage gain of 150).

Voltage gain is generally applied to the stage, and is equal to the difference in output voltage divided by a difference in input voltage.

Power gain is also generally applied to the stage, and is equal to the difference in output power divided by the difference in input power.

Except in a common-base amplifier, power gain is always higher than voltage gain, since power is based on the square of voltage (power = E^2/R), as well as the square of current (I^2R). Using the same values, the input power of the stage is 1×10^{-9}, the output power is 1.5×10^{-6}, and the power gain is 1500.

As in the case of current gain, both voltage gain and power gain are dependent on transistor characteristics and circuit values.

The function of a *voltage amplifier* is to receive an input signal consisting of a small voltage of definite waveform and produce an output signal consisting of a voltage with the same waveform, but much larger in amplitude. For example, as a radio wave cuts across the antenna of a radio receiver, the wave induces in that antenna a fluctuating voltage, usually on the order of microvolts. The voltage amplifier of the receiver amplifies this voltage to produce a similarly fluctuating voltage that is large enough to operate a *power amplifier*, which, in turn, operates the power-consuming loudspeaker.

Those transistors designed for voltage amplification usually have high betas with little current-carrying capability. On the other hand, power amplifier transistors have large current-carrying capacity, but relatively low betas. Typically, there is at least one (and usually two) voltage-amplifier stage ahead of the power-amplifier stage. This permits a low voltage input signal (say from an antenna, tape head, or industrial transducer) to operate a power-consuming device (radio or stereo loudspeaker, or industrial servo motor). If the power involved exceeds about 1 W, the transistor must be operated with *heat sinks*.

1-5. COMMON-BASE AMPLIFIER

The common-base circuit is shown in Fig. 1-4, where the power-supply connections for both *NPN* and *PNP* transistors are illustrated. In this configuration, the base is common to both the input and output circuits. Although the base is frequently called the grounded element, the base is not necessarily connected to ground. Battery B_1 forward biases the emitter–base junction. Battery B_2 reverse biases the collector–base junction. In a practical circuit, the voltage of B_2 is generally larger than that of B_1.

The input signal is applied between the base and emitter, and the output signal appears between the base and collector. This provides the lowest possible input impedance and a very high output impedance. The output signal is in phase with the input.

Capacitor C_1 is used to block out any dc components in the input signal. The input signal voltage is developed across resistor R_1, which also serves to

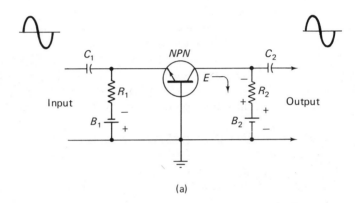

(a)

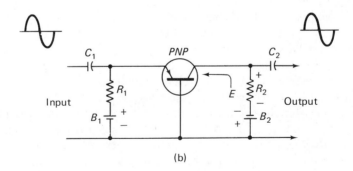

(b)

Typical Characteristics

Input resistance 30 - 150 Ω
Output resistance 300 - 500 kΩ
Voltage gain 500 - 1500
Current gain less than 1
Power gain 20 - 30 dB

Fig. 1-4   Basic common-base amplifier stage

provide a closed circuit for current flow in the emitter–base circuit. The flow of collector current produces a voltage drop, with the indicated polarity, across R_2. The ac component of this voltage is the output signal of the amplifier. Capacitor C_2 is the coupling capacitor, which blocks out the steady dc component and passes the ac component.

The phase relationship between the input and output signals is shown by the sinewaves appearing in Fig. 1-4. In the *NPN* amplifier (Fig. 1-4a), as a negative input voltage increases the forward bias of the emitter–base junction, the emitter and collector currents are increased, producing a corresponding increase in the voltage drop across R_2. This larger voltage drop makes the top of R_2 more negative with respect to its bottom. Thus, as the input signal goes negative, so does the output signal.

As the input signal goes positive, the forward bias for the emitter–base junction is reduced, resulting in a decrease of collector current and the voltage drop across R_2. The top of R_2 becomes less negative and so does the output signal. Thus, as the input signal goes positive, so does the output signal.

In the *PNP* amplifier (Fig. 1-4b), a similar action takes place, although with reversed polarities. Thus, where the common-base configuration is used, the input and output signals are in phase.

In a practical amplifier, both junctions are biased from a single power supply (or battery). There are several ways of supplying the bias, as is discussed in Sec. 1-7. A typical method of supplying bias to a common-base amplifier stage from a single supply is shown in Fig. 1-5. The circuit is a common-base amplifier using an *NPN* transistor. Battery V_{CC} furnishes the bias for both junctions.

The collector is connected to the positive terminal of battery V_{CC}, placing it at the most positive potential. The emitter is connected to the negative terminal of V_{CC}, placing it at the most negative potential. Resistors R_1 and R_2 form a voltage divider across V_{CC}. Current flows through this divider, causing a voltage drop across R_1 with the indicated polarity. Since the polarity of this voltage drop is opposed to the polarity of V_{CC}, the potential of the base is less positive than the potential of the collector although more positive than the potential of the emitter.

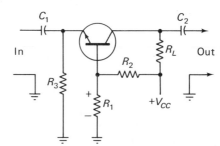

Fig. 1-5 Practical common-base amplifier operated from a single power supply

If a *PNP* transistor is used, the action is the same, except that the polarities of the battery and the voltage drop across R_1 are reversed.

The forward-biased emitter–base junction produces low impedance, whereas the reverse-biased collector–base junction produces high impedance. Typically, the input resistance of the common-base amplifier stage is between 30 and 150 Ω, and the output resistance is between 300 and 500 kΩ. However, it is possible to set input and output resistance of an amplifier stage by proper selection of resistors (typically, R_3 and R_L, of Fig. 1-5).

Typically, the current gain, voltage gain, and power gain of common-base amplifiers are 0.9 to 0.99, 500 to 1500, and 20 to 30 dB, respectively.

To summarize the common-base amplifier, the input signal is applied between the base and emitter, and the output signal appears between base and collector. This provides an extremely low input impedance and a very high output impedance. The output signal is in phase with the input. The common-base amplifier produces high voltage gains and modest power gains, even though there is no current gain. This is possible because of the resistance gain, as described in Sec. 1-4.1.

As an example, assume that the input resistance is 100 Ω, the output resistance is 15 kΩ, the current gain is 0.9 (less than 1), and the input signal is 1 mV. With 1 mV across 100 Ω, there is a 10-μA current change. With a current "gain" of 0.9, the output current is 9 μA. This 9-μA current passes through a 15-kΩ output resistance to produce a voltage change of 135 mV. Thus, a 1-mV input signal produces a 135-mV output signal (a voltage gain of 135).

1-6. COMMON-COLLECTOR AMPLIFIER

The common-collector circuit is shown in Fig. 1-6, where the power-supply connections for both *NPN* and *PNP* transistors are illustrated. The common-collector circuit is also known as an *emitter-follower* circuit, since the output is taken from the emitter resistance, and the output follows the input (in phase relationship).

Battery B_1 forward biases the emitter–base junction. Battery B_2 reverse biases the collector–base junction. The input signal is applied at the base, and the output signal appears at the emitter. This provides a high input impedance and a very low output impedance. The output signal is in phase with the input.

Capacitor C_1 is used to block out any dc components in the input signal. The input signal voltage is developed across resistor R_1. The flow of emitter current produces a voltage drop, with the indicated polarity, across R_2. The ac component of this voltage is the output signal of the amplifier. Capacitor C_2 is the coupling capacitor, which blocks out the steady dc component and passes the ac component.

The phase relationship between the input and output signals is shown by

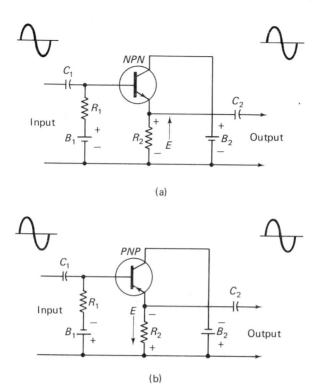

(a)

(b)

Typical characteristics

Input resistance 20 - 500 kΩ
Output resistance 50 - 1000 Ω
Voltage gain less than 1
Current gain 25 - 50
Power gain 10 - 20 dB

Fig. 1-6   Basic common-collector (emitter-follower) amplifier stage

the sinewaves appearing in Fig. 1-6. If the input voltage increases the forward bias of the emitter–base junction, the total emitter current is increased. This increased emitter current produces an increased voltage drop across R_2. In the *NPN* amplifier (Fig. 1-6a) this larger voltage drop makes the top of R_2 more positive with respect to its bottom. Thus, the output voltage becomes more positive.

If the input voltage decreases the forward bias of the emitter–base junction, the total emitter current is decreased, producing a corresponding decrease in the voltage drop across R_2. As a result, in the *NPN* amplifier this means that the top of R_2 becomes less positive with respect to its bottom, and the output voltage becomes less positive.

In the *PNP* amplifier (Fig. 1-6b) a similar action takes place, but with reversed polarities. Thus, where the common-collector configuration is used, as the input voltage rises, so does the output voltage. As the input voltage falls, the output voltage falls in like degree. Thus, the input and output signals are in phase.

In a practical amplifier both junctions are biased from a single power supply (or battery). There are several ways of supplying the bias. A typical method of supplying bias to a common-collector amplifier stage from a single supply is shown in Fig. 1-7. The circuit is a common-collector *NPN* transistor. The battery V_{CC} makes the collector positive relative to the base, thus reverse biasing the collector–base junction. Resistors R_1 and R_2 form a voltage divider across V_{CC}, which forward biases the emitter–base junction. Where a *PNP* transistor is used, the polarity of the battery is reversed, but the action is the same as before.

Typically, the input resistance of the common-collector amplifier stage is between 20 and 500 kΩ, and the output resistance is between 50 and 1000 Ω. However, it is possible to set input and output resistances of an amplifier stage by proper selection of resistors (typically, R_1 and R_L of Fig. 1-7).

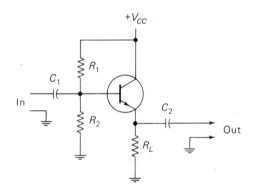

Fig. 1-7   Practical common-collector (emitter-follower) amplifier operated from a single power supply

Typically, the current gain, voltage gain, and power gain of common-collector amplifiers are 25 to 50, less than 1, and 10 to 20 dB, respectively.

To summarize the common-collector amplifier, the input signal is applied to the base, and the output signal appears at the emitter. This provides extremely high input impedance and low output impedance. The output signal is in phase with the input. The common-collector amplifier produces modest current and power gains, even though there is no voltage gain. In general, the common-collector circuit current gain (and hence the power gain) is limited by the current gain (beta) of the transistor.

1-7. AMPLIFIER BIAS NETWORKS

All solid-state amplifiers require some form of bias. As a minimum, the collector–base junction of any solid-state amplifier must be reverse biased. That is, current should not flow between collector and base. Any collector–base current that does flow is a result of leakage or break-down. Breakdown must be avoided by proper design. Leakage, usually listed as I_{CBO}, is an undesirable (but almost always present) condition that must be reckoned with in practical use. (The C and B of the subscript indi-cate that the current flows between the collector and base. The O indicates that it is measured with the emitter disconnected, or open.)

Under no-signal conditions, the emitter–base circuit of a solid-state ampli-fier can be forward biased, reverse biased, or zero biased (no bias). However, emitter–base current must flow under some condition of operation. For example, some current flows all the time in class A and class AB amplifiers. In class B and C amplifiers, current flows only in the presence of an operating signal. With any class of operation, the emitter–base circuit must be biased so that current can flow under some conditions. (Classes of amplifiers are discussed in Sec. 1-8.)

The desired bias is accomplished by applying voltages to the corresponding transistor elements through bias networks, usually composed of resistors. The following paragraphs describe several basic bias networks. These circuits (or variations of them) represent most of the bias methods used in solid-state amplifiers.

The bias networks (or the resistors used to form them) serve more than one purpose. Typically, the bias network resistors (1) set the operating point, (2) stabilize the circuit at the operating point, and (3) set the approximate input–output impedances of the circuit.

The basic purpose of the bias network is to establish collector–base–emitter voltage and current relationships *at the operating point of the circuit.* (The operating point is also known as the *quiescent point, Q point, no-signal point, idle point,* or *static point.*) Since transistors rarely operate at this Q point, the basic bias networks are generally used as a reference or starting point for design. The actual circuit configuration and (especially) the bias network values are generally selected on the basis of dynamic circuit con-ditions (desired output voltage swing, expected input signal level, etc.).

Once the desired operating point is established, the next function of the bias network is to stabilize the amplifier circuit at this point. Although there are many bias networks, each with its own advantages and disadvantages, one major factor must be considered for any network. The basic bias network must maintain the desired current relationships in the presence of temperature and power-supply changes, and possible transistor replacement. In some cases, frequency changes and changes caused by component aging must also be offset by the bias network. This process is generally referred to as *bias stabilization.*

Two undesirable conditions can result when adequate bias stabilization is not provided. First, any changes in temperature, power-supply voltage, and, possibly, frequency produce changes in collector–emitter and/or base–emitter current. For example, an increase in temperature or power-supply voltage will increase currents. In turn, this will *shift the operating point* of the amplifier circuit. As is discussed in later paragraphs, a shift in operating point can produce distortion and a change in frequency response, as well as other undesired effects. In any event, the amplifier circuit is no longer at the optimum operating point for which it is designed.

The other undesirable effect of inadequate bias stabilization has to do with power dissipation limits of the transistor. When a transistor is operated at or near its maximum power dissipation limits, the transistor is subject to *thermal runaway*. When current passes through a transistor junction, heat is generated. If not all of this heat is dissipated by the case or heat sink (often an impossibility), the junction temperature rises. This, in turn, causes more current to flow through the junction, even though the voltage, circuit values, and so on, remain the same. In turn, this causes the junction temperature to increase even further, with a corresponding increase in current flow. If the heat is not dissipated by some means, the transistor will burn out and be destroyed.

Adequate bias stabilization will prevent any drastic change in junction currents, despite changes in temperature, voltage, and so on. Thus, proper bias stabilization will maintain the amplifier circuit at the desired operating point (within practical limits), and will prevent thermal runaway.

The resistors used in bias networks also have the function of setting the input and output impedances of the amplifier circuit. From a theoretical standpoint, the input–output impedances of a circuit are set by a wide range of factors (transistor beta, transistor input–output capacitance, etc.). However, for practical purposes the input–output impedances of a resistance-coupled amplifier (operating at frequencies up to about 100 kHz) are set by the bias network resistors. For example, the output impedance of a common-base or common-emitter amplifier is approximately equal to the collector resistor (between the collector and power supply).

1-7.1. Bias-Stabilization Techniques

There are several methods for providing bias stabilization of solid-state amplifiers. All these methods involve a form of *negative feedback* or *inverse feedback*. That is, any change in transistor currents produces a corresponding voltage or current change that *tends to offset* the initial change. There are two basic methods for producing inverse or negative feedback: *inverse-voltage feedback* and *inverse-current feedback* (also known as *emitter feedback*).

Inverse-voltage feedback. Figure 1-8 illustrates a typical inverse-voltage bias network. The emitter–base junction is forward biased by the voltage at

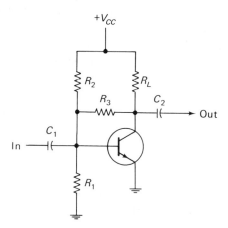

Fig. 1-8 Typical bias network
with inverse-voltage feedback

the junction of R_1 and R_2. The base–collector junction is reverse biased by
the differential between voltages at the collector and base. Generally, the
collector of a resistance-coupled amplifier is at a voltage about one half that
of the supply. For example, if the supply V_{CC} is 20 V, the collector is set at
about 10 V (at the Q point). This allows for the maximum output voltage
swing. The base is at a voltage about 0.5 to 1 V greater than the emitter.
Since the emitter is grounded in the circuit of Fig. 1-8, the base is set to about
0.5 to 1 V by R_1 and R_2.

Resistor R_3 is connected between the collector and base. Since the collector
voltage is positive (for an *NPN* transistor), a portion of this voltage is fed
back to the base to aid the forward bias. The normal (or Q point) forward
bias on the emitter–base junction is the result of all the voltages between the
emitter and base.

Should the collector current increase (due to a temperature rise, power-
supply increase, change in frequency, or any other cause), a larger voltage
drop is produced across R_L. As a result, the voltage on the collector decreases,
reducing the voltage fed back to the base through R_3. This reduces the emitter–
base forward bias, reducing the emitter current and lowering the collector
current to its normal value. Should there be an initial increase in collector
current (for any reason), an opposite action takes place, and the collector current is raised to its normal (Q point) value.

*Note that any form of negative or inverse feedback in an amplifier tends to
oppose all changes, even those produced by the signal being amplified.* Thus,
inverse or negative feedback tends to reduce and stabilize gain, as well as
undesired change. This principle of stabilizing gain by means of feedback
is used in virtually all types of amplifiers, as is discussed throughout remaining chapters.

Inverse-current feedback (emitter feedback). Current feedback is more
commonly used than voltage feedback in solid-state amplifiers. This is because

transistors are primarily current-operated devices, rather than voltage-operated devices.

Figure 1-9 shows a typical inverse-current (emitter-feedback) bias network using an *NPN* transistor. Other bias networks using the same principle are

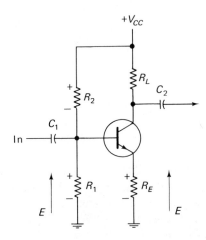

Fig. 1-9 Typical bias network with inverse-current feedback (emitter-feedback)

discussed in the following sections. However, the use of an emitter-feedback resistance in any bias circuit can be summed up as follows:

Base current (and, consequently, collector current) depends upon the *differential in voltage* between base and emitter. If the differential voltage is lowered, less base current (and, consequently, less collector current) will flow. The opposite is true when the differential is increased. All current flowing through the collector (ignoring collector–base leakage, I_{CBO}) also flows through the emitter resistor. The voltage drop across the emitter resistor is therefore dependent (in part) on the collector current.

Should the collector current increase (for any reason), emitter current, and the voltage drop across the emitter resistor, will also increase. This negative feedback tends to decrease the differential between base and emitter, thus lowering the base current. In turn, the lower base current tends to decrease the collector current, and offset the initial collector-current increase.

1-7.2. Some Representative Amplifier Bias Networks

Figures 1-10 through 1-16 illustrate typical bias schemes used in solid-state amplifiers. All the bias networks use inverse-current feedback (emitter feedback) of some form. Note that the bias network resistors in addition to providing the required voltage relationships for emitter, base, and collector also set the *approximate input and output impedances* of the amplifier circuit, as shown by the equations on the illustrations. Likewise, the *approximate voltage and current gains are set by resistance ratios* in many of the circuits. The equations on the illustrations hold true for operation up to about 100 kHz (and higher in many instances). As operating frequencies increase beyond about 100 kHz, other factors enter into gain and impedance relationships. These factors are discussed is later chapters.

In the circuit of Fig. 1-10, the value of R_B is selected to provide a given base current at the Q point. For an *NPN* silicon transistor, it can be assumed that the base voltage will be 0.5 to 1 V more positive than the emitter. (The base of *PNP* silicon transistors will be 0.5 to 1 V more negative than the emitter.) If germanium transistors are used, the voltage difference will be about 0.2 to 0.5 V. The emitter voltage is dependent upon the drop across R_E. As shown by the equations, the value of R_E is dependent upon a tradeoff between stability and gain. An increase in the value of R_E, in relation to R_L, increases stability but decreases gain. Typically, the value of R_E is less than 1 kΩ, and is generally not greater than 0.2 times R_L. An R_E that is 0.1 times R_L can be considered as typical.

The output impedance is approximately equal to R_L, whereas the input impedance is approximately equal to R_E times beta of the transistor. Since input impedance is dependent upon transistor beta, the input impedance is subject to wide variation from transistor to transistor, and with changes in frequency. Generally, this is considered as a disadvantage.

The circuit voltage gain is approximately equal to the ratio of R_L/R_E times beta, whereas the current gain is approximately equal to beta. Again, this produces wide variation in gain. The circuit of Fig. 1-10 offers the greatest possible voltage gain, but the least stability, of all the basic bias circuits described here.

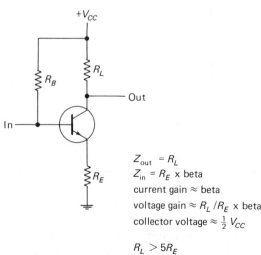

$$Z_{out} = R_L$$
$$Z_{in} = R_E \times beta$$
current gain ≈ beta
voltage gain ≈ R_L/R_E × beta
collector voltage ≈ $\frac{1}{2} V_{CC}$

Fig. 1-10   Bias network with maximum gain and minimum stability

$$R_L > 5R_E$$
$$E$$
$$R_L \approx 10R_E$$

In the circuit of Fig. 1-11, the values and characteristics are essentially the same as for the circuit of Fig. 1-10. However, the stability of the Fig. 1-11 circuit is increased. The increase in stability is brought about by connecting base resistance R_B to the collector, rather than to the power supply. If collector current increases for any reason, the drop across R_L also increases, lowering the voltage at the collector. This lowers the base voltage and current, thus reducing the collector current. The feedback effect is combined with that produced by the emitter resistor R_E to offset any variation in collector current. However, because of the increased stability produced by the feedback, gain for the Fig. 1-11 circuit is slightly less than for the Fig. 1-10 circuit.

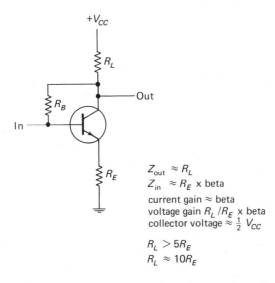

$$Z_{out} \approx R_L$$
$$Z_{in} \approx R_E \times \text{beta}$$
current gain $\approx$ beta
voltage gain $R_L/R_E \times$ beta
collector voltage $\approx \frac{1}{2} V_{CC}$

$$R_L > 5R_E$$
$$R_L \approx 10R_E$$

Fig. 1-11   Bias network with maximum gain and improved stability

In the circuit of Fig. 1-12, the output impedance is approximately equal to R_L, whereas the input impedance is approximately equal to R_B. In theory, the input impedance is equal to R_B in parallel with R_E times (beta $+$ 1). However, unless beta is very low, the R_E times (beta $+$ 1) factor will be much greater than R_B. Thus, the value of R_B (or slightly less) can be considered as the stage or circuit input impedance.

Since the input impedance is approximately equal to R_B, and the output impedance is equal to R_L, the approximate current gain of the stage is equal to the ratio of R_L/R_B. Of course, this assumes that the transistor beta is greater than the R_L/R_B ratio. For example, if the transistor beta is 20, the R_L/R_B ratio must be somewhat less than 20. The voltage gain of the Fig. 1-12 circuit is approximately equal to the ratio R_L/R_E.

The circuit of Fig. 1-12 offers more stability than the other bias networks, but with a tradeoff of lower gain and lower input impedance.

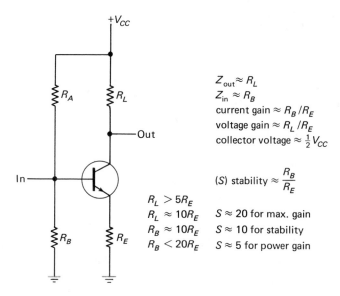

$Z_{out} \approx R_L$
$Z_{in} \approx R_B$
current gain $\approx R_B / R_E$
voltage gain $\approx R_L / R_E$
collector voltage $\approx \frac{1}{2} V_{CC}$

(S) stability $\approx \dfrac{R_B}{R_E}$

$R_L > 5R_E$
$R_L \approx 10R_E$ $S \approx 20$ for max. gain
$R_B \approx 10R_E$ $S \approx 10$ for stability
$R_B < 20R_E$ $S \approx 5$ for power gain

Fig. 1-12   Bias network with improved stability. Circuit characteristics are dependent upon circuit values

The value of R_L is determined by the desired collector voltage and current, or by an arbitrary need for a given output impedance. The value of R_E is dependent upon a tradeoff between stability and gain. An increase in the value of R_E, in relation to R_L, increases stability and decreases voltage gain. An increase in the value of R_E, in relation to R_B, increases stability and decreases current gain. The value of R_E is typically less than 1 kΩ, and should not be greater than 0.2 times R_L. An R_E that is 0.1 times R_L can be considered as typical.

The value of R_B is dependent upon tradeoffs between the value of R_E, current gain, stability, and the desired input impedance. As a basic rule, R_B is about 10 times R_E. A higher value of R_B will increase current gain and decrease stability. Input impedance of the circuit will be approximately equal to R_B (actually slightly less). Therefore, if the input impedance is of special importance in the circuit, the value of R_B is often selected on that basis. This may require a specific value of R_E to maintain stability and current gain relationships. Of course, any change in R_E results in a change of the voltage gain (assuming that R_L remains the same). The value of R_A is selected to provide a given base voltage at the Q point.

In addition to stability, the major advantage of the Fig. 1-12 circuit is that the input and output impedances, as well as voltage and current gain, are not dependent upon transistor beta. Instead, circuit characteristics are dependent upon circuit values.

In the circuit of Fig. 1-13 the basic characteristics are the same as for the

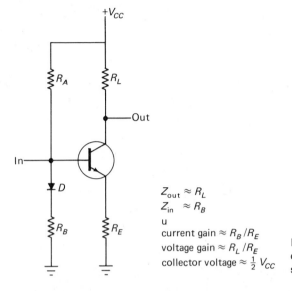

$Z_{out} \approx R_L$

$Z_{in} \approx R_B$

u

current gain $\approx R_B / R_E$

voltage gain $\approx R_L / R_E$

collector voltage $\approx \frac{1}{2} V_{CC}$

Fig. 1-13 Bias network with diode for improved temperature stability

Fig. 1-12 circuit, except that temperature stability is increased. The increase in temperature stability is brought about by connecting diode D between the base and R_B. Diode D (always forward biased) is of the same material (silicon or germanium) as the base–emitter junction, and is maintained at the same temperature. In practice, diode D is mounted near the transistor so that the base–emitter junction and diode D are at the same temperature. Thus, the voltage drops across diode D and the base–emitter junction are the same, and remain the same with changes in temperature.

For example, should temperature rise, the emitter and collector currents of the transistor also tend to rise. The same temperature rise causes the forward resistance of diode D to decrease. As a result, the current flowing through R_A increases, causing an increase in the voltage drop across R_A. Since this voltage drop tends to reverse bias the emitter–base junction, the net forward bias for that junction is reduced, and the emitter–collector currents are reduced to normal (or tend to reduce to normal). Should the temperature fall, the action is in the reverse direction, and the emitter–collector currents are raised toward their normal values.

In the circuit of Fig. 1-14 the basic characteristics are the same as for the Fig. 1-12 circuit, except that the *negative temperature coefficient* (NTC) characteristics of a *thermistor* are used to provide temperature compensation. The resistance of a thermistor decreases with increases in temperature, and vice versa. For best results, the thermistor is mounted near the transistor, so that both devices are at the same temperature.

In the circuit of Fig. 1-14 the thermistor is used to vary the emitter voltage with temperature variations to minimize the effects of these variations upon

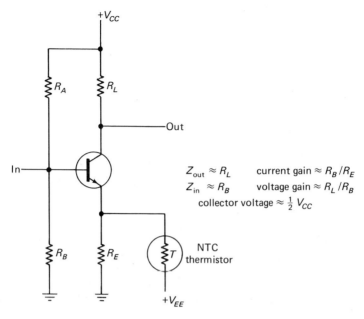

$$Z_{out} \approx R_L \qquad \text{current gain} \approx R_B/R_E$$
$$Z_{in} \approx R_B \qquad \text{voltage gain} \approx R_L/R_B$$
$$\text{collector voltage} \approx \tfrac{1}{2} V_{CC}$$

Fig. 1-14   Bias network with thermistor for improved temperature stabililty

the emitter current. Resistors R_A and R_B form a voltage divider to apply a portion of the supply voltage V_{CC} in a direction to forward bias the emitter–base junction. Resistor R_E and the thermistor form a second voltage divider across V_{EE}. The direction of the voltage drop across R_E is such as to reverse bias the base–emitter junction. However, since the forward bias applied to the base is larger than the reverse bias applied to the emitter, the net result is that the emitter–base junction is forward biased.

Should the temperature rise, the emitter and collector currents tend to rise. The same rise in temperature reduces the resistance of the thermistor. This action permits more current to flow through the voltage divider. The increase in current flow increases the voltage drop across R_E, thus increasing the reverse bias being applied to the emitter–base junction. As a result, the net forward bias of the junction is reduced, thus reducing the emitter and collector currents toward their normal values. Should the temperature decrease, the action is reversed, thus preventing a decrease in emitter and collector currents.

In the circuit of Fig. 1-15 the basic characteristics are the same as for the Fig. 1-12 circuit. However, the circuit of Fig. 1-15 is used in those special applications that require a negative and positive voltage, each with respect to ground, to control base current. The values for R_L, R_E, and R_B in Fig. 1-15 are the same as for the Fig. 1-12 circuit. However, the value of R_A in

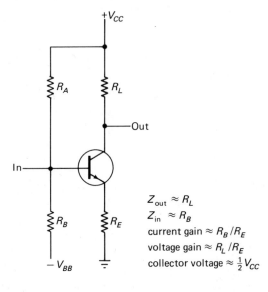

$$Z_{out} \approx R_L$$
$$Z_{in} \approx R_B$$
current gain $\approx R_B/R_E$
voltage gain $\approx R_L/R_E$
collector voltage $\approx \frac{1}{2}V_{CC}$

Fig. 1-15 Bias network with positive and negative supply voltages for control of base current

the Fig. 1-15 circuit is different, due to the large amount of current through R_B.

In the circuit of Fig. 1-16 the characteristics are somewhat different than in all the previous bias networks. The Fig. 1-16 circuit is used in those special applications where it is necessary to supply collector–emitter current from both a positive and negative source. Since the transistor is *NPN*, the collector is connected to the positive source through R_L, while the emitter is connected to the negative source through R_E. If both sources are approximately equal, there is generally no voltage gain. The collector and emitter currents are approximately equal (ignoring the base current). Therefore, if R_L drops the

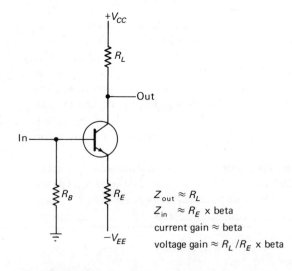

$$Z_{out} \approx R_L$$
$$Z_{in} \approx R_E \times beta$$
current gain $\approx$ beta
voltage gain $\approx R_L/R_E \times$ beta

Fig. 1-16 Bias network with positive and negative supply for emitter–collector currents

positive source to half (say from 20 to 10 V), then R_E would be approximately twice the resistance of R_L. This will produce a voltage loss, all other factors being equal.

1-8. AMPLIFIER CLASSIFICATIONS BASED ON OPERATING POINT

As discussed, amplifiers are often classified as to *operating point*, or the amount of current flow under no-signal conditions. The following is a brief summary of the four basic operating-point classifications.

Note that in all four classifications, the base–collector junction is always reverse biased at the operating point, as well as under all signal conditions. Thus, no base–collector current flows (with the possible exception of reverse leakage current I_{CBO}). On the other hand, the base–emitter junction is biased such that base–emitter current will flow under certain conditions, and possibly under all conditions. When base–emitter current flows, emitter–collector current also flows.

1-8.1. Class A Amplifier

In the class A amplifier, the base–emitter bias and the input voltage are such that the transistor operates only over the *linear portion* of the *characteristic curve*. Such a curve, representing the relationship between base voltage (input) and collector current (output), is shown in Fig. 1-17. At no point of the input signal cycle does the base become so positive or negative as to cause the transistor to operate at the nonlinear portion of the curve. The transistor collector current is never cut off, nor does the transistor ever reach saturation.

The main advantage of the class A amplifier is the relative lack of distortion. The output waveform follows that of the input waveform, except in amplified form. However, with any class of amplifier there is some distortion, as is discussed in later sections.

The main disadvantages of class A amplifiers are their relative inefficiency (low power output for a high power input dissipated by the transistor), and their inability to handle large signal voltage swings. Rarely is a class A amplifier over about 35 per cent efficient. Thus, if the power input to a class A amplifier is 1 W (generally, the maximum power dissipation capability of a single transistor), the output will be less than 0.3 W. The peak-to-peak output signal voltage swing of a class A amplifier is limited to something less than the total supply voltage. Since the output voltage must swing both positive and negative, the peak output is less than one half the supply voltage. For example, assume that the supply voltage is 20 V and the amplifier is biased so that the Q-point collector voltage is one half the supply, or 10 V.

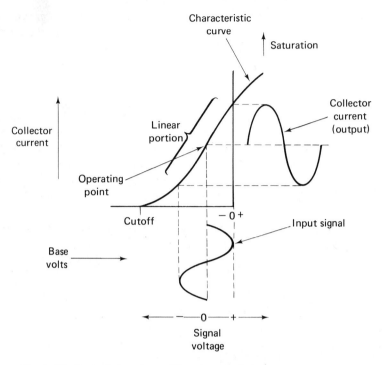

Fig. 1-17   Typical class A amplifier characteristic curve

(Such a Q point is generally typical for a class A amplifier.) Under these conditions, the output voltage swing cannot exceed ± 10 V. If distortion must be kept to a minimum, the output will usually be on the order of ± 5 V, so as to keep the transistor on the linear portion of the characteristic curve. (In most cases, the curve becomes nonlinear near the cutoff and saturation points.) However, this can be determined only from an actual test of the amplifier circuit, as described in later sections.

The input voltage swing of a class A amplifier is limited by the output voltage swing capability and the voltage amplification factor. For example, if the output is limited to ± 10 V, and the voltage amplification factor is 100, the input is limited to ± 0.1 V (100 mV).

Because of these limitations, class A amplifiers are generally used as voltage amplifiers, rather than power amplifiers. Typically, a class A amplifier stage is used ahead of a power amplifier stage.

1-8.2. Class B Amplifier

If the base–emitter bias is changed so that the operating point coincides with the transistor cutoff point, we obtain class B amplification.

For an *NPN* transistor this means making the base more negative than for class A operation. (For *PNP* transistors, class B is obtained by making the base more positive than for class A.) Either way, the base–emitter reverse bias is increased for class B operation.

As shown in Fig. 1-18, when the input signal voltage is zero, there is no flow of collector current. During the positive half-cycle of the signal voltage (Fig. 1-18 is for an *NPN* transistor), the collector current rises to its peak and then falls back to zero in step with the variations of that half-cycle. During the negative half-cycle of the signal voltage, there is no collector current since the base–emitter reverse bias is at all times greater than the cutoff voltage of the transistor. Hence, collector current flows only during half the input signal cycle.

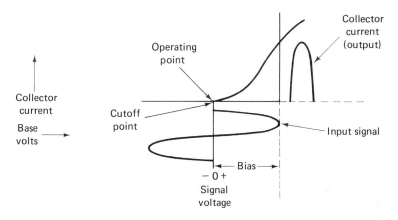

Fig. 1-18   Typical class B amplifier characteristic curve

If a single transistor is operated as class B, there will be considerable distortion. This is because the waveform of the resulting collector current resembles that of the positive half-cycle of the input signal and, consequently, does not resemble the complete waveform of the input. As is discussed in later sections, it is possible to use two transistors, one for each half-cycle of the input signal, and by combining the outputs of these transistors in *push–pull* to reconstruct an output whose waveform resembles the full waveform of the input.

The peak output voltage swing of a class B amplifier is slightly less than the supply voltage. Since the output appears only on half-cycles, it is possible to operate class B amplifiers at a higher current (or power) rating than class A, all other factors being equal. For example, if a transistor is capable of 0.3-W dissipation (without damage) as class A, the same transistor can be operated at 0.6 W, class B, since the transistor is conducting collector current only half the time. (This is a theoretical example. In practice, there are factors which limit class B power dissipation to something less than twice that of class A. These factors are discussed in later sections.)

Also note that the peak output of a class B amplifier is equivalent to the peak-to-peak output of a class A amplifier. Thus, if two transistors are connected in push–pull and operated as class B, the output voltage can be twice that of class A.

Because of these voltage and power factors, class B amplifiers are generally used as power amplifiers, rather than voltage amplifiers. Typically, two push–pull transistors are operated in class B, preceded by a single class A amplifier stage. The class A stage provides voltage amplification, whereas the class B stage produces the necessary power amplification.

1-8.3. Class AB Amplifier

Class B is the most efficient operating mode for audio amplifiers, since it draws the least amount of current. That is, the transistors are cut off at the Q point and draw collector current only in the presence of an input signal. Class B operation can, however, result in a form of distortion known as *crossover distortion*. The effects of crossover distortion can be seen by comparing the input and output waveforms on Fig. 1-19. In true class B operation, the transistor *remains cut off* at very low signal inputs (because transistors have very low current gain at cutoff) and turns on abruptly with a large signal. For example, a silicon transistor does not have appreciable collector current flow until the base–emitter junction is forward biased by about 0.5 V. Assuming that the input signal starts at 0 V, there will be little or no collector current flow (and thus no change in the output voltage) during the time that the input signal is going from 0 to 0.5 V. When the

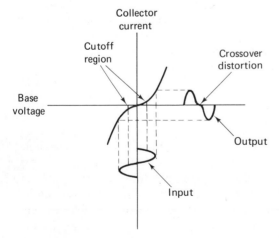

Fig. 1-19 Example of crossover distortion

input reaches 0.5 V, the collector current increases rapidly and follows the input signal in a linear fashion.

Crossover distortion can be minimized by operating the stage as class AB (or somewhere between B and AB). That is, the transistors are forward biased just enough for a *small amount* of collector current to flow at the Q point. For a typical silicon transistor, the forward bias is just below 0.5 V for class AB. Therefore, some collector current is flowing at the lowest signal levels, and there is no abrupt change in current gain. Class AB is less efficient than class B, since more current must be used. Generally, class AB is only used in push–pull circuits.

1-8.4. Class C Amplifier

If the transistor is reverse biased considerably below the cutoff point, we obtain a class C amplifier. As shown in Fig. 1-20, during the positive half-cycle of the input signal, the signal voltage starts from zero, rises to the positive peak value, and falls back to zero. (Figure 1-20 is for an *NPN* transistor.) Note that *a portion of the input signal* causes the base–emitter junction to be forward biased. As a result, there is a flow of collector current for a portion of one half the input cycle. The negative half-cycle of the input signal lies well below the cutoff point of the transistor.

Collector current flows only during that portion of the positive half-cycle of the input signal between the cutoff point and the peak. The resulting collector current is a pulse, the duration of which is considerably less than a half-cycle of the input signal.

Obviously, the waveform of the output signal cannot resemble that of the input signal. Nor can this resemblance be restored by the push–pull method

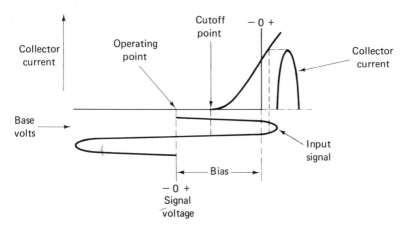

Fig. 1-20   Typical class C amplifier characteristic curve

mentioned in the discussion of class B and AB amplifiers. For this reason, class C is limited to those applications where distortion is of no concern. Generally, class C operation is limited to use in radio-frequency (RF) amplifiers.

1-9. AMPLIFIER DISTORTION

For the purposes of this book, distortion is defined as that condition when the output signal of an amplifier is not identical to the wave-form of the input signal. A small amount of distortion is generally present in all amplifiers. However, amplifiers are usually designed to keep such distortion within acceptable limits. In some special cases, amplifiers contain circuits that introduce a form of distortion. However, this is generally to offset or compensate for distortion already present in the signal.

There are many specific types and causes of distortion in amplifiers. These include crossover distortion (Sec. 1-8), intermodulation distortion, and harmonic distortion. However, there are three general types of distortion found in amplifiers: *amplitude distortion*, *frequency distortion*, and *phase distortion*. Any of these, either separately or in combination, may be present in an amplifier of any type.

1-9.1. Amplitude Distortion

Amplitude distortion arises within the transistor itself and is the result of operating the transistor over the nonlinear portion of its charac-teristic curve. The usual remedy is to use a base–emitter bias that places the operating point well within the linear portion of the curve, preferably at the center of the linear portion. In addition, care must be taken so that the amplitude of the input signal is small enough that its positive and negative half-cycles do not drive the transistor beyond the linear portion of the curve. Often, the combination of low input signals and placement of the operating point to keep the output linear (and thus minimize distortion) means that gain must be sacrificed. An overdriven amplifier, used to get maximum gain, almost always results in some amplitude distortion.

1-9.2. Frequency Distortion

Frequency distortion arises from the fact that the input signal rarely, if ever, is a single frequency. Instead, the input signal usually contains components of several frequencies, making the signal waveform somewhat complex.

In addition to a transistor, a solid-state amplifier circuit is composed of resistors, capacitors (unless the amplifier is direct coupled; refer to Chapter

4), and possibly inductances (coils and transformers). Capacitors and inductances have reactance. Since reactance is a function of frequency, the different frequencies of the signal encounter different reactances. Thus, the high and low frequencies of the signal can be impeded in *different degrees*. This produces distortion of the signal waveform from the original.

For example, assume that the input signal is a complex waveform composed of three frequencies; 10, 100, and 1000 Hz, all of the same amplitude. The reactance of coupling capacitors between stages will be different for each of the three frequencies. Capacitive reactance increases with a decrease in frequency. Thus, the 10-Hz signal will be attenuated more than the 100-Hz signal, and much more than the 1000-Hz signal. Even though the transistor amplifies all three frequencies equally, they will no longer be of equal amplitude at the output, and the output waveform will be different (distorted) from the input.

To minimize the effect of frequency distortion, amplifiers are usually designed to eliminate unwanted capacitance and inductance. Likewise, compensating components may be introduced into the circuit.

1-9.3. Phase Distortion

The fact that the input signal contains components of different frequencies is also responsible for phase distortion in an amplifier. When an alternating current (such as an ac signal) flows through a capacitor or an inductor, the current encounters a shift of phase. The degree of this phase shift is a function of frequency. Hence, the high- and low-frequency components of the signal are phase shifted in different degrees. These different phase shifts cause a distortion of the signal waveform.

As in the case of frequency distortion, the phase distortion in an amplifier may be minimized by proper design to eliminate unwanted capacitance and inductance.

1-10. DECIBEL MEASUREMENTS

When working with amplifiers, we are frequently interested in the relationship between the power input and the power output. Likewise, we may be interested in the ratio between voltage input and voltage output. The decibel (dB) is the unit that has been widely adopted in amplifier work to express logarithmically the ratio between two power or voltage levels (and less commonly the ratio between two current levels). A decibel is one tenth of a bel. (The bel is too large for most practical applications.)

Although power, voltage, or current amplification or the magnitude of a particular power, voltage, or current relative to a given reference value can be expressed as an ordinary ratio, the decibel has been adopted for two reasons:

(1) the decibel is a convenient unit to use for all types of amplifiers, and (2) the decibel is related to reaction of the human ear, and is thus well suited for use with audio amplifiers.

This latter use can best be understood when we examine audio or sound power. The human ear does not hear sounds in their direct power ratio. Thus, we can listen to ordinary conversation quite comfortably, and yet be able to hear thunder (which is taken to be 100,000 times louder than conversation) without damage to our ears. This is because the response of the human ear to sound waves is *approximately proportional to the logarithm of the energy* of the sound wave, and is not proportional to the energy itself.

The common logarithm ($\log_{10}$) of a number is the number of times 10 must be multiplied by itself to equal that number. Thus, the logarithm of 100 (that is, 10×10, or 10^2) is 2. Likewise, the logarithm of 100,000 (10^5) is 5. In mathematics, we write this relationship as

$$\log_{10} 100,000 = 5.$$

In comparing two powers, we could use the unit bel, which is the logarithm of the ratio of the two powers. For example, in comparing the power of ordinary conversation with that of thunder, the increase in sound is equal to

$$\log_{10} \frac{\text{power of thunder}}{\text{power of conversation}} \quad \text{or} \quad \log_{10} \frac{100,000}{1}.$$

Using the more convenient decibel, the increase in sound from ordinary conversation to thunder is equal to

$$10 \log_{10} \frac{100,000}{1} \quad \text{or} \quad 50 \text{ decibels (50 dB).}$$

For convenience, the same method is used in measuring the increase in power of amplifiers, whether the amplifiers are used with audio frequencies or not. Thus, the increase in power of any amplifier can be expressed as

$$\text{gain in dB} = 10 \log_{10} \frac{\text{power output}}{\text{power input}}.$$

This relationship can also be expressed as

$$\text{gain in dB} = 10 \log \frac{P_2}{P_1}.$$

Usually, P_2 represents power output and P_1 represents power input. Therefore, if P_2 is greater than P_1, there is a power gain expressed in positive decibels ($+$dB). With P_1 greater than P_2, there is a power loss expressed in negative decibels ($-$dB). Whichever is the case, the ratio of the two powers (P_1 and P_2) is taken, and the logarithm of this ratio is multiplied by 10. From this, we can obtain the following:

$$\text{power ratio of } 10 = 10\text{-dB gain}$$
$$\text{power ratio of } 100 = 20\text{-dB gain}$$
$$\text{power ratio of } 1000 = 30\text{-dB gain}$$

and so on.

Doubling the power produces a gain of $+3$ dB. Thus, if the volume control of an amplifier is turned up so that the power rises from 4 to 8 W, the gain is up $+3$ dB. If, conversely, the power output is reduced from 4 to 2 W, the gain is down -3 dB. Again, if the original 4 W is increased to 8 W, the gain is up $+3$ dB. Increasing the power output to 16 W produces another gain of $+3$ dB, and the total gain is $+6$ dB. At 40 W the power has been increased 10 times, and the total gain is $+10$ dB, and so on.

There is another convenience in using decibels. When several amplifier stage are connected to work into one another (stages connected in cascade), the amplifications are multiplied. For example, if three stages, each having a power gain of 10, are connected, there is a total gain of $10 \times 10 \times 10$, or 1000. In the decibel system, the *decibel gains are added*. Thus, the decibel gain is $+10 +10 +10$, or $+30$ dB. Similarly, if two amplifiers or stages are connected, one of which has a gain of $+30$ dB and the other a loss of -10 dB, the net result is $+30 -10$, or $+20$ dB.

The decibel system is also used to compare the *voltage input and output* of an amplifier. (Decibels can be used to express current ratios. However, this is generally not practical in amplifiers.) When voltages (or currents) are involved, the decibel is a function of

$$20 \log \frac{\text{output voltage}}{\text{input voltage}}, \qquad 20 \log \frac{\text{output current}}{\text{input current}}.$$

The ratio of two voltages (or currents) is taken, and the logarithm of this ratio is multiplied by 20.

It is important to note that although power ratios are independent of source and load impedance values *voltage and current ratios in these equations hold true only when the source and load impedances are equal.*

In circuits where input and output impedances differ, voltage and current ratios are calculated as follows:

$$20 \log \frac{E_1 \sqrt{R_2}}{E_2 \sqrt{R_1}}, \qquad 20 \log \frac{I_1 \sqrt{R_1}}{I_2 \sqrt{R_2}}$$

where R_1 is the source impedance and R_2 the load impedance. ($E_1 \sqrt{R_2}$ and $I_1 \sqrt{R_1}$ are always higher in value than $E_2 \sqrt{R_1}$ and $I_2 \sqrt{R_2}$.)

As is true for the power relationship, if the voltage output is greater than the input, there is a decibel gain (+dB). If the output is less than the input, there is a voltage loss (−dB).

Note that doubling the voltage produces a gain of +6 dB. Conversely, if the voltage is halved, there is a loss of −6 dB. To get the net effect of several voltage amplifiers working together, add the decibel gains (or losses) of each.

When an amplifier has a power gain of +20 dB, this has no meaning in actual power output. Instead, it means that the power output is 100 times as great as the power input. For this reason, decibels are often used with specific reference levels.

The most common reference levels in use are the *volume unit*, or VU, and the *decibel meter*, or dBm.

When the volume unit is used, it is assumed that the zero level is equal to 0.001 W (1 mW) across a 600-Ω impedance. Therefore,

$$VU = 10 \log \frac{P_2}{0.001} = 10 \log \frac{P_2}{10^{-3}} = 10 \log 10^3 P_2$$

where P_2 is the output power.

Both the decibel meter and volume unit have the same zero level base. A decibel-meter scale is (generally) used when the signal is a sinewave (normally 1 kHz), whereas the volume unit is used for complex audio waveforms.

1-11. FIELD-EFFECT TRANSISTOR AMPLIFIERS

Before going into specific considerations for field-effect-transistor (FET) amplifier circuits, we shall discuss basic FET operating characteristics in this chapter. Such a review is necessary, since FET characteristics are unique when compared to any other type of transistor. (For example, a FET is often biased at the *zero temperature coefficient point* when used as an amplifier. This is an operating point where the FET drain–source current will not vary with temperature.) Likewise, the characteristics shown on FET datasheets do not correspond to those of conventional two-junction (*NPN–PNP*) transistors. It is necessary to analyze these datasheet characteristics, as they apply to amplifier circuits. FET types and operating modes are discussed fully in the author's *Practical Semiconductor Databook for Electronic Engineers and Technicians* (Prentice-Hall, Inc., Englewood Cliffs, N.J., 1970). The following paragraphs summarize this information.

1-11.1. Advantages and Disadvantages of FETs

The FET has several advantages over a conventional transistor in amplifier applications. The FET is relatively free of noise, and is more resistant to the degrading effects of nuclear radiation because carrier lifetime effects are comparatively unimportant to FET operation. The FET is inherently more resistant to burnout than a conventional two-junction (or bipolar) transistor.

There are additional advantages for certain amplifier considerations. For example, the high input impedance (typically several megohms) is very useful in impedance transformations and where the amplifier must be matched to a high input impedance signal source. Since the FET is a voltage-controlled device, in contrast to current-controlled two-junction transistors, the FET can readily be "self-biased." This frequently makes for a more simple circuit than is possible with a bipolar transistor. The FET also has a nonlinear region of operation, but this is generally of small value for amplifiers, except where automatic gain control is used.

The junction field-effect transistor (or JFET, Sec. 1-11.2) has a very high output resistance, making it useful as a constant current source. Figure 1-21 illustrates some JFET key parameters and their relative magnitudes as compared to vacuum tubes, bipolar transistors, and MOSFETs.

Characteristics	Vacuum Tube	JFET	MOSFET	Bipolar
Input impedance	High	High	Very high	Low
Noise	Low	Low	Unpredictable	Low
Warm-up time	Long	Short	Short	Short
Size	Large	Small	Small	Small
Power consumption	Large	Small	Small	Small
Aging	Noticeable	Not noticeable	Noticeable	Not noticeable
Bias voltage temp coefficient	Low, not predictable	Low predictable	High, not predictable	Low predictable
Typical gate/grid current	1 nA	0.1 nA	10 pA	—
Gate/grid current change with temp.	High unpredictable	Medium predictable	Low unpredictable	—
Reliability	Low	High	High	High
Sensitivity to overload	Very good	Good	Poor	Good

Courtesy Motorola

Fig. 1-21   Comparison of FET characteristics

When compared with bipolar transistors, the chief shortcoming of the FET is its relatively small gain–bandwidth product. Although the JFET is free from carrier-transit-time limitations, parasitic capacitances limit the FET at higher frequencies. This aspect is discussed further in Chapter 3.

1-11.2. Types of FETs and Modes of Operation

There are two types of FETs in common use, junction (JFET) and metal oxide silicon (MOSFET). As the names imply, a JFET uses the characteristics of a reverse-biased *junction* to control the drain–source current, whereas with a MOSFET the gate is a metal deposit on an oxide layer. The

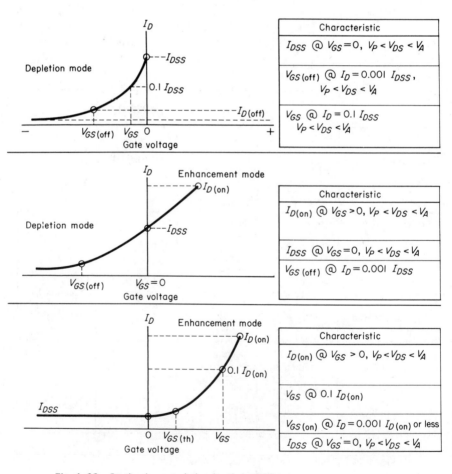

Fig. 1-22   Static characteristics for three FET types

gate is insulated from the source and drain. Because of this *insulated gate*, the MOSFET is often referred to as an insulated-gate FET, or IGFET. The terms MOSFET and IGFET are used interchangeably throughout this book.

Both JFETs and MOSFETs operate on the principle of a *channel* current controlled by an electric field. The control mechanism for the two are different, resulting in considerably different characteristics. The main difference between the two is in the gate characteristics. The input of the JFET acts like a reverse-biased diode, whereas the input of a MOSFET is similar to a small capacitor.

In addition to the two basic types, there are two fundamental modes of operation for FETs—*depletion* and *enhancement*. These modes are illustrated

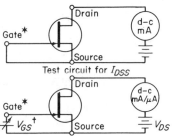

Description
Zero–gate–voltage drain current Represents maximum drain current
Gate voltage necessary to reduce I_D to some specified negligible value at the recommended V_{DS}, i.e. cutoff
Gate voltage for a specified value of I_D between I_{DSS} and I_{DS} at cutoff–normally 0.1 I_{DSS}

Test circuit for I_{DSS}

Test circuit for V_{GS} and $V_{GS\,(off)}$

*Gates internally connected
†Adjust for desired I_D

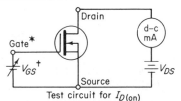

Description
An arbitrary current value (usually near max rated current) that locates a point in the enhancement operating mode
Zero–gate–voltage drain current
Voltage necessary to reduce I_D to some specified negligible value at the recomended V_{DS}, i.e. cutoff

Test circuit for $I_{D\,(on)}$

*Gates internally connected
†Adjust for desired I_D, normally near max-rated I_D

Test circuits for I_{DSS} and $V_{GSS\,(off)}$ same as type A

Description
An arbitrary current value (usually near max rated current) that locates a point in the enhancement operating mode
Gate-source voltage for a specified drain current of 0.1
Gate cutoff or turn-on voltage
Leakage drain current

$I_{D\,(on)}$ test circuit same as for type B

V_{GS} test circuit same as for $I_{D\,(on)}$

$V_{GS\,(th)}$ test circuit same as $V_{GS\,(off)}$ in type A, except reverse V_{GS} battery polarity

I_{DSS} test circuit same as for type A

Courtesy Motorola

in Fig. 1-22, which shows the transfer characteristics and basic test circuits for each mode.

In the depletion mode, maximum drain current (I_{DSS}) flows when the gate–source voltage (V_{GS}) is zero, and decreases for increasing V_{GS}.

Enhancement mode is just the opposite in that minimum drain current flows at $V_{GS} = 0$. With enhancement mode, the drain current increases with increasing V_{GS}.

Field-effect transistors designated as type A operate in the depletion mode only. Type B FETs operate in both depletion and enhancement modes. Type C FETs operate in the enhancement mode only.

The test circuits in Fig. 1-22 show the biasing for N-channel FETs. Note that V_{DS} is always positive for the three N-channel types. In the useful range of operation, V_{GS} is negative for a type A FET, positive for type C, and either polarity for type B. For a P-channel FET, all polarities must be reversed.

1-11.3. Basic FET Operating Regions

The FET has three distinct characteristic regions, only two of which are operational. Figure 1-23a, the output transfer characteristics, illustrates the different regions. Below the pinch-off voltage V_P, the FET operates in the *ohmic* or *resistance region*. The ohmic region is not generally used for amplifiers, except in special cases. Above the pinch-off voltage up to the drain–source breakdown voltage $V_{(BR)DSS}$, the FET operates in the *constant-current region*, which is the region most used for amplifier circuits. The third region, above the breakdown voltage, is the *avalanche region*, where the FET is not operated in practical circuits.

The drain–source resistance r_{DS} at any point on these curves is given by the slope of the curve at that point. Above the pinch-off voltage, changes in the drain–source voltage V_{DS} result in small changes in drain current I_D. This produces a very high drain–source resistance, and is characteristic of a constant-current source. Also, the actual operating drain current is variable, and is dependent on the gate–source voltage. This results in a voltage-controlled current source.

The I_D–V_{GS} curve, shown in Fig. 1-23b and found on a typical FET datasheet, illustrates how the drain current varies with changes in gate–source voltage. For depletion FETs, the drain current decreases as the gate–source voltage is increased. For enhancement devices, the drain current is enhanced or increased as the gate–source voltage is increased.

If the FET is operated with a drain–source voltage below the pinch-off voltage, the slope of the curves varies considerably as the gate–source voltage is varied. This is shown in Fig. 1-23c. Since the slope varies, the drain–source resistance varies. This is considered as operation in the ohmic region.

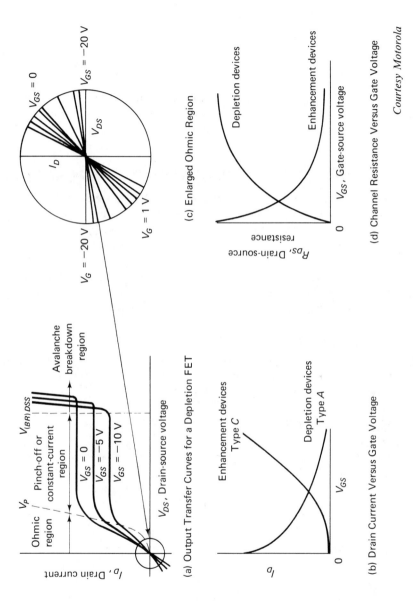

(a) Output Transfer Curves for a Depletion FET

(b) Drain Current Versus Gate Voltage

(c) Enlarged Ohmic Region

(d) Channel Resistance Versus Gate Voltage

Courtesy Motorola

Fig. 1-23   Characteristics of FET operating regions

In effect, the drain–source channel is a voltage-variable or voltage-controlled resistor. As shown in Fig. 1-23d, the drain–source resistance decreases with increasing gate–source voltage for enhancement FETs. For depletion FETs, the converse is true.

Note that the curves near the origin (Fig. 1-23c) are relatively symmetrical. This means that ac as well as dc signals can be handled. In other words, the drain–source channel is bilateral, not unilateral.

Note that a FET is generally operated in the pinch-off region for linear devices (which includes most amplifiers), whereas the ohmic region is used only for voltage-variable resistor applications.

1-11.4. Zero-Temperature-Coefficient Point

An important characteristic of all FETs is their ability to operate at a zero-temperature-coefficient point ($0TC$). This means that if the gate–source is biased at a specific voltage and is held constant, the drain current will not vary with changes in temperature. This characteristic makes for very stable amplifier circuits.

The I_D–V_{GS} curves of Fig. 1-24 show that the various curves at different temperatures intersect at a common point. If the FET is operated at this value of I_D and V_{GS}, (shown as I_{DZ} and V_{GSZ}), zero-temperature-coefficient ($0TC$) operation will result.

The $0TC$ point varies from one FET to another, and is dependent upon I_{DSS}, the zero gate voltage drain current, and V_P. The equations shown in Fig. 1-24 provide good approximations of the zero-temperature-coefficient point. For example, if the pinch-off voltage V_P is 1 V, the $0TC$ mode is obtained if the gate–source voltage V_{GS} is 0.37 V ($1 - 0.63 = 0.37$).

Typically, JFETs show the $0TC$ characteristic over a wide range of tem-

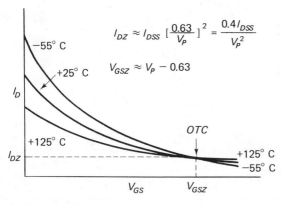

$$I_{DZ} \approx I_{DSS} \left[\frac{0.63}{V_P} \right]^2 = \frac{0.4 I_{DSS}}{V_P^2}$$

$$V_{GSZ} \approx V_P - 0.63$$

Courtesy Motorola

Fig. 1-24   Zero temperature co-efficient of FETs

peratures, approximately 150°C. MOSFETs are limited to a much narrower range, approximately 50°C.

It is sometimes assumed that the forward transadmittance (Y_{fs} or Y_{21}) of the FET does not vary with temperature, particularly if the FET is biased at the $0TC$ point. However, this is not correct. The transadmittance of the FET is the slope of the I_D–V_{GS} curve. The curve of Fig. 1-24 shows that the slope varies with temperature at every point on the curve.

Figures 1-25 and 1-26 illustrate the temperature coefficients for a typical JFET.

Keep in mind that *it is not always practical to operate a FET at the zero-temperature-coefficient point.* For example, assume that the required V_{GS} to produce $0TC$ is 0.37 V, and the FET is to operate as an amplifier with 0.5-V input signals. A part of the input signal will be clipped. Or, assume that the circuit is to be self-biased with a source resistor (Sec. 1-11.5). An increase in bias resistance to produce $0TC$ could reduce gain.

Practical methods for finding $0TC$ of FETs. The values of I_D and V_{GS} that produce $0TC$ can be found using datasheet curves or by equations, as shown in Fig. 1-24. However, these values are typical approximations.

A more practical method for determining I_{DZ} requires a soldering tool, coolant (a can of freon), and a curve tracer (such as the Tektronix Type 575). By placing a 1000-Ω resistor across the base and emitter terminals of the curve-tracer test socket, the constant-current base drive is converted to a relatively constant voltage for driving the FET gate. The curve tracer is then adjusted to display the I_D–V_{DS} output family of curves (Fig. 1-23a). By alternately bringing the soldering tool near the FET, and spraying the FET

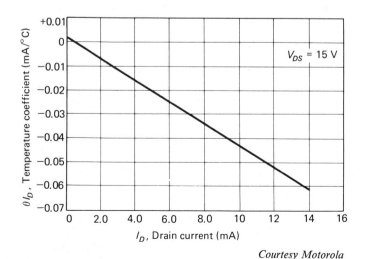

Courtesy Motorola

Fig. 1-25   Drain-current temperature coefficient versus drain current

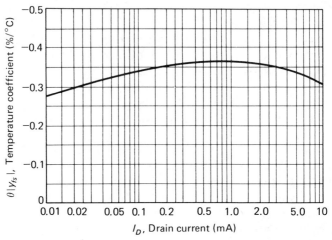

Fig. 1-26 Forward-transadmittance temperature coefficient versus drain current

with freon, the *voltage step of V_{GS} that remains motionless in the presence of temperature changes* can be observed on the curve tracer. The I_D at this voltage step is I_{DZ}.

Typically, FETs with an I_{DSS} of about 10 to 20 mA will have an I_{DZ} of about 0.5 mA. Usually, I_{DZ} increases as I_{DSS} increases (but not always, and not in proportion). For example, the I_{DZ} of 300-mA FETs is often on the order of 1 mA.

1-11.5. Bias Methods for FETs

In linear amplifier circuit applications, the FET is biased by an external supply, by self-bias, or by a combination of these two techniques. This applies to all FETs, whether biased at the $0TC$ point or at some other operating point.

Figure 1-27 shows the familiar common source–drain characteristic curves of a JFET (as they might appear on a typical curve tracer). For a constant level of drain–source voltage V_{DS}, drain current I_D can be plotted versus gate–source voltage V_{GS} as shown in Fig. 1-27b. This latter curve is generally referred to as a transfer characteristic.

From a practical standpoint, the curves of Fig. 1-27 show the amount of current that flows through the FET for a given gate–source voltage. For example, either curve shows that if a -1-V bias is applied between gate and source, approximately 10 mA will flow. If the supply voltage (drain–source

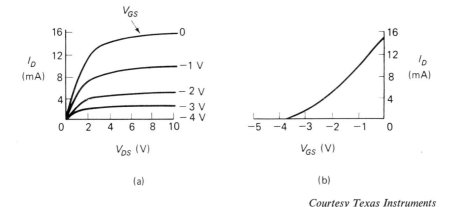

Courtesy Texas Instruments

Fig. 1-27   FET characteristic curves; common-source, drain characteristics (a); transfer characteristics (b)

voltage) is 10 V, and a 500-Ω resistor is connected between the drain and
supply, there is a 5-V drop across the resistor. Of course, this reduces the
drain–source voltage down to 5 V, and possibly changes the characteristics.
(In the example of Fig. 1-27a, there is very little change in characteristics.)

The following paragraphs show how similar curves can be used to find the
correct value of bias for a given quiescent (no-signal) operating point for a
FET. Keep in mind that the following paragraphs provide basic or theoretical
methods for finding bias values. In Chapter 2 we shall discuss step-by-step
procedures for finding FET bias values, using actual datasheet information.

External bias for FETs. Figure 1-28 shows the FET biased by an external
voltage source. The input portion of this circuit is redrawn in Fig. 1-28b so
that a graphical analysis may be used to determine the quiescent drain current.
The graphical analysis consists of plotting the (voltage-current) V–I characteristics looking into the source terminal, and the V–I characteristics looking
into the supply-voltage terminal. When the source terminal is connected to
V_{OUT}, currents I_1 and I_2 are equal. Consequently, the quiescent level of source
(and drain) current is determined by the point of intersection of the V–I
plots. For example, approximately 9 mA of current (I_{D1}) flows when V_{OUT}
(now the gate–source voltage) is 1.5 V. Figure 1-28c shows this graphical
analysis.

Note that two I_2 curves are given. These two curves illustrate a typical
spread of transfer characteristics among FETs of the same family or type.
For example, with the same V_{OUT} of 1.5 V, the lower I_2 curve shows that the
current is approximately 1 mA (I_{D2}). Thus, if it is desirable that I_D be maintained at some level, a form of self-bias must be used.

Self-bias for FETs. Self-bias of a FET amplifier will reduce (but not

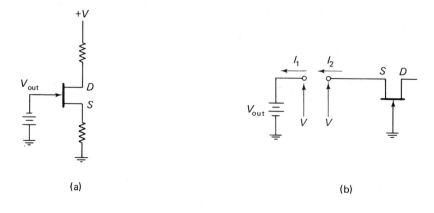

(a)   (b)

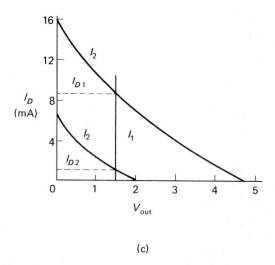

(c)

Courtesy Texas Instruments

Fig. 1-28   FET biased by external supply voltage (a) ; equivalent circuit
for bias network (b) ; graphical analysis (c)

eliminate) variation in quiescent levels of I_D. Figure 1-29a shows the use of
a source resistor R_S to develop a gate–source reverse-bias voltage. As I_D
increases, V_{GS} becomes more negative, thus tending to prevent an increase
in I_D. The input portion of the self-bias circuit is redrawn in Fig. 1-29b and is
analyzed graphically in Fig. 1-29c. The V–I characteristics for the resistor is a
straight line, having a slope equal to the reciprocal of R_S, or $1/R_S$. This line
intersects the two I_2 plots at points P_1 and P_2. Quiescent levels of I_D are
somewhat closer for the circuit of Fig. 1-29 than for the circuit of Fig. 1-28.

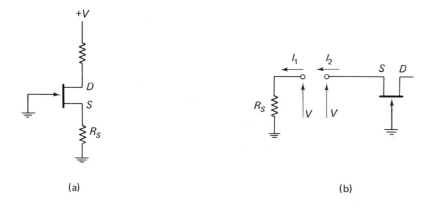

(a)   (b)

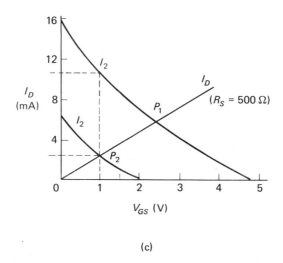

(c)

Courtesy Texas Instruments

Fig. 1-29   Self-bias for FET (a) ; equivalent circuit for bias network (b) ;
graphical analysis (c)

However, it is still possible to have a wide variation in I_D. For example, I_D could vary from about 2.5 to 11 mA with 1 V of V_{GS}. Thus, if it is quite essential that I_D be maintained within narrow limits, to achieve a given amplifier characteristic, then both fixed bias and self-bias must be used.

Combined fixed bias and self-bias. Assume that it is desired to limit I_D to a range between 3 and 7 mA (points A and B of Fig. 1-30). Although this cannot be accomplished with self-bias alone (Fig. 1-29), Fig. 1-30 shows two circuits that will restrict I_D to the desired range of values. A representation

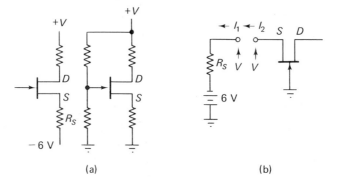

(a)   (b)

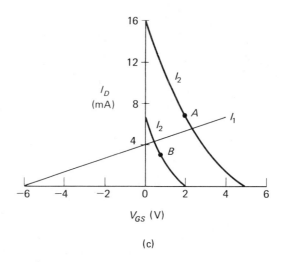

(c)

Courtesy Texas Instruments

Fig. 1-30   Circuits incorporating a combination of fixed and self-bias
(a) ; equivalent circuit for either bias network (b) ; graphical analysis (c)

for the input portion of these two combination circuits is given in Fig. 1-30b.
The graphical analysis in Fig. 1-30c shows that, by proper selection of power
supply and resistance values, I_D can be bounded by points A and B. The
step-by-step procedures for finding the values are described in Chapter 2.

Constant-current bias. Figure 1-31a shows a bias network for a differential
amplifier. Transistor Q_3 is biased as a constant-current generator in order to
improve the common-mode rejection ratio of the differential amplifier (Q_1–
Q_2). (Differential amplifiers are discussed in Chapter 5.) Figure 1-31b shows
the bias network, and Fig. 1-31c shows a graphical analysis of the bias cir-

cuit. Note that the two transfer curves are shifted to the left by the amount of negative voltage appearing at the gate terminal. For example, the top curve of Fig. 1-30 shows a maximum gate voltage of about -10 V. This same curve is shifted 15 V to the left in Fig. 1-31, and intersects the voltage axis at about 10 V. The -15-V voltage is developed across resistors R_1 and R_2 in Fig. 1-31.

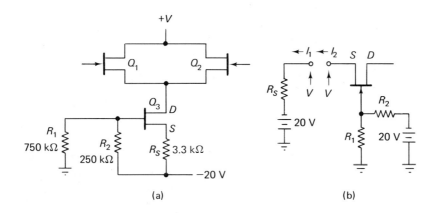

(a)

(b)

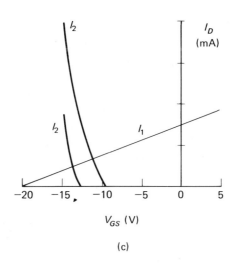

(c)

Courtesy Texas Instruments

Fig. 1-31   Bias circuit for differential amplifier (a) ; equivalent circuit for bias network to Q_3 (b) ; graphical analysis (c)

2. AUDIO-FREQUENCY
AMPLIFIERS

The average human ear is able to hear sounds ranging up to about 15,000 *vibrations per second.* Thus, corresponding electrical signals (ranging in *frequency* up to about 15 kHz) are known as *audio-frequency* (or AF) signals. However, the frequency range from 15 to 20 kHz is also considered as part of the audio range, since some humans can hear sounds at that frequency. For the purposes of this book, we shall define audio frequencies as any frequency up to about 20 kHz.

2-1. FREQUENCY LIMITATIONS OF AMPLIFIER COMPONENTS

Were it not for reactance, a transistor (by itself) should be capable of operating at any frequency from zero (direct current) on up. That is, the top frequency limit would be set only by the transit time of electrons across the transistor junctions. However, there are limitations placed on the operating frequency of any amplifier by the transistor characteristics. Likewise, the other components (capacitors, resistors, inductors, etc.) used in the amplifier circuit also limit the operating frequency. In this section, we shall see how each of the components affects operation of amplifiers.

Every electronic component has some impedance, and is thus *frequency sensitive.* That is, the component will not attenuate (or pass) signals of all frequencies equally. Even a simple length of wire has impedance. Wire, being a conductor, has some resistance. If alternating current is passed through the wire, there is some inductive reactance. If the wire is near another conductor (or metal chassis), there is some capacitance between the two conductors, and thus some capacitive reactance. The reactance and resistance combine to produce impedance, which, in turn, varies with frequency.

Of course, many of the impedances presented by components are of little practical concern. On the other hand, certain of the impedances have a very pronounced effect on amplifier design and operation. There are four major components used in audio-amplifier designs: transistors, resistors, capacitors, and inductances (coils and transformers). Let us examine how the impedances and reactances of these components affect audio-amplifier operation.

Transistor frequency limitations. All transistors have some capacitance between the junctions (emitter–base, collector–base). If any of the elements is common or ground, the remaining elements have some capacitance to ground. For example, in a common-emitter amplifier, there is some capacitance from base to ground (across the input) and collector to ground (across the output). Likewise, there is capacitance from collector to base (which forms a feedback path from output to input). This is shown in Fig. 2-1.

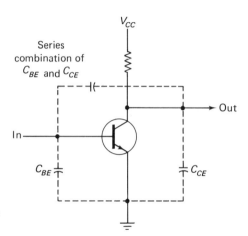

Fig. 2-1   Capacitances associated with transistor elements

Capacitive reactance decreases with an increase in frequency, and vice versa. A capacitance in series with a conductor presents less attenuation to the signal as frequency increases. A capacitance across a conductor (for example, in parallel or shunt from the conductor to ground) acts as a short to signals of increasing frequency. Consider a common-emitter amplifier where transistor capacitances are across the input and output. As frequency increases, the capacitive reactance drops, producing a short across the input and output, and increases attenuation of the signal. At some frequency, the attenuation equals the transistor amplification, so there is no gain. At higher frequencies, the attenuation exceeds amplification, and there is a loss, even though the transistor may still continue to operate.

From a practical standpoint, the input and output capacitances of transistors have little effect at audio frequencies. Most modern transistors will operate well beyond the AF range, and will generally produce equal (or flat) *frequency response*. That is, all signals up to about 20 kHz (or higher) are amplified by the same amount. (Refer to Chapter 7 for a further discussion of frequency response.) However, with most transistors, as signal frequencies increase into the RF range, amplification begins to drop. (Radio-frequency amplifiers are discussed in Chapter 3.)

All transistor have some *inductance* in their leads. This produces inductive reactance in series with the transistor elements. Inductive reactance increases with frequency. In the AF range the inductive reactance is of little concern. However, at radio frequencies the inductive reactance can produce considerable attenuation, as is discussed in Chapter 3.

Resistor frequency limitations. At audio frequencies, resistors offer relatively few problems, since resistors attenuate signals equally. Only at very high frequencies, where the resistor leads and body could produce some kind of reactance, is there any particular concern about frequency limits imposed by resistors. However, resistors do produce *voltage drops*. These voltage drops can be a problem when considering interstage coupling methods (described in Sec. 2-2), and when used in conjunction with coupling capacitors.

Capacitance frequency limitations. Capacitors have three major uses in audio amplifiers: bypass, decoupling, and coupling.

Bypass capacitors are used to provide a signal path around high resistances. For example, if the power supply of an audio amplifier does not have a filter capacitor, or a battery is used, the collector–emitter current must pass through a high resistance. This can impede the ac component of the signal. A bypass capacitor provides a signal path, as shown in Fig. 2-2.

When several stages of amplification are connected, they all join at one point, the common power supply. In multistage amplifiers there is the possibility of one stage feeding back through the power supply to a previous

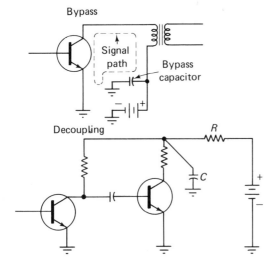

Fig. 2-2 Examples of bypass and decoupling capacitors

stage, thus causing interference with the signal. To avoid this feedback, one or more of these stages may be *decoupled* from the power supply. Figure 2-2 also shows a typical decoupling network. Resistor R is placed in series between the load resistors of the stages and the power supply. Hence, it offers a high-resistance path for the ac signal component to the power supply. Capacitor C, on the other hand, offers a low shunt reactance to this component and thus decouples (bypasses) the component to ground.

Actually, the functions of bypass and decoupling capacitors are the same, and the terms are interchanged. In either case, the main concern is that the *reactance be low at the lowest frequency involved.* This requires a capacitor of a given value, which increases as frequency decreases. For example, assume that the lowest frequency involved in 100 Hz, and the minimum required reactance is 100 Ω. This requires a capacitance value of about 10 μF. If the required frequency is decreased to 10 Hz, the capacitance value must be raised to 160 μF to keep the reactance below 100 Ω.

Coupling capacitors are used at the input and output of each stage to block direct current. For example, if a coupling capacitor is not used between two transistor stages, the collector of the first stage is connected directly to the base of the following stage, and both elements are at the same bias voltage. While it is possible to operate transistors in this way, direct coupling does increase problems, as is described in Chapter 4.

The values of coupling capacitors are dependent upon the *low-frequency limit* at which the amplifier is to operate, and on the resistances with which the capacitors operate. As frequency increases, capacitive reactance decreases and the coupling capacitors become (in effect) a short to the signal. Therefore, the high-frequency limit need not be considered in audio circuits. Figure 2-3 shows how a high-pass filter is formed by coupling capacitors. Capacitor C_1 forms a high-pass RC filter with R_B. Capacitor C_2 forms another high-pass filter with the input resistance of the following stage (or the load). The input voltage is applied across the capacitor and resistor in series. The output voltage is taken across the resistance. The relation of input voltage to output voltage is

$$\text{output voltage} = \text{input voltage} \times \frac{R}{Z}$$

where R is the dc resistance value, and Z is the impedance obtained by the vector combination of series capacitive reactance and dc resistance.

When the reactance drops to approximately one half the resistance, the output will drop to about 90 per cent of the input (or approximately a 1-dB loss). Using the 1-dB loss as the low-frequency cutoff point, the value of C_1 or C_2 can be found by

$$\text{capacitance} = \frac{1}{3.2FR}$$

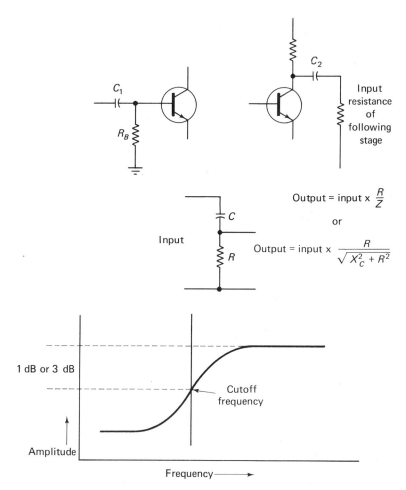

Fig. 2-3 Formation of high-pass (low-cut) RC filter by coupling capacitors and related resistances

where capacitance is in farads, F is the low-frequency limit in hertz, and R is resistance in ohms.

If a 3-dB loss at the low-frequency cutoff point can be tolerated, the value of C_1 or C_2 can be found by

$$\text{capacitance} = \frac{1}{6.2FR}$$

Inductance frequency limits. Both coils and transformers are used in audio amplifiers. As discussed in Sec. 2-2, coils are sometimes used in place of the collector resistor as a load. This permits the collector to be operated at a higher voltage. Likewise, transformers are used for coupling between stages. This provides impedance matching, as discussed in Sec. 2-2.

The inductive reactance of coils and transformers increases with frequency. At the high end of the audio range, the attenuation produced by this increased reactance is usually sufficient to impair operation of the amplifier. At the low end of the AF range, the reactance of a typical transformer drops to a few ohms. This low impedance acts as a short across the line and attenuates the signal. Thus, coils and transformers tend to attenuate signals at both the high end and low ends of the AF range.

Stray impedances. As discussed, any conductor (wiring, terminals, etc.) can have resistance, reactance, and impedance. Thus, care must be used in the routing of wires and placement of terminals to minimize the effects of this *stray* impedance. Likewise, the effects of stray impedances can alter the characteristics of components. A classic example of this is stray capacitance, which is added to the input and output capacitances of transistors. The effects of stray impedances are usually not critical at audio frequencies. However, as discussed in Chapter 3, the effects of stray impedance on amplifiers operating at radio frequencies can be of considerable importance.

2-2. COUPLING METHODS

All amplifiers require some form of coupling. Even a single-stage audio amplifier must be coupled to the input and output devices. If more than one stage is involved, there must be *interstage coupling*. Amplifiers are often classified as to coupling method. For example, the four basic coupling methods are *capacitor* (*or capacitance*) *coupling*, *inductive coupling*, *direct coupling*, and *transformer coupling*. All four methods require resistance and could be called *resistance-coupled* amplifiers. However, the term resistance coupled is generally used to indicate that the amplifier does not have inductances or transformers between stages, and that the input and/or output impedance is formed by a resistance. Capacitor coupling is often called *resistance–capacitance* (or *RC*) coupling.

In this section, we shall see how the different methods affect operation of practical audio amplifiers. Figure 2-4 shows the four coupling methods.

With *direct coupling*, Fig. 2-4a, the collector of one transistor is connected directly to the base of the following transistor. The outstanding characteristic of a direct-coupled amplifier is the ability to amplify direct current and low-frequency signals. Because of the special nature of direct-coupled amplifiers, they are discussed in a separate chapter (Chapter 4).

With *capacitor coupling*, or *RC coupling*, Fig. 2-4b, the coupling is accomplished by means of the load resistor R_{L1} of stage 1, the base resistor R_{B2} of stage 2, and the coupling capacitor C_2. The original signal is acted upon by stage 1 and appears in amplified form as the voltage drop across R_{L1}.

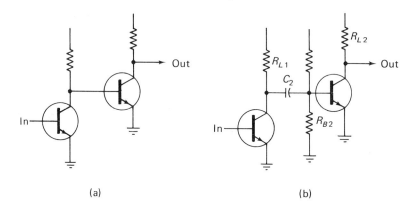

(a)   (b)

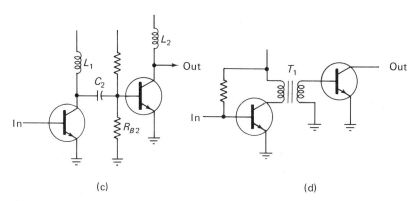

(c)   (d)

Fig. 2-4   Four basic types of coupling used in audio amplifiers

The dc component of the amplified signal is blocked by C_2, which passes the ac component to the input section of stage 2 for further amplification. If necessary, more stages may be coupled to the output of stage 2 for further amplification of the signal. The main advantage of capacitor or RC coupling is that the amplifier will amplify uniformly over nearly the entire audio range, since resistor values are independent of frequency changes. However, as discussed in Sec. 2-1, RC coupled amplifiers do have a low-frequency limit imposed by reactance of the capacitor (which increases as frequency decreases). Reactance–capacitance coupling is also small, light, inexpensive, and produces no magnetic field to interfere with the signal. One disadvantage of the RC coupling method is that the supply voltage is dropped (usually to one half) by the load resistance. Thus, the collectors must operate at a reduced voltage.

With *inductive* or *impedance coupling*, Fig. 2-4c, the load resistors R_{L1} and R_{L2} are replaced by inductors L_1 and L_2. The advantage of impedance coupling over resistance coupling is due to the fact that the ohmic resistance of the load inductor is less than that of the load resistor. Thus, for a power supply of given voltage, there is a higher collector voltage. However, impedance coupling also suffers from a number of disadvantages. Impedance coupling is larger, heavier, and costlier than resistance coupling. To prevent the magnetic field of the inductor from affecting the signal, the inductor turns are wound upon a closed, iron core and usually are shielded further. The main disadvantage of impedance coupling is frequency discrimination.

With impedance coupling at very low frequencies, the gain is low due to the capacitive reactance of the coupling capacitor, just as in the RC-coupled amplifier. The gain increases with frequency, leveling off at the middle frequencies of the audio range. (However, the frequency spread of this level portion is not as great as for the RC amplifier.)

With impedance coupling at very high frequencies, the gain drops because of the increased reactance. Impedance coupling is rarely, if ever, used at frequencies above the audio range.

With *transformer coupling*, Fig. 2-4d, the transformer T_1 serves several purposes. As the fluctuating collector current of the first stage flows through the primary winding of T_1, the current induces an alternating voltage with similar waveform in the secondary of T_1. This voltage forms the input signal to the second stage. Since the secondary of T_1 conveys the ac component of the signal directly to the base of the second stage, there is no need for a coupling capacitor. Also, since the secondary winding furnishes a base return path, there is no need for a base resistance.

Compared to the RC-coupled amplifier, the transformer-coupled amplifier has essentially the same advantages and disadvantages as the impedance-coupled amplifier. The transistor collectors can be operated at higher voltages. The impedances are set by the transformer primary and secondary windings. However, transformers are frequency sensitive (impedance changes with frequency). Therefore, the frequency range of transformer-coupled amplifiers is limited.

The inductances and transformers used in AF work are generally of the iron-core type. If air-core transformers are used at audio frequencies, the inductive reactance (and the impedance) will be so small as to be ineffective. At frequencies above the audio range (or at the high end), the reactance of iron-core inductances and transformers is so large that signals cannot pass (or are greatly attenuated). Therefore, air-core transformers and inductances are used for higher-frequency amplifiers, as is discussed in Chapter 3.

Coupling transformers also provide for impedance matching between stages. Because the transistor is a current-operated device, impedance matching between the output of one stage to the input of the next is desirable for maximum transfer of power. This can be accomplished by making the primary and secondary transformer windings of different impedance. Typically, the input impedance of a transistor stage is less than the output impedance. Thus, the secondary impedance of an interstage transformer is typically lower than the primary impedance. When two common-emitter stages are impedance matched, the overall gain is greater than when identical stages are resistance coupled.

Transformer-coupling is also effective when the final amplifier output must be fed to a low-impedance load. For example, the impedance of a typical loudspeaker is in the order of 4 to 16 Ω, whereas the output impedance of a transistor stage is several hundred (or thousand) ohms. A transformer at the output of an audio amplifier can offset the obviously undesired effects of such a mismatch.

2-2.1. Effects of Coupling on Audio-Amplifier Frequency Response

A simplified *frequency-response graph* or curve is illustrated in Fig. 2-5. (A more comprehensive graph, as well as the procedures for producing such graphs, is discussed in Chapter 7.) The graph of Fig. 2-5 is provided here to illustrate the effects of coupling methods on amplifier frequency response. The response is measured by the gain of the amplifier at various frequencies in the audio range.

Note that the gain falls off at the very low frequencies. In an *RC*-coupled amplifier, this drop in gain (generally referred to as *rolloff*) at the low end is due to the capacitive reactance of the coupling capacitor. Since the coupling capacitor is between the output of the first stage and the input to the second stage, the signal is attenuated by the voltage drop across the capacitor. Hence, the lower the frequency, the larger the capacitive reactance, and the smaller

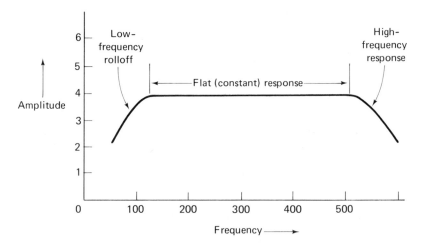

Fig. 2-5 Simplified frequency-response graph

is the signal input to the second stage. In impedance-coupled or transformer-coupled amplifiers, the low-frequency rolloff is caused by the very low inductive reactance, which acts as a short across the signal path. In effect, the low reactance bypasses some of the signal to ground.

As shown in Fig. 2-5, the gain also falls off at the higher frequencies. In *RC*-coupled amplifiers, this high-frequency rolloff is due to the output capacitance of the first stage, the input capacitance of the second stage, and the stray capacitance furnished by the coupling network. These capacitances act to bypass some of the signal to ground. The higher the frequency, the smaller the capacitive reactance becomes, and the greater the amount of signal so bypassed. Hence, the overall gain falls, At frequencies between these two extremes, the gain remains fairly constant. In impedance-coupled or transformer-coupled amplifiers, the high-frequency rolloff is due to the large inductive reactance that attenuates the signal.

In sum, then, resistance coupling produces the lowest gain, transformer coupling the highest. As a general rule of thumb, three stages of *RC*-coupled amplification will produce approximately the same gain as two stages of *comparable* transformer-coupled amplification. On the other hand, *RC*-coupling produces the least frequency distortion of the signal. Transformer coupling has the added advantage of providing an impedance match at the amplifier input and output.

2-3. HOW TRANSISTOR RATINGS AFFECT AMPLIFIER CHARACTERISTICS

The characteristics of solid-state amplifier circuits are directly related to the ratings or capabilities of the transistors involved. For example, the current gain of an individual amplifier stage can be no greater than the maximum possible current gain of the transistor in that stage. Whether the problem is one of amplifier design or amplifier test and troubleshooting, it is always helpful (and often necessary) to know the characteristics of the transistors involved. These characteristics are given in transistor manufacturer's *datasheets*.

Most of the amplifier-related characteristics for a particular transistor can be obtained from the datasheet. There are some exceptions to this rule. For extreme high-frequency work (Chapter 3) and in digital work where switching characteristics are of particular importance, it may be necessary to test the transistor under simulated operating conditions. In any event, it is always necessary to interpret datasheets. Each manufacturer has its own system of datasheets. It would be impractical to discuss all datasheet formats here. Instead, we shall discuss typical information found on datasheets, and see how this information affects amplifier circuits.

The subject of transistor ratings is discussed thoroughly in the author's *Practical Semiconductor Databook for Electronic Engineers and Technicians* (Prentice-Hall, Inc., Englewood Cliffs, N.J., 1970); and *Handbook of Simplified Solid-State Circuit Design* (Prentice-Hall, Inc., Englewood Cliffs, N. J., 1971). The following is a summary of these discussions, with particular emphasis on the relationship between transistor ratings and amplifier circuit operation.

2-3.1. Two-Junction Transistor Ratings

Most two-junction transistor datasheets list certain *absolute maximum ratings* that apply to voltage, current, power, and temperature range.

The *maximum collector voltage* is usually listed as V_{CBO}, which is a test voltage, rather than an operating design voltage. (V_{CBO} usually indicates collector-base breakdown voltage, with the emitter circuit open. However, transistors do not operate in this way in amplifiers. For a common-emitter amplifier, V_{CEO} or BV_{CEO} is the dominant breakdown voltage which is always less than V_{CBO}.)

Except for RF amplifiers, most transistors are operated with their collector at some voltage value *less* than the source voltage (V_{CC}). For example, in a typical class A amplifier circuit, the collector voltage is one half the source voltage at the normal operating point. However, the collector voltage will rise to or near the source voltage when the transistor is at or near cutoff. For this reason, the collector of any transistor amplifier should never be connected to a source higher than the maximum voltage rating, even through a resistance.

If an electronic power supply is used for a transistor amplifier, always allow for some variation in source voltage. Of course, a battery power source will not deliver more than its rated voltage.

Maximum voltage is affected by temperature, and is listed on datasheets at some particular temperature, usually 25°C. Typically, breakdown will occur at a lower voltage when temperature is increased.

The *maximum base–emitter voltage* is usually listed as V_{EBO}. Again, this is a test voltage rather than an operating or design voltage. Typically, the voltage drop across the base–emitter junction is 0.2 to 0.3 V for germanium, and 0.5 to 0.6 V for silicon transistors. Lower base–emitter voltages (typical of class B and C amplifiers) produce less current drain, and lower no-signal (Q-point) power dissipation. Higher base–emitter voltages may result in operation on a more linear portion of the transistor characteristics (typical of class A amplifiers). In practical amplifiers of all classes, it is often necessary to select a bias (base–emitter voltage) on the basis of input signal, rather than on some arbitrary point of the transistor's characteristic curve. Keep in mind that the input signal to an amplifier stage can come from an external source or a previous stage, or can be a form of feedback. Thus, when analyzing any amplifier stage, always consider any input signal that may be applied to the base–emitter junction, in addition to the normal operating bias.

Collector current, usually listed as I_C, will increase with temperature. For that reason, it is unwise to operate any transistor at or near its maximum current rating. Of course, if you could be absolutely certain that the transistor is dissipating any temperature increases (a practical impossibility), the circuit could operate near the maximum current rating.

In practical amplifiers it is the power dissipated in the collector circuit (rather than a given current) that is of major concern. For example, assume that the collector is operated at 50 V and 25 mA. This will result in a power dissipation of over 1 W, somewhat above the typical small-signal transistors found in amplifier stages.

The power dissipation capabilities of a transistor in any amplifier are closely associated with the temperature range. For example, a typical power dissipation of 150 mW at 25°C must be *derated* to 50 mW at 125°C, indicating a derating of 1 mW for each degree (°C) increase in ambient temperature.

Small-signal characteristics, such as current transfer ratio, input impedance, reverse voltage transfer ratio, output admittance, power gain, and noise figure, can be defined as those where the ac signal is small compared to the dc bias. For example, h_{fe} or *forward current transfer ratio* (also known as ac beta or dynamic beta) is properly measured by noting the change in collector alternating current for a given change in base alternating current with the collector voltage held constant and without regard to static base and collector currents.

Small-signal characteristics do not provide a truly sound basis for practical amplifier circuit analysis. As discussed in Chapter 1, and in other related chapters, the performance of a transistor in a working amplifier circuit can be set by the circuit component values (within obvious limits, of course).

The small-signal characteristics found on datasheets are sometimes confusing. For example, not all manufacturers list the same small-signal characteristics on their datasheets. To further complicate matters, manufacturers call the same characteristic by different names (or even use the same name to identify different characteristics). In any event, the small-signal characteristics listed in datasheets are based on a set of fixed operating conditions. If the conditions change (as they must in any practical amplifier), the characteristics will change. For example, beta changes drastically with temperature, frequency, and operating point. With this in mind, use small-signal characteristics as a starting point for amplifier circuit analysis, not as hard and fast rules.

High-frequency characteristics, such as frequency cutoff, collector-to-base capacity, and power gain, are not especially important in audio amplifiers.

However, as discussed in Chapter 3, such characteristics are essential in analyzing RF networks found in RF amplifiers. The networks used in RF amplifiers provide the dual function of frequency selection (tank circuit) and impedance matching between the transistor and a load.

Direct-current characteristics, such as collector breakdown voltage, collector cutoff current, and collector saturation resistance, are important in the design of basic amplifier bias circuits (Sec. 1-7). However, the dc ratings of a transistor do not have too critical an effect on the operation of audio amplifiers. The dc characteristics shown on datasheets are primarily test values, rather than design or operating parameters. The important points to remember regarding such dc characteristics are that they serve as starting points for bias circuit design, and they will change with temperature in any practical amplifier circuit.

Switching characteristics, such as delay time, rise time, storage time, and fall time, are important in amplifiers used to handle *pulse* signals and in switching amplifiers. The switching time of a transistor (which includes the sum of all four times) is defined in the test and troubleshooting information of Chapter 7.

The time factors of delay, storage, rise, and fall determine the operating limits for switching amplifiers. For example, if a transistor amplifier stage with a 20-ns rise time is used to pass a 15-ns pulse, the pulse would be hopelessly distorted. Likewise, if the 1.5-μs delay were added to a 1-μs pulse, an absolute minimum of 2.5 μs would be required before the next pulse could occur. This would mean a maximum pulse repetition frequency (prf) of 400 kHz ($1/2.5^{-6} = 400,000$ Hz).

2-3.1.1. Determining Transistor Rating at Different Frequencies

One of the great weaknesses in using transistor datasheet ratings to analyze amplifier circuits is that datasheets specify most ratings at some given frequency. In practical applications it is convenient to know the ratings over the entire frequency range of the amplifier. Transistor gain (either alpha or beta) is a classic example of this requirement. If a common-emitter transistor circuit is to provide a given gain operating from 10 Hz to 20 kHz, the transistor beta must be greater than the required circuit gain across the same frequency range.

Alpha or beta (or both) are found on most datasheets for transistors used in the audio range. Alpha is often listed as h_{fb}, or common-base ac forward current gain. Beta is listed as h_{fe}, or common-emitter ac forward current gain.

Common-base ratings. h_{fbo} (the value of h_{fb} at 1 kHz) will remain constant as frequency is increased, until a top limit is reached. After the top limit, h_{fb} begins to drop rapidly. The frequency at which a significant drop in h_{fb}

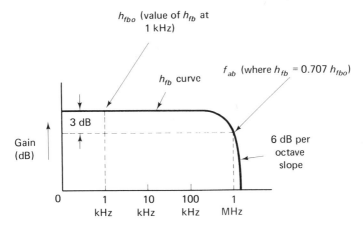

Fig. 2-6   Curve of h_{fb} versus frequency for a transistor with an f_{ab} of 1 MHz

occurs provides a basis for comparison of the expected frequency performance (amplification) of different transistors. This frequency is known as f_{ab} and is defined as that frequency at which h_{fb} is 3 dB below h_{fbo}.

A curve of h_{fb} versus frequency for a transistor with an f_{ab} of 1 MHz is shown in Fig. 2-6. This curve has the following significant characteristics:

At frequencies below f_{ab}, h_{fb} is nearly constant and approximate to h_{fbo}. h_{fb} begins to decrease significantly in the region of f_{ab}.

Above f_{ab}, the rate of decrease of h_{fb} (with increasing frequency) approaches 6 dB per octave. (The term 6 dB per octave means that the gain drops by 6 dB each time frequency is doubled. This is the same as a 20-dB drop each time the frequency is increased by a factor of 10, which is referred to as 20 dB per decade.)

The curve of common-base current gain versus frequency for any transistor has the same characteristics, and the same general appearance as the curve of Fig. 2-6, although not necessarily the same frequency range.

Common-emitter ratings. The common-emitter rating that corresponds to f_{ab} is f_{ae}, the common-emitter current gain cutoff frequency. This f_{ae} is the frequency at which h_{fe} (beta) has decreased 3 dB below h_{feo} (the value of h_{fe} at 1 kHz). A typical curve of h_{fe} versus frequency for a transistor with an f_{ae} of 100 kHz is shown in Fig. 2-7. The curve of Fig. 2-7 has the same significant characteristics described for Fig. 2-6. That is, h_{fe} is considered to be decreasing at a rate of 6 dB per octave at f_{ae}.

These characteristics allow such a curve to be constructed for a particular transistor by knowing only h_{feo} and f_{ae}. With the curve constructed, h_{fe} (at any frequency) can be determined. Furthermore, if f_{ae} is not known, a curve could also be constructed if h_{feo} and h_{fe} *at any frequency above* f_{ae} were

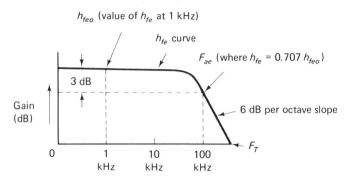

Fig. 2-7 Curve of h_{fe} versus frequency for a transistor with an f_{ae} of 100 kHz

known. Thus, to find h_{fe} at any frequency, it is necessary to know only h_{feo} (which is available on most datasheets) and either f_{ae} or h_{fe} at some frequency greater than f_{ae}.

Gain–bandwidth product, or f_T, is sometimes specified on datasheets instead of f_{ae}. This f_T is the frequency at which gain drops to unity (0 dB; no amplification, but no loss). f_T can be approximated when f_{ae} is multiplied by h_{feo} ($f_T = f_{ae} \times h_{feo}$).

On those datasheets where h_{fe} is specified at some frequency greater than f_{ae}, f_T can be approximated when the specified frequency is multiplied by the specified h_{fe}.

It should be noted that f_T is a common-emitter rating, and should not be used with common-base calculations.

It should also be noted that common emitter f_T is approximately equal to the common-base parameter of f_{ab}. Usually, f_T is slightly less than f_{ab}.

Maximum operating frequency. Although common-emitter current gain is equal to 1 at f_T, there may still be considerable power gain at f_T due to different input and output impedance levels (as described in Sec. 1-4.1). Thus, f_T is not necessarily the highest useful frequency for a transistor operating as a common-emitter amplifier. An additional rating, the maximum frequency of oscillation (f_{max}), is found.

The term f_{max} is the frequency at which *common-emitter power gain* is equal to 1. A plot of common-emitter power gain versus frequency has the same characteristics as the voltage-gain plot (Fig. 2-7).

f_{max} may be found by measuring power gain at some frequency on the 6 dB per octave (slope) portion of the *power gain versus frequency*, and multiplying the *square root of the power gain* (in magnitude) by the frequency of measurement.

The problem here is that datasheets do not always specify if the power-gain figure is on the slope of the power gain versus frequency curve. However, there are some clues that can be used to estimate the location of the power-gain figure on the curve.

If two power-gain figures are given, the high-frequency figure can be considered to be on the slope, and can be used to find f_{max}. For example, if the datasheet shows a power gain of 35 dB (magnitude of 50) at 1 kHz and 17 dB at 5 MHz, the 17-dB figure (5 MHz) can be used to find f_{max} by

$$f_{max} = \text{frequency } \sqrt{\text{power gain in magnitude}}$$
$$f_{max} = 5\sqrt{50}$$
$$f_{max} = 35 \text{ MHz} \quad \text{(approximately)}$$

If a power-gain figure is given for a frequency higher than f_{ae}, the gain figure can be considered as being on the slope, and can be used to find f_{max}.

Conversion between ratings. One problem with datasheets is the mixing of amplification or gain-frequency ratings. For example, the datasheet of a 2N337 shows a "typical" current transfer ratio h_{fe} (or beta) of 55 (magnitude), and a "typical" alpha cutoff frequency of 30 MHz. (The datasheet lists alpha cutoff frequency as f_{hfb}, rather than f_{ab}.) This means that if the transistor is used as a common-emitter amplifier the gain is 55, up to some unknown frequency. Or, if the transistor is used as a common-base amplifier, the gain will drop 3 dB from some unknown level at 30 MHz.

Since the mixing of ratings is not uncommon on most two-junction transistor datasheets, it is necessary to convert from one rating to another. Since the majority of amplifiers are common emitter, it is usually necessary to convert from common base to common emitter, but the reverse can also be true. The following rules summarize the conversion process:

$$\text{beta} = \frac{\text{alpha}}{1 - \text{alpha}}, \qquad h_{feo} = \frac{h_{feo}}{1 + h_{feo}}$$

$$\text{alpha} = \frac{\text{beta}}{1 + \text{beta}}, \qquad f_{ae} = K(1 - h_{fbo})f_{ab}$$

$$h_{feo} = \frac{h_{fbo}}{1 - h_{fbo}}, \qquad f_{ab} = \frac{f_{ae}}{K(1 - h_{fbo})}$$

where K refers to a phase shift factor and ranges between 0.8 and 0.9 for most transistors (MADT transistors have a K of about 0.6).

To find h_{fe} at a particular frequency:

h_{fe} is approximately equal to h_{feo} when the frequency of interest is less than f_{ae}.

h_{fe} is approximately equal to $0.7h_{feo}$ when the frequency of interest is near f_{ae}.

h_{fb} decreases at a rate of 6 dB per octave and is approximately equal to f_T/frequency when the frequency of interest is above f_{ae}.

h_{fe} is equal to 1 (unity, no amplification, no loss) at f_T.

To find f_T:

When h_{feo} and f_{ae} are given, $f_T = h_{feo} \times f_{ae}$.

When h_{fbo} and f_{ab} are given, $f_T = h_{fbo} \times f_{ab} \times K$.

As an example of how these calculations can be put to a practical use, assume that a transistor is to be used in a common-emitter audio amplifier, and that a gain of magnitude 55 must be no more than 3 dB down at some frequency higher than 20 kHz. That is, f_{ae} must be 20 kHz or greater. Furthermore, assume that the datasheet shows an h_{fbo} of 55 (magnitude), and an f_{ab} (common base) of 200 kHz. Assuming a phase shift constant K of 0.9, is the transistor capable of providing the desired gain at 20 kHz, and higher?

The first step is to convert the common emitter h_{feo} into common base h_{fbo} as follows:

$$h_{fbo} = \frac{h_{feo}}{1 + h_{feo}} \quad \text{or} \quad \frac{55}{56} = 0.982$$

Then find f_{ae} from the known h_{fbo} and f_{ab} as follows:

$$f_{ae} - K(1 - h_{fbo})f_{ab} = 0.9 \times 0.018 \times 200 = 32.4 \text{ kHz}$$

2-3.1.2. Interpreting Transistor Characteristic Curves with Load Lines

Many datasheets show two-junction transistor characteristics by means of curves that are reproductions of displays obtained with an oscilloscope-type curve tracer. The collector voltage–current curves shown in Fig. 2-8 are typical. Such curves are obtained by applying a series of stepped base currents, then sweeping the collector voltage over a given range. Several curves are made in rapid succession at different base currents. Current gain (beta) can be found by noting the *difference* in collector current for a given *change* in base current, while maintaining a fixed collector voltage.

Load lines can be superimposed on the datasheet curves to provide an indication of how the transistor will perform in an amplifier circuit. For example, assume that a collector load of 1000 Ω was used with the transistor displayed in Fig. 2-8. When the collector current reaches 20 mA (base current approximately 0.8 mA), the collector voltage drops to zero. Likewise, when the collector current drops to zero, collector voltage rises to 20 V (base current zero). If a load line is connected between these two extreme points (marked A and B on Fig. 2-8), the instantaneous collector voltage and collector current can be obtained for any base current along the line. Approximate transistor gain can be estimated from the load line. For example, with a base current change from 0.2 to 0.4 mA (a 0.2-mA change), the collector

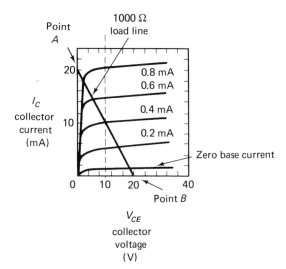

Fig. 2-8 Typical load line drawn
on characteristic curves of common-emitter transistor

current changes from about 4 to 9 mA (a 5-mA change). This represents an
approximate current gain of 25.

Using datasheet curves and load lines to find transistor characteristics has
some drawbacks. The datasheet curves are "typical" for a transistor of a
given type, and represent an average gain (or, in some cases, the *minimum*
gain). Also, as discussed, transistor gain is temperature and frequency
dependent. Therefore, the selection of load values, bias values, and operating
point on the basis of static gain curves is subject to error. If the transistor's
beta shifts, the operating point must shift, requiring different bias and load
values.

As is discussed throughout this book, the problem of variable gain in
transistor amplifiers can be overcome by means of *feedback to stabilize the
gain*. With feedback, amplifier characteristics are primarily dependent upon
the relationship of circuit values, rather than on transistor gain characteristics.

2-3.1.3. How Temperature Affects Transistors in Amplifier
Circuits

There are several basic temperature-related problems when
two-junction transistors are used in amplifier circuits.

Most important, datasheets specify transistor characteristics at a given
temperature. Many of these characteristics will change with temperature.
Since amplifiers rarely operate at the exact temperature shown on the datasheet, it is important to know the characteristics at the actual operating
temperature. To compound the problem, changes in characteristics can

affect transistor temperature. For example, an increase in current gain or power dissipation will result in a temperature increase. This can produce *thermal runaway*, as described in Sec. 1-7.

In addition to knowing the effects of temperature on characteristics, it is important to know that *heat sinks* or transistor mounting can be used to offset the effects of temperature. For example, if a transistor is used with a heat sink, or is mounted on a metal chassis that acts as a heat sink, an increase in temperature (from any cause) can be dissipated into the surrounding air.

The following rules of thumb can be applied to the effects of temperature on two-junction transistor characteristics:

Collector leakage, or I_{CBO}, increases with temperature. Typically, I_{CBO} doubles with every 10°C increase in temperature for germanium transistors, and doubles with every 15°C increase in temperature for silicon transistors. Also, always consider the possible effects of a different collector voltage when approximating I_{CBO} at temperatures other than those on the datasheet, since collector leakage increases with voltage applied to the collector.

Current gain, or h_{fe}, increases with temperature. Typically, current gain doubles when the temperature is raised from 25 to 100°C for germanium transistors. With silicon, current gain doubles when the temperature is raised from 25 to 175°C. In any amplifier, do not exceed 100°C for germanium transistors or 200°C for silicon.

The power dissipation capabilities of a transistor must be carefully considered, especially when designing an amplifier circuit. Of course, in small-signal circuits the power dissipation is usually less than 1 W, and heat sinks are not needed. In such amplifiers, the only concern is that the rated power dissipation (as shown on the datasheet) not be exceeded. For typical small-signal amplifier circuits, do not exceed 90 to 95 per cent of the maximum power dissipation.

Maximum power dissipation is specified on datasheets in many ways. Some manufacturers provide *safe-area curves* for temperature and/or power dissipation. Others specify *maximum power dissipation* in relation to a given ambient temperature, or a given case temperature, Still others specify a *maximum junction temperature*, or a *maximum case temperature*.

Transistors designed for power applications usually have some form of *thermal resistance* specified to indicate the power dissipation capability. Thermal resistance can be defined as *the increase in temperature of the transistor junction (with respect to some reference) divided by the power dissipated* (°C/W). In power transistors, thermal resistance is normally measured from the transistor junction to the case. This results in the term θJC. On those transistors where the case is bolted directly to the mounting surface, the terms θMB (thermal resistance to mounting base) or θMF (thermal resistance to mounting flange) are used. If a transistor is *not mounted* on a heat sink or chassis, the thermal resistance from case-to-ambient air, or θCA, is so large in relation to that from junction-to-case (or mount) that the total thermal resistance from junction to ambient air, or θJA, is primarily the result of the θCA term.

A typical TO-5 style case has a 150°C/W rating, whereas the power-type TO-3 case has a 30°C/W rating. From this it will be seen that the TO-3 case (used in final power audio-amplifier stages) has a small temperature increase (for a given wattage) in comparison to the TO-5. That is, the heavy-duty cases will dissipate the heat into the ambient air.

Assume that a germanium transistor (maximum 100°C safe temperature) with a TO-5 case is to be used, and it is desired to find the absolute maximum power dissipation (without a heat sink and at an ambient of 25°C). The maximum allowable increase is 75°C (100 − 25). Maximum power dissipation is the maximum allowable temperature increase divided by degrees Celsius per watt, or 75/150 = 0.5 W. Of course, the current gain will double when the temperature is increased from 25 to 100°C. Assuming that the collector voltage remains constant, the dissipation will double. Thus, to be safe, the maximum power dissipation should be calculated at one half the 25°C value, or 0.5 W $\times \frac{1}{2}$, or 0.25 W.

Transistors operated at about 1 W, or higher, are usually used with an external heat sink. Sometimes, the chassis or mounting area serves as the heat sink. In other cases, a heat sink is attached to the case. Either way, the primary purpose of the heat sink is to increase the effective heat-dissipation area of the case, and provide a low-heat-resistance path from case to ambient.

Power-transistor datasheets specify the θJA that must be combined with the heat-sink thermal resistance to find the *total power-dissipation capability*. Commercial heat sinks are rated in terms of thermal resistance (usually as °C/W). When heat sinks are used, some form of electrical insulation must be provided between the case and heat sink (unless a grounded-collector circuit is used). Because good electrical insulators usually are also good thermal insulators, it is difficult to provide electrical insulation without introducing some thermal resistance between case and heat sink. Usually the insulation is provided by washers (between transistor and chassis) or cups (between transistor and heat sink). These washers introduce additional thermal resistance, typically in the order of 0.25 to 0.5°C/W. The use of a zinc oxide filled silicon compound (such as Dow Corning #340 or Wakefield #120) between the washer or cup and transistor or chassis helps to decrease thermal resistance.

Note that any insulation between collector and the chassis (as is produced by the washer between the case and heat sink) will also result in capacitance between the two metals. This capacitance is usually no problem at audio frequencies, but must be considered in RF amplifier circuits (Chapter 3).

Steady-state power dissipation. In any amplifier operated under steady-state conditions, typical of audio amplifiers used to amplify voice or music signals, the no-signal (or Q-point) dc collector voltage and current can be used to find the *approximate* power dissipation. That is, power dissipation equals dc collector voltage times dc collector current. Using this factor, and ignoring such factors as collector–base leakage, emitter–base current, and the like, the maximum power-dissipation capability of a transistor in any amplifier circuit can be calculated as follows: Assume a maximum junction temperature of 200°C (typical for a silicon power transistor), a junction-to-case thermal resistance θJC of 2°C/W, a heat sink with a thermal resistance of 3°C/W (including the washer), and an ambient temperature of 25°C.

First find θJA, which is equal to the sum of the transistor and heat sink resistances, or $2 + 3 = 5$.

Then find the maximum permitted power dissipation, which is equal to

$$\frac{\text{max junction temp (200°)} - \text{ambient temp (25°C)}}{\theta JA(5)}$$

or 35 W (maximum).

If the same transistor were to be used *without a heat sink*, but under the same conditions and with a TO-3 case (30°C/W), the maximum power is calculated as follows: $\theta CA(30) + \theta JC(2) = 32$.

$$\frac{200 - 25}{32} = 5 \text{ W} \quad \text{(approx)}$$

If *maximum case temperature* is specified, rather than maximum junction temperature, subtract the ambient temperature from the maximum permitted case temperature; then divide the case temperature by the heat-sink thermal resistance. For example, assume a maximum case temperature of 130°C, an ambient of 25°C, and a 3°C/W heat sink. Then, $130 - 25 = 105$°C, and $105/3 = 35$ W maximum power.

Pulse power dissipation. In amplifiers operating with pulse signals, the maximum permitted power dissipation is greater than with steady-state operation. For a *single, nonrepetitive* pulse, the transient thermal resistance must be calculated. Generally, this transient factor is given in the form of a *power multiplier*, related to a given pulse width and case temperature. If the case temperature is increased, the factor must be derated.

For example, assume that a transistor is used to amplify 1-ms pulses, the transistor has a maximum junction temperature of 200°C, the case temperature under these conditions is 130°C, the ambient temperature is 25°C, and that the maximum steady-state power is 35 W (as in the previous example). Furthermore, assume that the datasheet specifies a power multiplier of 3 for 1-ms pulses with a case temperature of 25°C. The derating factor would be

$$\frac{130°C - 25°C}{200°C - 25°C} = \frac{105}{175} = \frac{3}{5}$$

To find the maximum single, nonrepetitive pulse power:

$$\text{max power} = \text{multiplier} \times (1 - \text{derating factor}) \times \text{steady-state power}$$
$$= 3 \times (1 - \tfrac{2}{5}) \times 35 = 42 \text{ W}$$

These calculations for single-pulse operation are based on the assumption that the heat-sink capacity is large enough to prevent the heat-sink temperature from rising between pulses.

For repetitive pulses, both the case and heat-sink temperatures will rise. This can be accounted for as follows:

$$\text{max permitted power} = \frac{\text{power multiplier} \times \text{max junction} - \text{ambient}}{\theta JC + [\text{power multiplier} \times \% \text{ duty cycle} \times \theta JA]}$$

Assume that it is desired to find the maximum permitted power of the same transistor described for the steady-state amplifier, but now used to amplify 1-ms pulses repeated at 100-Hz intervals. The conditions are power multiplier of 3, θJC of 2°C/W, θJA of 5°C/W (including heat sink), maximum junction temperature of 200°C, ambient temperature of 25°C, and a duty cycle of 10 per cent (1 ms on and 9 ms off):

$$\text{max permitted power} = \frac{3(200 - 25)}{2 + [3(0.1)5]} = 150 \text{ W}$$

The case temperature of a transistor operated by repetitive pulses is

$$\text{case temperature} = \text{peak pulse power} \times \% \text{ duty cycle} \times \theta JA + \text{ambient}$$

Peak pulse power is obtained by multiplying the collector voltage by the collector current (assuming that the transistor is operated in a grounded-emitter or grounded-base configuration, and that the transistor is switched full on and full off by the pulses).

For example, assume an ambient temperature of 25°C, a duty cycle of 10 per cent, a total θJA of 5°C/W, and peak pulses of 120 W (say a collector voltage of 60 and a collector current of 2 A). The case temperature would be

$$\text{case temperature (or } T_c) = 120 \times 0.1 \times 5 + 25 = 85°C$$

2-3.2. Field-Effect Transistor (FET) Ratings

Characteristics of the FET are unique when compared to any other type of transistor. For example, as discussed in Chapter 1, a FET is often biased at the zero-temperature-coefficient point, where the FET drain–source current will not vary with temperature. Likewise, the characteristics shown on FET datasheets do not correspond to those of conventional two-junction transistors. The following paragraphs summarize FET characteristics and datasheet ratings.

2-3.2.1. Datasheet Format for the FET

Typically, the first page of the datasheet gives the maximum ratings of the FET, mechanical dimensions, and pin layout, similar to a two-junction transistor datasheet. Obviously, if the maximum ratings are exceeded, abnormal circuit operation will occur, and the FET may be destoryed. For example, assume that 70 V is applied to the FET, which has a maximum drain–source rating of 60 V. Damage, if not total destruction, will surely result. All the precautions concerning maximum ratings for two-junction transistors also apply to FETs.

The second page of the datasheet presents the electrical characteristics and operating conditions under which they were measured. Page 2 shows both "on" and "off" characteristics, as well as "small-signal characteristics." Keep in mind that all the characteristics listed in any datasheet are based on a set of fixed operating conditions. If the conditions change (as they must in any practical amplifier), the characteristics will change.

2-3.2.2. Characteristics of the FET

The following characteristics appear on most FET datasheets.

Forward gate current $I_{G(f)}$ is the maximum recommended forward current through the gate terminal. This is a limiting factor in some amplifier applications, and is caused by a large forward bias current on the gate. When this condition occurs, the gate current must be limited or degeneration of the FET will occur. A resistor in series with the gate will limit the current, but its value will determine the variance of gate bias as affected by the gate leakage current.

Total device dissipation P_D is the maximum power that can be dissipated within the device at 25°C without exceeding the maximum allowable internal temperature (typically 200°C). The power is derated according to the *thermal resistance* (Sec. 2-3.1.3) value. For example, assume that the FET has a 200-mW rating at 25°C and a 1.33-mW/C° thermal resistance value. At 125°C the power dissipated in the FET must not exceed 77 mW (125 − 25 = 100°C; 100 × 1.33 = 133; 200 − 133 = 77 mW). Operation above the maximum value could damage the device.

Gate–source breakdown voltage $V_{(BR)GSS}$ is the breakdown voltage from gate to source, with the drain and source shorted. Under these conditions, the gate–channel junction also meets the breakdown specification, since the drain and source are connections to the channel. This means that the drain and source may be interchanged, for symmetrical devices, without fear of individual junction breakdown.

Gate–source cutoff voltage $V_{GS(off)}$ is defined as the gate-to-source voltage required to reduce the drain current to 0.01 (or preferably 0.001) of the minimum I_{DSS} value.

Pinch-off voltage V_P is essentially the same as $V_{GS(off)}$, only measured in a different manner. V_P is shown in Fig. 2-9 and is the drain-to-source voltage at which the drain current increases very little for an increase in drain-to-source voltage at $V_{GS} = 0$. Most equations used to describe the operation of a FET use V_P, but the value of $V_{GS(off)}$ can be used instead.

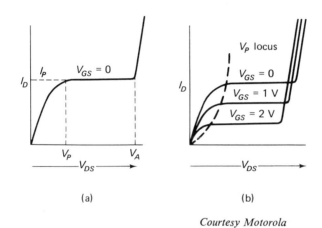

(a)   (b)

Courtesy Motorola

Fig. 2-9   FET output characteristics

An *approximate value* for $V_{GS(off)}$ can also be determined from the transfer characteristic curve shown in Fig. 2-10. $V_{GS(off)}$ is found by taking the *slope of the curve* at $V_{GS} = 0$ (where the curve intersects the I_D axis), and extending the slope to the V_{GS} axis. *Twice the value of this intercept is* $V_{GS(off)}$.

For example, using the curves of Fig. 2-10, V_{GS} minimum is about 2×1.25 V, or 2.5 V. V_{GS} maximum is about 2×1.8, or 3.6 V at 25°C. This technique can be used on any of the I_D–V_{GS} curves for other temperatures.

Gate leakage (reverse) current I_{GSS} is defined as the gate–channel leakage with the drain shorted to the source, and is a measure of the static short-circuit input impedance. Since gate-to-channel is a reverse-biased diode junction (for a JFET), I_{GSS} doubles (approximately) every 15°C increase in temperature, and is proportional to the square root of the applied voltage.

Zero gate voltage drain current I_{DSS} is defined as the drain-to-source current, with the gate shorted to the source, at a specified drain–source voltage. I_{DSS} is a basic parameter for JFETs, and is considered to be a figure of merit.

Gate–source voltage V_{GS} is a range of gate-to-source voltages with $0.1I_{DSS}$ drain current flowing. The specified drain-to-source voltage is the same as for I_{DSS}. This characteristic gives the min/max variation in V_{GS} for different FETs, for a given I_D and V_{DS}.

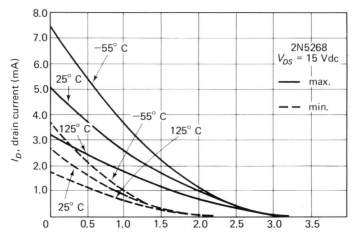

Fig. 2-10   Forward transfer characteristic curves for min/max I_{DSS} limits

Forward transadmittance Y_{fs} is defined as the magnitude of the common-source forward transfer admittance. Y_{fs} shows the relationship between input signal voltage and output signal current, and is the key dynamic characteristic for all FETs. All other factors being equal, an increase in Y_{fs} produces an increase in gain, when a FET is used in an audio-amplifier stage. Y_{fs} is specified at 1 kHz, with dc operating conditions the same as for I_{DSS}.

At 1 kHz, Y_{fs} is almost entirely real. Thus, $Y_{\prime}$ at 1 kHz $= Y_{fs}$. At higher frequencies, Y_{fs} includes the effects of gate-to-drain capacitance and may be misleadingly high. However, for FETs operating in the audio range, Y_{fs} is generally an accurate figure. For higher-frequency operation, the real part of transconductance $\text{Re}(Y_{fs})$, as discussed in later paragraphs, should be used.

Figure 2-11 shows Y_{fs} versus I_D with typical and minimum values shown. The curves end sharply where $I_D = I_{DSS}$ and, at this value, Y_{fs} is at its maximum.

Figure 2-12 shows a typical temperature coefficient of Y_{fs} versus drain current. For this curve, $V_{DS} = 15$ V, and V_{GS} is varied to obtain the variations of I_D.

Forward transconductance $\text{Re}(Y_{fs})$ is defined as the common-source forward transfer conductance (drain current versus gate voltage). For high-frequency applications, $\text{Re}(Y_{fs})$ is considered a figure of merit. The dc operating conditions are the same as for Y_{fs}, but the test frequency is 100

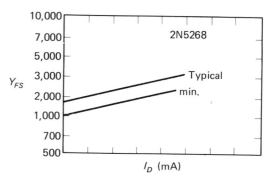

Fig. 2-11 Typical and minimum
forward admittance Y_{fs}

Courtesy Motorola

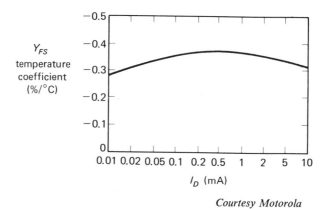

Courtesy Motorola

Fig. 2-12 Forward transadmittance coefficient versus drain current

MHz. All other factors being equal, an increase in Re(Y_{fs}) produces an increase in the voltage gain produced by a FET amplifier stage used at radio frequencies.

In comparing Re(Y_{fs}) with Y_{fs}, the minimum values of the two are quite close, considering the difference in frequency at which the measurements are made. At high frequencies, about 30 MHz and above, Y_{fs} will increase due to the effect of the gate–drain capacitance C_{Gd}, so that Y_{fs} will be misleadingly high.

Output admittance Y_{os} is defined as the magnitude of the common-source output admittance, and is measured at the same operating conditions and frequency as Y_{fs}. Since Y_{os} is a complex number at low frequencies, only the magnitude is specified. For higher frequencies, Y_{os} can be calculated using r_{oss}, as shown in Fig. 2-13.

The common-source output resistance r_{oss} is the real part of Y_{os}. Figure 2-14 shows the variation of r_{oss} versus I_D for several values of I_{DSS}. These are also measured at 1 kHz, $V_{DS} = 15$ V, and V_{GS} varied to obtain the different

73

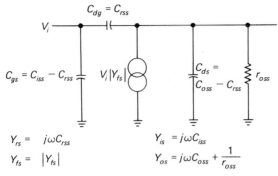

$$Y_{rs} = j\omega C_{rss}$$
$$Y_{fs} = |Y_{fs}|$$

$$Y_{is} = j\omega C_{iss}$$
$$Y_{os} = j\omega C_{oss} + \frac{1}{r_{oss}}$$

Courtesy Motorola

Fig. 2-13 Equivalent FET low-frequency circuit

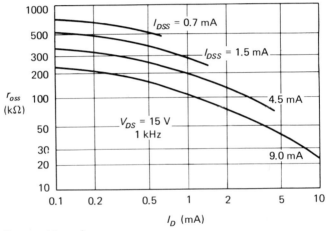

Courtesy Motorola

Fig. 2-14 Output resistance versus drain current

values of I_D. Note that the lower I_{DSS} FETs have a higher output resistance for the same drain current.

Input capacitance C_{iss} is the common-source input capacitance with the output shorted, and is used in place of Y_{iss}, the short-circuit input admittance. Y_{iss} is entirely capacitive at low frequencies, since the input is a reverse-biased silicon diode (for JFETs). The real parts of Y_{iss} could be calculated from I_{GSS}, but is negligible even at low frequencies.

Reverse transfer capacitance C_{rss} is defined as the common-source reverse transfer capacitance with the input shorted. C_{rss} is used in place of Y_{rs}, the short-circuit reverse transfer admittance, since Y_{sr} is almost entirely capacitive over the useful frequency range of the FET. Figure 2-15 shows C_{iss}, C_{rss}, and C_{oss} as functions of V_{DS}. These same ratings are shown on Fig. 2-13. The circuit of Fig. 2-13 is valid up to about 30 MHz, and shows how to

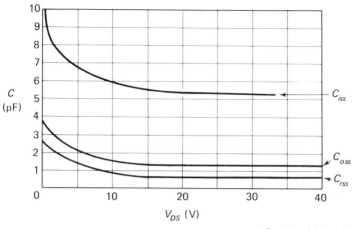

Courtesy Motorola

Fig. 2-15   Capacitance versus drain–source voltage

obtain all the short-circuit Y parameters from the data given in the data-sheet. Although all this information may not be necessary for simple linear FET amplifiers used at audio frequencies, the data can be of great value for RF amplifiers, as discussed in Chapter 3.

Common-source noise figure NF represents a ratio between input signal-to-noise and output signal-to-noise, and is measured in decibels. Short-circuit input noise voltage e_n is the equivalent short-circuit input noise, expressed in volts per root cycle.

NF, as specified, includes the effects of e_n and i_n, where i_n is the equivalent open-circuit input noise current. For a FET, the contribution of i_n is small compared to e_n. As shown in Fig. 2-16, NF attains its highest value for a small generator resistance and decreases for increasing generator resistance, indicating a large noise contribution from the noise-voltage generator. For this reason, NF and e_n are specified, and i_n is neglected. NF is independent of operating current and proportional to voltage. However, the voltage effects are slight over the normal operating range.

Figure 2-17 is a nomograph for converting noise figure to equivalent input noise voltage for different generator source impedances, R_s. This nomograph can be used with any FET. Since NF and e_n are frequency dependent, Fig. 2-17 must be used in conjunction with Fig. 2-18 (noise figure versus frequency at a specified source impedance) to determine e_n. For example, to find the input noise at 50 Hz and $R_s = 1$ MΩ, NF, from Fig. 2-18, is about 1.5 dB. Next, from Fig. 2-17, e_n is found to be 1.5×10^{-7} volts (on the 10^6 or 1-MΩ curve), or about 1.5 nV. If the stage has a voltage gain of 10, the output noise is 10×150 nV, or 1.5 μV of 50-Hz noise.

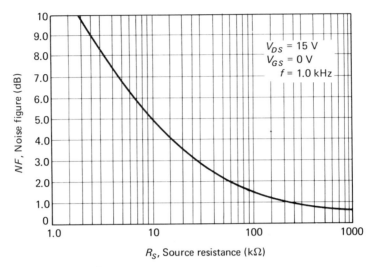

Courtesy Motorola

Fig. 2-16   Noise figure versus source resistance

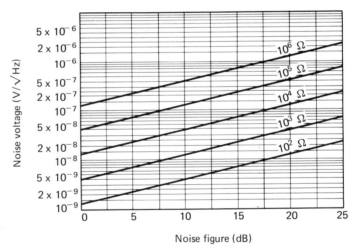

Courtesy Motorola

Fig. 2-17   Noise-figure conversion chart

Figure 2-16 shows *NF* versus generator resistance at a fixed frequency. The only point common to Figs. 2-18 and 2-16 is at $R_s = 1$ MΩ and $f = 1$ kHz. For other values of R_s and f, *NF* can be estimated by using both figures; then e_n is found, as before, from Fig. 2-17.

Zero temperature coefficient, 0*TC*. Although there is no single specification dealing with the 0*TC* effect of FETs (refer to Sec. 1-11), its importance to

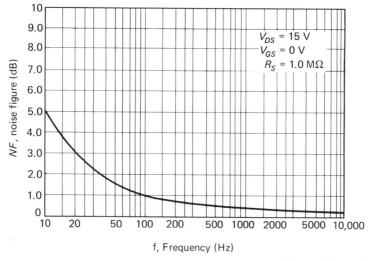

Courtesy Motorola

Fig. 2-18   Noise figure versus frequency

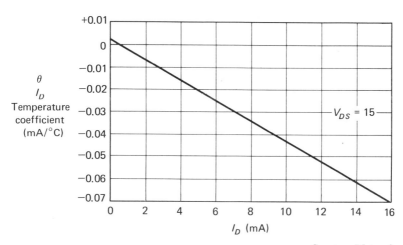

Courtesy Motorola

Fig. 2-19   Drain-current temperature coefficient versus drain current

amplifier circuits warrants discussion. Figure 2-19 shows the typical I_D temperature coefficient θI_D versus I_D for 2N5265 through 2N5270 FETs. All these FETs have a temperature coefficient of zero at approximately $I_D = 0.5$ mA.

The positive temperature coefficient occuring at small values of drain current is caused by a change in the width of the thermally generated depletion layer at the gate–channel junction. For the above FETs, the change is at a rate equivalent to -2.2 mV/°C across the gate–source junction. This is the same value and phenomenon as the forward-biased base–emitter junction of a two-junction transistor. The negative temperature coefficient occurs at high voltages of I_D due to a decrease in carrier mobility (increasing resistivity) to the channel. At some value of I_D these two effects cancel each other, and zero temperature coefficient ($0TC$) of I_D or V_{GS} occurs.

The rule-of-thumb equations for $0TC$ operation are

$$V_{GSZ} \approx V_P - 0.63 \tag{2-1}$$

$$I_{DZ} \approx I_{DSS} \left(\frac{0.63}{V_P}\right)^2 \tag{2-2}$$

$$\text{drift in mV/°C} \approx 2.2 \left(1 - \frac{\sqrt{I_D}}{I_{DZ}}\right) \tag{2-3}$$

where subscript Z indicates the $0TC$ point for the parameter.

Equation (2-3) is valued near, but not at, I_{DZ}. Equations (2-1) and (2-2) must both be satisfied for $0TC$ operation. Since V_P and I_{DSS} vary from device to device, an adjustment for V_{GS} or I_D is necessary. It is noted that resistor temperature coefficients, and the effects of I_{GSS} with temperature, must also be considered.

2-4. BASIC TWO-JUNCTION TRANSISTOR AMPLIFIER STAGE

Figure 2-20 is the working schematic of a basic, single-stage audio amplifier using a two-junction transistor. Note that the basic audio circuit is similar to the bias circuits discussed in Sec. 1-7 (Fig. 1-12), except that input and output coupling capacitors C_1 and C_2 are added. These capacitors prevent dc flow to and from external circuits. Note that a bypass capacitor C_3 is shown connected across the emitter resistor R_E. Capacitor C_3 is required only under certain conditions, as discussed in later paragraphs.

Input to the amplifier is applied between base and ground across R_B. Output is taken across the collector and ground. The input signal adds to, or subtracts from, the bias across R_B. Variations in bias voltage cause corresponding variations in base current, collector current, and the drop across collector resistor R_L. Therefore, the collector voltage (or circuit output) follows the input signal waveform, except that the output is inverted in phase. (If the input swings positive, the output swings negative, and vice versa.)

Variations in collector current also cause variations in emitter current. This results in a change of voltage drop across the emitter resistor R_E, and a change in the base–emitter bias relationship. As discussed in Chapter 1,

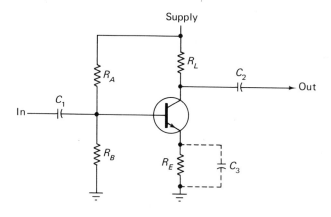

$Z_{in} \approx R_B$ $R_L > 5R_E$ $R_E \approx 100 - 1000\ \Omega$

$Z_{out} \approx R_L$

$I_{gain} \approx \dfrac{R_B}{R_E}$ $R_L \approx 10R_E$

$V_{gain} \approx \dfrac{R_L}{R_E}$ $R_B \approx 10R_E$

                              $R_B < 20R_E$

$V_{collector} \approx 0.5 \times$ supply (set by R_A)

$I_{gain} \approx$ stability $\approx \dfrac{R_B}{R_E}$ $C_1 \approx \dfrac{1}{3.2\ FR}$ for 1 dB

$S \approx 20$ for high gain

$S \approx 10$ for stability $C_2 \approx \dfrac{1}{6.2\ FR}$ for 3 dB

$S \approx 5$ for power gain

Fig. 2-20   Basic, single-stage, audio amplifier using a two-junction transistor

the change in bias that results from the voltage drop across R_E tends to cancel the initial bias change caused by the input signal, and serves as a form of negative feedback to increase stability (and limit gain). This form of emitter feedback (current feedback) is known as *stage feedback* or *local feedback*, since only one stage is involved. As discussed in later chapters, *overall feedback* or *loop feedback* is sometimes used where several stages are involved.

2-4.1. Circuit Analysis

The outstanding characteristic of the circuit in Fig. 2-20 is that circuit characteristics (gain, stability, impedance) are determined (primarily) by circuit values, rather than transistor characteristics (beta). The circuit is shown with an *NPN* transistor. The power supply polarity must be reversed if a *PNP* transistor is used.

The maximum peak-to-peak output voltage is set by the source voltage. For class A operation, the collector is operated at approximately one half the source voltage. This permits the maximum positive and negative swing of output voltage. Generally, the absolute maximum peak-to-peak output can be between 90 and 95 per cent of the source. For example, if the source is 20 V, the collector will operate at 10 V (Q point), and swing from about 1 to 19 V. However, there is less distortion if the output is one half to one third of the source. In any circuit, the maximum collector voltage rating of the transistor cannot be exceeded.

The input and/or output impedances are set by the values of R_B and R_L, as shown by the equations of Fig. 2-20. Maximum power transfer occurs when R_B and R_L match the impedances of the previous stage and following stage, respectively.

Stability versus gain tradeoff. As shown by the equations, maximum voltage gain occurs when R_L is made larger in relation to R_E. Likewise, current gain increases when R_B is made larger in relation to R_E. However, circuit stability is greatest when R_B and R_L are made smaller in relation to R_E. That is, the circuit gain will remain more constant in the presence of temperature, source voltage, or input signal changes when R_E is made larger in relation to R_B and R_L. Thus, there is a tradeoff between gain and stability.

In a practical amplifier, the circuit of Fig. 2-20 should be limited to a maximum current gain of 10 and a maximum voltage gain of 20. Higher gains are possible, but the stability is generally poor. Of course, even though gain is set by circuit values, the minimum ac beta of the transistor must be higher than the desired gain. For example, if the circuit values are chosen for a gain of 20, the minimum beta must be 20 across the entire frequency range.

The Q point is affected by the values of all four resistors in Fig. 2-20. However, the value of R_A is the *final determining factor for Q point*. That is, in a practical amplifier, the remaining resistor values are selected for the desired gain, impedance, and stability of the circuit; then the value of R_A is selected (or adjusted) to give a desired operating point.

The values of C_1 and C_2 are dependent upon the *low-frequency limit* at which the amplifier is to operate. As discussed in Sec. 2-1, C_1 forms a high-pass (or low-cut) *RC* filter with R_B. Capacitor C_2 forms another filter with the input resistance of the following stage (or the load). For a given resistance value, a lower frequency requires a larger capacitor value, as shown by the equations on Fig. 2-20. Of course, if the resistance can be made larger (with the same desired frequency), the capacitor value can be reduced. Since two-junction transistors are low-impedance, current-operated devices, the coupling capacitors in two-junction amplifiers are generally large. That is, the coupling capacitors are generally large in relation to those of FET (and vacuum-tube) amplifiers.

2-4.2. Emitter Bypass for Transistor Amplifier Stage

Figure 2-20 shows (in phantom) a bypass capacitor C_3 across emitter resistor R_E. This arrangement permits R_E to be removed from the circuit as far as the signal is concerned, but leaves R_E in the circuit (in regards to direct current). With R_E removed from the signal path, the voltage gain is approximately R_L divided by the dynamic resistance of the transistor, and the current gain is approximately ac beta of the transistor. Thus, the use of an emitter-bypass capacitor permits the highly temperature stable dc circuit to remain intact while providing a high signal gain.

An emitter-bypass capacitor also creates some problems. Transistor input impedance changes with frequency, and from transistor to transistor, as does beta. Therefore, current and voltage gains can only be approximated. When the emitter resistance is bypassed, the circuit input impedance is approximately beta times transistor input impedance, making circuit input impedance subject to variation and unpredictable.

The emitter bypass is generally used where maximum gain must be obtained from a single stage of amplification, and a stable gain is of little concern. The value of the emitter-bypass capacitor should be such that the *reactance is less than the transistor input impedance* at the lowest frequency of operation. This will effectively short the emitter (signal path) around R_E.

The value of C_3 can be found by

$$\text{capacitance} = \frac{1}{6.2FR}$$

where capacitance is in farads, F is the low-frequency limit in hertz, and R is the maximum input impedance of the transistor in ohms.

2-4.3. Basic Audio Amplifier with Partially Bypassed Emitter

Figure 2-21 shows a basic single-stage two-junction transistor used as an audio amplifier with a partially bypassed emitter resistor. This circuit is a compromise between the basic amplifier without bypass and the fully bypassed emitter. The dc characteristics of both the unbypassed and partially bypassed circuits are essentially the same. The circuit values (except C_3 and R_C) are calculated in the same way for both circuits. However, the voltage and current gains for a partially bypassed amplifier are greater than an unbypassed circuit, but less than for the fully bypassed circuit.

The value of R_C is chosen on the basis of voltage gain, even though current gain will be increased when voltage gain increases. R_C should be substantially smaller than R_E. Otherwise, there will be no advantage to the partially bypassed circuit. However, a smaller value for R_C will require a larger value for

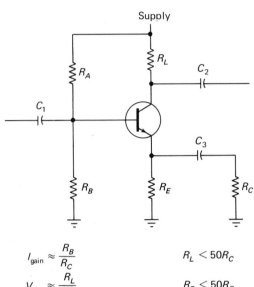

$$I_{gain} \approx \frac{R_B}{R_C}$$

$$V_{gain} \approx \frac{R_L}{R_C}$$

$$C_3 \approx \frac{1}{6.2FR_C}$$

$$R_L < 50R_C$$

$$R_B < 50R_C$$

Fig. 2-21 Basic single-stage two-junction transistor used as an audio amplifier with partially bypassed emitter resistor

C_3, since the C_3 value is dependent upon the R_C value and the desired low-frequency cutoff point.

With the circuit of Fig. 2-21, current gain and voltage gain are dependent upon the ratios of R_B/R_C and R_L/R_C, respectively. Thus, the value of R_E has little or no effect on circuit gain.

2-5. BASIC FET AMPLIFIER STAGE

As discussed, the characteristics of FETs are quite different from those of two-junction transistors. This makes for considerable differences in amplifier circuit characteristics. For this reason, we shall summarize FET amplifier characteristics before going into a basic FET amplifier stage.

The basic FET bias circuit is shown in Fig. 2-22. This circuit can be converted to a single FET amplifier stage by the addition of input and output coupling capacitors. The basic bias network is modified as necessary to produce a FET stage with desired characteristics (stage gain, input–output impedance, etc.).

The purpose of the basic bias circuit is to establish a given I_D, and to maintain that I_D (plus or minus some given tolerance) over a given temperature range. Generally, this is to keep the FET at the $0TC$ operating point. When the basic FET circuit is used to form a linear amplifier, the output

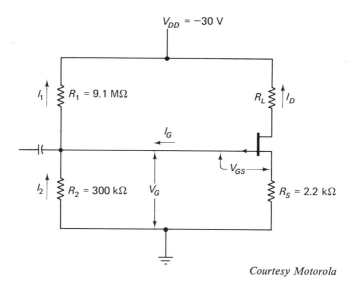

Courtesy Motorola

Fig. 2-22   Basic FET stage

(drain terminal) should be at one half the supply voltage, if maximum output voltage swing is wanted. Keep in mind that amplifiers rarely operate under static conditions. Thus, the basic bias circuit is used as a reference or starting point for design or analysis. The actual circuit configuration, and (especially) the bias circuit values, are selected on the basis of dynamic circuit conditions (desired output voltage swing, expected input signal level, etc.).

2-5.1. Selecting Values for Basic FET Bias Circuit

Assume that the circuit of Fig. 2-22 is to maintain I_D at 1 ± 0.25 mA, over a temperature range from -55 to $+125°C$, with a supply voltage V_{DD} of 30 V. The first step is to draw a $1/R_S$ load line on the FET transfer characteristics, as shown in Fig. 2-23. (Note that this illustration is similar to that of Fig. 1-30.) As shown by the equations, the value of R_S is set by the limits of V_{GS} and I_D. The value of $V_{GS(min)}$ is the point where the $I_{D(min)}$ of 0.75 mA crosses the high-temperature-limit curve of $+125°C$, or approximately 0.8 V. The value of $V_{GS(max)}$ is the point where the $I_{D(max)}$ of 1.25 mA crosses the low-temperature-limit curve of $-55°C$, or approximately 1.9 V. Using these values, the first trial value for R_S is

$$R_S = \frac{1.9 - 0.8}{1.25 - 0.75} = 2.2 \text{ k}\Omega$$

The fixed bias voltage V_G is determined from the intercept of the $1/R_S$ load line with the V_{GS} axis, and is computed using the same set of values, as

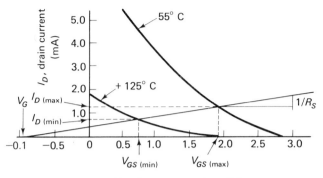

$$V_{GS}, \text{ Gate-source voltage (V)}$$

$$R_S = \frac{V_{GS\,(max)} - V_{GS\,(min)}}{I_{D\,(max)} - I_{D\,(min)}}$$

$$V_{RL} = V_{DD} - (1.5 \times V_{GS\,(off)}) - (I_{DC\,(max)} \times R_S)$$

$$R_2 = \frac{V_G R_1}{V_{DD} - V_G}$$

$$R_{L\,(max)} = \frac{V_{RL}}{I_{DC\,(max)}}$$

$$V_G = \frac{I_{D\,(min)}\,V_{GS\,(max)} - I_{D\,(max)}\,V_{GS\,(min)}}{I_{D\,(max)} - I_{D\,(min)}}$$

Courtesy Motorola

Fig. 2-23   The $1/R_S$ load line for basic FET stage

shown in the Fig. 2-24 equations:

$$V_G = \frac{0.75 \times 1.9 - 1.25 \times 0.8}{0.5} = 0.85 \text{ V}$$

The maximum value of R_1 is determined by the maximum gate reverse current, as specified on the datasheet. The variation in V_G versus temperature

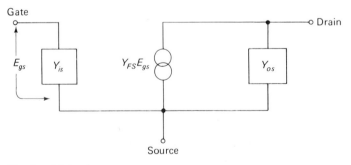

Courtesy Motorola

Fig. 2-24   Small-signal analysis of basic FET stage

will generally not be too great if a value for R_1 is chosen such that I_1 (Fig. 2-22) is at least six times greater than the maximum reverse current. Assume a maximum gate reverse current of 0.5 μA at the maximum temperature of 125°C, and the drop across R_1 is approximately 29 V (30 V — a nominal V_{GS} of 1 = 29 V). Six times 0.5 μA is 3 μA. A value of 9.1 MΩ for R_1 will produce approximately 3.2 μA of I_1.

The value of R_2 is found from a simple voltage-divider relationship, ignoring the effect of I_G, as shown in the Fig. 2-23 equations:

$$R_2 = \frac{0.85 \times 9.1 \times 10^6}{30 - 0.85} \approx 300 \text{ k}\Omega$$

The maximum value of R_L is determined by the voltage drops across R_L and $I_{D(max)}$. As shown by the Fig. 2-23 equations, the voltage drop across R_L (or V_{RL}) is

$$V_{RL} = 30 - (1.5 \times 3.6) - (1.25 \times 2.2) = 21.85 \text{ V}$$

Note that $I_{D(max)}$ is shown in Fig. 2-23, and $V_{GS(off)}$ is calculated using the technique of Sec. 2-3.2.2.

With V_{RL} established at 21.85 V, and an $I_{D(max)}$ of 1.25 mA, the maximum value of R_L is

$$R_{L(max)} = \frac{21.85}{1.25} = 17.48$$

Assuming a standard 1 per cent resistor, a value of 16.9 kΩ can be used.

2-5.2. Small-Signal Analysis of Basic FET Bias Circuit

Small-signal analysis of a FET amplifier stage is easily accomplished with reasonable accuracy by using a few simple equations. The FET model used for the analysis is shown in Fig. 2-24. This model differs from the previous model (Fig. 2-13) in that C_{rss} is omitted. By omitting all capacitance and using only the real parts of Y_{is}, Y_{fs}, and Y_{os}, the model and the accompanying equations are useful up to about 100 kHz. If capacitance effects are included, the equations are useful up to several megahertz.

Figures 2-25, 2-26, and 2-27 show schematics for common-source, common-drain (source-follower), and common-gate circuits. In addition, the *exact* and *approximate* equations for voltage gain, input impedance, and output impedance are included. (The equations for finding the resistance values of R_1, R_2, R_S, and R_L are discussed in Sec. 2-5.1.)

For the common-source circuit (Fig. 2-25), the omission of C_{rss} will affect the equations for input impedance and voltage gain. However, for low frequencies (below about 100 kHz) the error is minimal. If *only dc feedback*

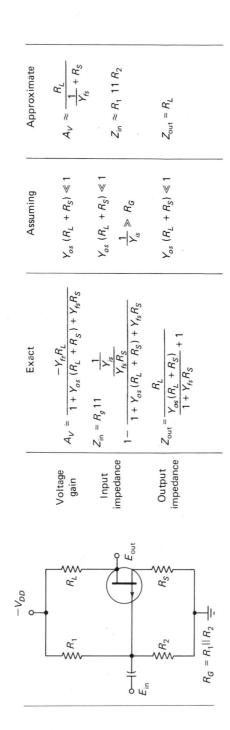

	Exact	Assuming	Approximate
Voltage gain	$A_V = \dfrac{-Y_{fs}R_L}{1+Y_{os}(R_L+R_S)+Y_{fs}R_S}$	$Y_{os}(R_L+R_S) \ll 1$	$A_V \approx \dfrac{R_L}{\frac{1}{Y_{fs}}+R_S}$
Input impedance	$Z_{in} = R_g \,\|\, \dfrac{\frac{1}{Y_{is}}}{Y_{fs}R_S\left(1-\dfrac{1}{1+Y_{os}(R_L+R_S)+Y_{fs}R_S}\right)}$	$Y_{os}(R_L+R_S) \ll 1$ $\dfrac{1}{Y_{is}} \gg R_G$	$Z_{in} \approx R_1 \,\|\, R_2$
Output impedance	$Z_{out} = \dfrac{R_L}{\dfrac{Y_{os}(R_L+R_S)}{1+Y_{fs}R_S}+1}$	$Y_{os}(R_L+R_S) \ll 1$	$Z_{out} = R_L$

Courtesy Motorola

Fig. 2-25   Basic common-source FET stage circuit and characteristics

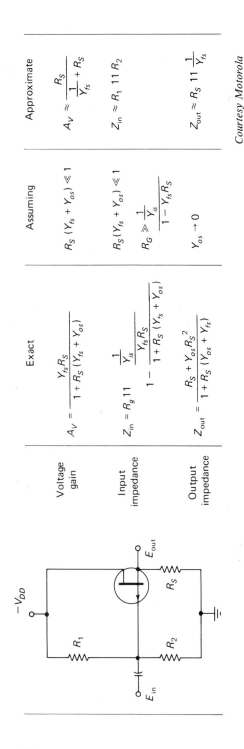

	Exact	Assuming	Approximate
Voltage gain	$A_V = \dfrac{Y_{fs} R_S}{1 + R_S(Y_{fs} + Y_{os})}$	$R_S(Y_{fs} + Y_{os}) \ll 1$	$A_V \approx \dfrac{R_S}{\dfrac{1}{Y_{fs}} + R_S}$
Input impedance	$Z_{in} = R_g \,\|\, \dfrac{\dfrac{1}{Y_{is}}}{1 - \dfrac{Y_{fs} R_S}{1 + R_S(Y_{fs} + Y_{os})}}$	$R_S(Y_{fs} + Y_{os}) \ll 1$ $R_G \gg \dfrac{\dfrac{1}{Y_{is}}}{1 - Y_{fs} R_S}$	$Z_{in} \approx R_1 \,\|\, R_2$
Output impedance	$Z_{out} = \dfrac{R_S + Y_{os} R_S^2}{1 + R_S(Y_{os} + Y_{fs})}$	$Y_{os} \to 0$	$Z_{out} \approx R_S \,\|\, \dfrac{1}{Y_{fs}}$

Courtesy Motorola

Fig. 2-26 Basic common-drain (source-follower) FET stage circuit and characteristics

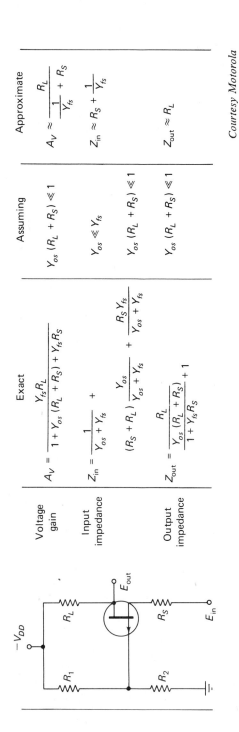

	Exact	Assuming	Approximate
Voltage gain	$A_V = \dfrac{Y_{fs}R_L}{1 + Y_{os}(R_L + R_S) + Y_{fs}R_S}$	$Y_{os}(R_L + R_S) \ll 1$	$A_V \approx \dfrac{R_L}{\dfrac{1}{Y_{fs}} + R_S}$
Input impedance	$Z_{in} = \dfrac{1}{Y_{os} + Y_{fs}} + (R_S + R_L)\dfrac{Y_{os}}{Y_{os} + Y_{fs}} + \dfrac{R_S Y_{fs}}{Y_{os} + Y_{fs}}$	$Y_{os} \ll Y_{fs}$ $Y_{os}(R_L + R_S) \ll 1$	$Z_{in} \approx R_S + \dfrac{1}{Y_{fs}}$
Output impedance	$Z_{out} = \dfrac{R_L}{\dfrac{Y_{os}(R_L + R_S)}{1 + Y_{fs}R_S} + 1}$	$Y_{os}(R_L + R_S) \ll 1$	$Z_{out} \approx R_L$

Courtesy Motorola

Fig. 2-27   Basic common-gate FET stage circuit and characteristics

is required, then the source resistor R_S is bypassed. With a capacitor across R_S, the effects on the design equations are to set R_S at zero. Under these conditions, voltage gain is the product of R_L and Y_{fs}.

With R_S not bypassed, the circuit characteristics are virtually independent of FET parameters (with the exception of Y_{fs}). Instead, the circuit characteristics (impedances, gain, etc.) are dependent upon R_L and R_S. By using precision resistors with close temperature coefficients, the common-source circuit can be made very stable over a wide temperature range.

The common-drain (source-follower) configuration (Fig. 2-26) is a very useful basic circuit. Some of its properties are a voltage gain always less than unity with no phase inversion, low output impedance (essentially set by the value of R_S), high input impedance, large signal swing, and active impedance transformation.

The common-gate stage (Fig. 2-27) offers impedance transformation opposite to that of the source follower. Common-gate produces low input impedance and high output impedance. The voltage gain is the same as for the common source, except that there is no phase inversion.

The circuits of Figs. 2-25, 2-26, and 2-27 are for *P*-channel JFETs. Reverse the polarity for *N*-channel FETs. MOSFET (or IGFET) devices may not conform exactly to the relations shown, but are sufficiently close for analysis of FET amplifier circuits.

2-5.3. Single-Stage Common-Source FET Amplifier

Figure 2-28 shows a basic single-stage FET amplifier. Note that the basic amplifier circuit is similar to the basic common-source circuit of Figs. 2-22 and 2-25, except that input and output coupling capacitors C_1 and C_2 are added to prevent dc flow to and from external circuits. Bypass capacitor C_3, connected across R_S, is required only under certain conditions.

Input to the amplifier is applied between gate and ground across R_2. Output is taken across the drain and ground. The input signal adds to, or subtracts from, the bias voltage across R_2. Variations in bias voltage cause corresponding variations in I_D, and the voltage drop across R_L. Therefore, the drain voltage (or circuit output) follows the input signal waveform, except that the output is inverted in phase. (If the input swings positive, the output swings negative, and vice versa.)

Variations in I_D also cause variations in voltage drop across R_S, and a change in the gate–source bias relationship. The change in bias that results from the voltage drop across R_S tends to cancel the initial bias change caused by the input signal, and serves as a form of *negative feedback* to increase stability (and limit gain). This gate–source feedback is known as *stage* or *local feedback*. As in the case of two-junction transistors, where several stages are involved, *overall* or *loop feedback* is sometimes used.

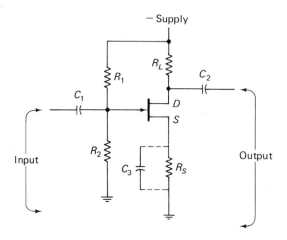

$$A_V \approx \frac{R_L}{\frac{1}{Y_{fs}} + R_S} \approx \frac{R_L}{R_S}$$

Drain voltage =
0.5 x supply

$$Z_{in} \approx R_1 \parallel R_2 \approx R_2 \qquad C_1 \approx \frac{1}{3.2\, FR_2} \text{ for 1 dB}$$

$$Z_{out} \approx R_L \qquad\qquad C_1 \approx \frac{1}{6.2\, FR_2} \text{ for 3 dB}$$

Fig. 2-28 Basic common-source
FET amplifier stage

2-5.3.1. Circuit Analysis

The outstanding characteristic of the circuit in Fig. 2-28 is that circuit gain, stability, impedance, and so on, are determined (primarily) by circuit values, rather than FET characteristics. The circuit shown is a *P*-channel FET. The power supply polarity must be reversed if an *N*-channel FET is used.

The maximum peak-to-peak output voltage is set by the supply voltage. For class A operation, the drain is operated at approximately one half the supply voltage. This permits the maximum positive and negative swing of output voltage. Generally, the absolute maximum peak-to-peak output can be between 90 and 95 per cent of the supply. For example, if the supply is 20 V, the drain will operate at 10 V (*Q* point), and swing from about 1 to 19 V. However, there is less distortion if the output is one half to one third of the supply. In any circuit the maximum drain–source voltage V_{DS} of the FET cannot be exceeded.

The input and/or output impedances are set by the resistance values (R_1, R_2, and R_L) as shown in Fig. 2-28. However, there are certain limitations for R_2 and R_L imposed by tradeoffs (for gain, impedance match, zero-temperature-coefficient operating point, etc.).

For example, the output impedance is set by R_L. If R_L is increased to match a given impedance, the gain will increase (all other factors remaining equal). However, an increase in R_L will lower the drain voltage Q point, since the same amount of I_D will flow through R_L and produce a larger voltage drop. This reduces the possible output voltage swing. A reduction in R_L produces the opposite effect, increasing the drain voltage Q point, but still reducing output voltage swing.

When R_1 is much larger than R_2 (which is generally the case), the input impedance of the circuit is set by the value of R_2. If R_2 is increased (or decreased) far from the value found in Sec. 2-5.1, the no-signal I_D point will change. If the value of I_D is chosen for $0TC$, and R_2 is changed drastically, the drain current will change, and the FET is no longer operating at the $0TC$ point. Generally, this is an undesirable condition. Typically, the common-source FET circuit is chosen for its high input impedance, thus presenting a low current drain to the preceding circuit. If the FET stage must provide a low input impedance, the common-gate circuit of Fig. 2-27 is generally preferred.

The values of C_1 and C_2 are dependent upon the *low-frequency limit* at which the amplifier is to operate. As discussed in Sec. 2-1, C_1 forms a high-pass (or low-cut) RC filter with R_B. Capacitor C_2 forms another filter with the input resistance of the following stage (or the load). For a given resistance values, a lower frequency requires a larger capacitor value, as shown by the equations on Fig. 2-28. Of course, if the resistance can be made larger (with the same desired frequency), the capacitor value can be reduced. Since FETs are high-impedance voltage-operated devices, the coupling capacitors in FET amplifiers are generally small (in relation to those of two-junction transistors).

Sufficient feedback. The design of a FET amplifier stage can be checked by noting if there is sufficient feedback. Such a condition occurs when the calculated gain is at least 75 per cent of the R_L/R_S ratio. If so, there is sufficient feedback to be of practical value. As an example, assume that R_L is 16.9 kΩ, R_S is 2.2 kΩ, and Y_{fs} is 2000 μmhos. The ratio of R_L/R_S is slightly over 7.6, with the gain slightly over 6.

$$\frac{R_L}{R_S} = \frac{16.9}{2.2} = 7.6 + ; \qquad \text{gain} \ \frac{16.9\text{k}\Omega}{(1/2000 \ \mu\text{mho}) + 2.2\text{k}\Omega} = \frac{16.9\text{k}\Omega}{2.7\text{k}\Omega} = 6+$$

Since 75 per cent of 7.6 is 5.7, the gain of 6 is greater, and there is sufficient feedback. Under these conditions the design should be stable.

2-5.4. Source Bypass for FET Amplifier Stage

Figure 2-28 shows (in phantom) a bypass capacitor C_3 across source resistor R_S. This arrangement permits R_S to be removed from the circuit as far as the signal is concerned, but leaves R_S in the circuit (in regards to direct current). With R_S removed from the signal path, the voltage gain is approximately equal to $Y_{fs} \times R_L$. Thus, the use of a bypass capacitor permits a temperature-stable dc circuit to remain intact, while providing a high signal gain.

A source resistance bypass capacitor also creates some problems. The Y_{fs} changes with frequency, and from FET to FET. Thus, circuit gain can only be approximated. The source bypass is generally used where maximum gain must be obtained from a single stage of amplification and a stable gain is of little concern. The value of C_3 can be found by

$$\text{capacitance} = \frac{1}{6.2F(R_S \times 0.2)}$$

where capacitance is in microfarads, F is low-frequency limit in hertz, and R_S is in megohms.

2-5.5. Basic FET Amplifier with Partially Bypassed Source

Figure 2-29 is the working schematic of a basic single-stage FET amplifier with a partially bypassed source resistor. This design is a compromise between the basic design without bypass and the fully bypassed

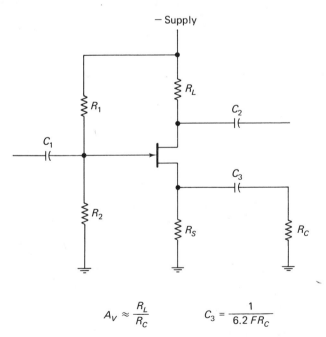

$$A_V \approx \frac{R_L}{R_C} \qquad C_3 = \frac{1}{6.2\,FR_C}$$

Fig. 2-29 Basic FET audio amplifier with partially bypassed source resistor

source. The dc characteristics of both the unbypassed and partially bypassed
circuits are essentially the same. All circuit values (except C_3 and R_C) can be
calculated in the same way for both circuits.

As shown in Fig. 2-29, the voltage gain for a partially bypassed FET
amplifier is greater than the unbypassed circuit, but less than for the fully
bypassed circuit. However, the gain can be set to an approximate value by
selection of circuit values, unlike the fully bypassed circuit where gain is
entirely subject to variations in Y_{fs}.

The value of R_C is chosen on the basis of desired voltage gain. R_C should be
substantially smaller than R_S. Otherwise, there will be no advantage to the
partially bypassed design. As shown by the equations, voltage gain is approximately equal to R_L/R_C. This holds true unless both Y_{fs} and R_C are very low
(where $1/Y_{fs}$ is about equal to R_C). In such a case, a more accurate gain
approximation is

$$\frac{R_L}{(1/Y_{fs}) + R_C}$$

2-5.6. Single-Stage Common-Drain FET Amplifier (Source Follower)

Figure 2-30 is the working schematic of a basic single-stage
FET source-follower (common-drain) circuit. Note that this circuit is similar
to that of Fig. 2-26, except that input and output coupling capacitors C_1
and C_2 are added to prevent dc flow to and from external circuits.

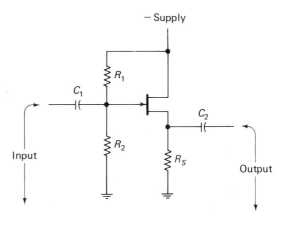

$$A_V \approx \frac{R_S}{\dfrac{1}{Y_{fs}} + R_S} \approx 0.6 \times \text{input}$$

$$Z_{in} \approx R_1 \,\|\, R_2 \approx R_2$$

$$Z_{out} \approx R_S \,\|\, \frac{1}{Y_{fs}}$$

Fig. 2-30 Basic common-drain
(source-follower) FET amplifier
stage

Input to the source follower is applied between gate and ground across R_2. Output is taken across the source and ground. The input signal adds to, or subtracts from, the bias voltage across R_2. Variations in bias voltage cause corresponding variations in I_D and the voltage drop across R_S. Therefore, the source voltage (or circuit output) follows the input signal waveform, and remains in phase.

Variations in voltage drop across R_s change the gate–source bias relationship, and tend to cancel the initial bias change caused by the input signal. This serves as a form of negative feedback to increase stability (and limit gain).

The circuit of Fig. 2-30 is used primarily where high input impedance and low output impedance (with no phase inversion) are required, but no gain is needed. The source follower is the FET equivalent of the two-junction transistor emitter follower and the vacuum-tube cathode follower.

2-5.6.1. Circuit Analysis

The Q-point voltage at the circuit output (source terminal) is set by I_D under no-signal conditions and the value of R_S. Since R_S is typically small, the Q-point voltage is quite low in comparison to the common-source amplifier. In turn, the maximum allowable peak-to-peak output voltage is also low. For example, if the source is at 1 V with no signal, the maximum possible peak-to-peak output is less than 1 V. Of course, the value of R_S can be increased as necessary to permit a higher output.

The input and/or output impedances are set by the resistance values (R_1, R_2, and R_S). However, there are certain limitations for R_S imposed by trade-offs for impedance match and output Q point.

For example, the output impedance is the parallel resistance combination of R_S and $1/Y_{fs}$. If R_S is made very small (less than 10 times) in relation to $1/Y_{fs}$, the output impedance is approximately equal to R_S. A low value of R_S decreases the source (output) voltage Q point, thus reducing output voltage swing. If R_S is made large in relation to $1/Y_{fs}$, the output impedance is approximately equal to $1/Y_{fs}$, and is subject to variation with frequency, and from FET to FET.

There is no voltage gain for a source follower. Typically, the output voltage is about 0.6 times the input voltage, depending upon the ratio of $1/Y_{fs}$ to R_S. However, the source follower is capable of current gain, and thus *power gain*. For example, assume that 1 V is applied at the input and 0.6 V is taken from the output. Furthermore, assume that the input impedance is 300 kΩ, and the output impedance is 300 Ω. The input power is approximately 0.0033 mW, while the output power is about 1.2 mW, indicating a power gain of about 350.

2-5.7. Single-Stage Common-Gate FET Amplifier

Figure 2-31 is the working schematic of a basic single-stage FET common-gate amplifier. Note that the basic circuit is similar to that of Fig. 2-27, except that C_1 and C_2 are added to prevent dc flow to and from external circuits.

Input is applied at the source, across a portion of R_S. Typically, the value of R_{S1} is equal to R_{S2}, although some circuits divide the resistance value unequally. The total value of $R_S(R_{S1} + R_{S2})$ must be considered when calculating the dc characteristics of the circuit. Output is taken across the drain and ground. The input signal adds to or subtracts from the bias voltage across R_S. Variations in bias voltage cause corresponding variations in I_D and the voltage drop across R_L. The drain voltage (or circuit output) follows the input signal, in phase.

Variations in voltage drop across R_S change the gate–source bias relationship. This change in bias tends to cancel the initial bias change caused by the input signal, and serves as a form of negative feedback to increase stability (and limit gain).

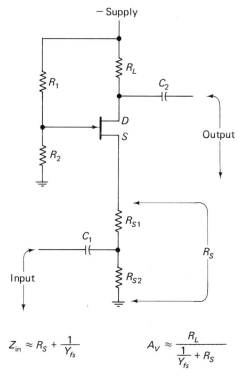

$$Z_{in} \approx R_S + \frac{1}{Y_{fs}} \qquad A_V \approx \frac{R_L}{\frac{1}{Y_{fs}} + R_S}$$

Fig. 2-31   Basic common-gate
FET amplifier stage $Z_{out} \approx R_L$

The circuit of Fig. 2-31 is used primarily where low input impedance and high output impedance (with no phase inversion) are required. Gain is determined (primarily) by circuit values, rather than FET characteristics. The common-gate amplifier is the FET equivalent of the two-junction transistor common-base amplifier and the vacuum-tube common-grid amplifier.

2-5.7.1. Circuit Analysis

Although the input and/or output impedances are set by the resistance values, the input impedance of the circuit is dependent upon the reciprocal of the Y_{fs} ($1/Y_{fs}$) factor. This is true unless the value of R_S is many times (at least 10) that of $1/Y_{fs}$.

Capacitor C_1 forms a high-pass RC filter with R_{S2}. Capacitor C_2 forms another high-pass filter with the input resistance of the following stage (or the load). Using a 1-dB loss as the low-frequency cutoff point, the value of C_1 is approximately

$$\frac{1}{3.2FR_{S2}}$$

where capacitance is in microfarads, F is the low-frequency limit in hertz, and R_{S2} is in megohms.

2-5.8. Basic FET Amplifier Without Fixed Bias

Figure 2-32 shows a basic single-stage FET amplifier without fixed bias. Capacitors C_1 and C_2 prevent dc flow to and from external circuits. The resistor R_1 provides a path for bias and signal voltages between gate and source.

Input is applied between gate and ground across R_1 Output is taken across the drain and ground. The input signal adds to, or subtracts from, the bias voltage across R_1. Variations in bias voltage cause corresponding variations in I_D and the voltage drop across R_L. Thus, the drain voltage (or circuit output) follows the input signal waveform, except that the output is inverted in phase.

Variations in I_D also cause variation in voltage drop across R_S and a change in the gate–source bias relationship. The change in bias that results from the voltage drop across R_S tends to cancel the initial bias change caused by the input signal, and serves as a form of negative feedback to increase stability (and limit gain).

The major difference in the circuit of Fig. 2-32 and the FET amplifier with fixed bias is that the amount of I_D at the Q point is set entirely by the value of R_S. It may not be possible to achieve a desired I_D with a practical value of

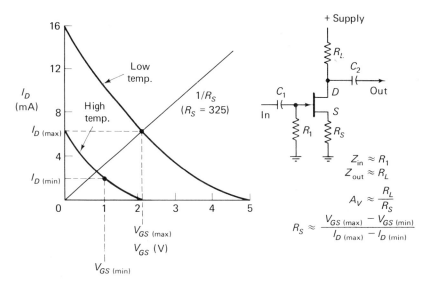

Fig. 2-32   Basic FET amplifier without fixed bias

R_S. Thus, it may not be possible to operate at the zero-temperature-coefficient point. If this is of less importance than minimizing the number of circuit components (elimination of one resistor), the circuit of Fig. 2-32 can be used in place of the fixed-bias FET amplifier.

2-5.8.1. Circuit Analysis

The input and/or output impedances are set by the resistance values of R_1, R_S, and R_L. However, there are certain limitations for the resistance values imposed by tradeoffs (for gain, impedance match, operating point, etc.).

For example, the value of R_S sets the amount of bias, and thus the amount of I_D. At the same time, the ratio of R_L/R_S sets the amount of gain. Going further, the value of R_L sets the output impedance. If R_S is changed to change the I_D, both the gain and Q point will be changed. If R_L is changed to match a given impedance, both the gain and Q point will change.

The input impedance is set by R_1. A change in R_1 will have little effect on gain, operating point, or output impedance. However, R_1 forms a high-pass RC filter with C_1. A decrease in R_1 requires a corresponding increase in C_1 to accommodate the same low-frequency cutoff point. As a general rule, the value of R_1 is high (in the megohm range). This minimizes current drain on the stage ahead of the FET.

2-5.9. Basic FET Amplifier with Zero Bias

Figure 2-33 is the working schematic of a basic single-stage FET amplifier operating at zero bias and without feedback. Capacitors C_1 and C_2 prevent dc flow to and from external circuits. Resistor R_1 provides a path for signal voltages between gate and source.

Input to the amplifier is applied between gate and ground across R_1. Output is taken across the drain and ground. Variations in gate voltage cause corresponding variations in I_D and the voltage drop across R_L. Thus, the drain voltage (or circuit output) follows the input signal waveform, except that the output is inverted in phase.

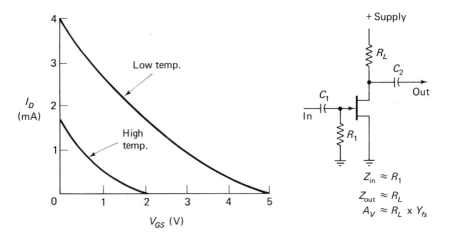

Fig. 2-33   Basic FET amplifier with zero bias

Any FET will have some value of I_D at zero V_{GS}. If the FET has characteristics similar to those of Fig. 2-33, the I_D will vary between about 1.75 and 4 mA, depending upon temperature, and from FET to FET. Therefore, with a zero-bias circuit it is impossible to set the Q point I_D at any particular value. Likewise, drain voltage Q point is subject to considerable variation. Since there is no source resistor, there is no negative feedback. Thus, there is no means to control this variation in I_D. For these reasons, the zero-bias circuit it used where circuit stability is of no particular concern.

2-5.9.1. Circuit Analysis

The input and/or output impedances are set by the values of R_1 and R_2. However, as in the case of other circuits, there are tradeoffs for gain, impedance match, operating point, and so forth.

For example, the value of R_L sets the amount of gain (with a stable Y_{fs}) and the drain voltage operating point (with a stable I_D). Thus, a change in Y_{fs} (which is usually accompanied by a change in I_D) causes a change in gain (and probably a shift in Q-point voltage). At best, the zero-bias voltage is unstable, even though the input and output impedances remain fairly constant. In analyzing the zero-bias circuit, both the *minimum* and *maximum* values of I_D must be considered, as well as the minimum and maximum Y_{fs}.

2-6. MULTISTAGE AMPLIFIERS

When stable voltage gains greater than about 20 are required, and it is not practical to bypass the emitter (or source) resistor of a single stage, two or more transistor amplifier stages can be used in *cascade* (where the output of one transistor is fed to the input of a second transistor). In theory, any number of two-junction or FET amplifiers can be connected in cascade to increase voltage gain. In practice, the number of stages is usually limited to three. The overall gain of the amplifier is equal (approximately) to the cumulative gain of each stage, multiplied by the gain of the adjacent stage.

As an example, if each stage of a three-stage amplifier has a gain of 10, the overall gain is (approximately) 1000 ($10 \times 10 \times 10$). Since it is possible to design a very stable single stage with a gain of 10, and adequately stable stages with gains of 15 to 20, a three-stage amplifier could provide gains in the 1000 to 8000 range. Generally, this is more than enough voltage gain for most practical applications. Using the 8000 figure, a $1\text{-}\mu\text{V}$ input signal (say from a low-voltage transducer or delicate electronic device) can be raised to the 8-mV range, while maintaining stability in the presence of temperature and power supply variations.

2-6.1. Basic Considerations for Multistage Amplifiers

Any of the single-stage amplifiers described in previous sections of this chapter could be connected to form a two-stage or three-stage *voltage* amplifier. For example, the basic stage (without emitter or source bypass) can be connected to two like stages in cascade. The result is a highly temperature stable voltage amplifier. Since each stage has its own feedback, the gain is precisely controlled and very stable.

It is also possible to mix stages to achieve some given design goal. For example, a three-stage amplifier can be designed using a highly stable, unbypassed amplifier for the first stage, and two bypassed amplifiers for the remaining stages. Assuming a gain of 10 for the unbypassed stage, and gains of 30 for the bypassed stages, this results in an overall gain of 9000. Of course,

with the bypassed stages the gain is dependent upon the transistor characteristics (dynamic impedance, h_{fe}, Y_{fs}, etc.), and is therefore unpredictable.
However, once the gain is established for a given amplifier, the gain should
remain fairly stable.

Since design of a multistage, capacitor-coupled voltage amplifier is essentially the same as for individual stages, no specific circuit example is given.
In practical terms, each stage is analyzed and designed as described in previous sections of this chapter. However, a few precautions must be considered.

Distortion and clipping. As in the case with any high-gain amplifier, the
possibility of *overdriving* a multistage solid-state amplifier is always present.
If the maximum input signal is known, check this value against the overall
gain and the maximum allowable output signal swing.

As an example, assume an overall gain of 1000 and a supply voltage of
20 V. Typically, this implies a 10-V Q point (for the output collector or drain)
and a 20-V (peak-to-peak) output swing (from 0 to 20 V). In practice, a
swing from about 1 to 19 V is more realistic. Either way, a 20-mV (P-P)
input signal, multiplied by a gain of 1000, will drive the final output to its
limits, and possibly into distortion or clipping.

Feedback. When each stage of a multistage amplifier has its own feedback
(local or stage feedback), the most precise control of gain is obtained. However, such feedback is often unnecessary. Instead, overall feedback (or loop
feedback) can be used, where part of the output from one stage is fed back
to the input of a previous stage. Usually, such feedback is through a resistance (to set the amount of the feedback), and the feedback is from the final
stage to the first stage. However, it is possible to use feedback from one stage
to the next (second stage to first stage, third stage to second stage, etc.).

Feedback phase inversion. There is a problem of phase inversion when using
loop or overall feedback. In a common-emitter or common-source amplifier,
the phase is inverted from input to output. If feedback is between two stages,
the phase is inverted twice, resulting in *positive feedback*. This usually produces oscillation. In any event, positive feedback will not stabilize gain. The
phase inversion problem can be overcome, when stages are involved, by
connecting the output collector (or drain) of the second stage back to the
emitter (or source) of the first stage. This will produce the desired *negative
feedback*.

As an example, if the base (or gate) of the first stage is swinging positive,
the collector (or drain) of that stage will swing negative, as will the base (or
gate) of the second stage. The collector (or drain) of the second stage will
swing positive, and this positive swing can be fed back to the emitter (or
source) of the first stage. A positive input at the emitter (or source) has the
same effect as a negative at the base (or gate). Thus, negative feedback is
obtained.

Low-frequency cutoff. Unless direct coupling is used (as described in Chapter 4), coupling capacitors must be used between stages, as well as at the input and output. Such capacitors form a low-pass RC filter with the base-to-ground (or gate-to-ground) resistance. Thus, each stage has its own low-pass filter. In multistage amplifiers the *effects of these filters are cumulative*.

As an example, if each filter causes a 1-dB drop at some given cutoff frequency, and there are three filters (one at the input and two between stages), the result is a 3-dB drop at that frequency in the final output. If this cannot be tolerated, the RC relationship must be redesigned. In practical terms, this means increasing the value of C, since a change in R will usually produce some undesired shift in operating point or other circuit characteristic.

2-6.2. Direct-Coupled and Differential Multistage Amplifiers

One method of eliminating the RC filter problem created by coupling capacitors is to use *direct coupling*. This eliminates the interstage coupling capacitor, as well as some of the interstage resistances. In addition to the direct-coupled amplifier, there are a number of transformerless multistage circuits used to provide voltage, current, and even power amplification at audio frequencies. The circuits for the most important of these are the Darlington-pair configuration (compound), the phase inverter or splitter, the emitter-coupled amplifier, the transformerless series-output amplifier, the quasi-complementary amplifier, and the full-complementary amplifier. Since all these circuits involve some form of direct coupling, they are all discussed in Chapter 4, along with an analysis of FET multistage dc amplifiers. Due to their highly specialized nature, differential amplifiers used at audio frequencies (as well as any other frequency) are discussed in Chapter 5.

2-7. MULTISTAGE AUDIO AMPLIFIERS WITH TRANSFORMER COUPLING

Audio amplifier stages can be coupled by means of transformers. Iron-core transformers are used in the AF range, particularly when power amplification is required. (As discussed in Chapter 3, air-core, or open-coil, transformers are used at higher frequencies.)

As is the case with any coupling method, transformers impose certain problems and have certain advantages. These factors can be traded off to meet a specific design need.

Inductive reactance. One major problem with transformers is the inductive reactance created by the transformer windings. (Inductive reactance increases with frequency.) At high frequencies (beyond about 20 kHz) the inductive reactance of iron-core transformers becomes so high that signals cannot pass, or are badly attenuated. For this reason, iron-core transformers are not used at higher frequencies. At low frequencies the reactance drops to near zero, even with iron cores. Since transformers are placed across (or in shunt with) the amplifier circuits, the transformer winding acts as a short at low frequencies. Therefore, if an amplifier must operate at very low frequencies (10 to 20 Hz), transformers of all types are usually avoided. For operation above the audio range, transformers are used less frequently (or air-core transformers are used).

Size and weight. For a transformer to handle any large amounts of power or current, the winding wire must be large. Also, the iron core must be large. Both of these add up to bulk weight. Except in certain cases, the added weight of transformers defeats the purpose of compact, lightweight, solid-state equipment.

Short-circuit burnout. Another problem with transformers is the danger of a short-circuited output, resulting in excessive (simultaneous) voltage and current in the transistor collector. There is very little voltage drop across a transformer winding in a transistor collector circuit (compared to the drop across the load resistor in an *RC* or direct-coupled amplifier). Therefore, if a short circuit in the output (say due to a short across the load) causes heavy current flow, the transistor can be damaged.

Impedance matching. One of the major advantages to transformer coupling is the impedance-matching capability. The output impedance of a typical *RC* amplifier is on the order of several hundred (or thousand) ohms (generally set by the output collector resistor value). In audio systems (particularly for voice and music reproduction) this large output impedance must be matched to 4-, 8-, and 16-Ω loudspeaker systems. The severe mismatch results in power loss. With a transformer, the primary winding can be designed (or selected) to match the transistor circuit output impedance with the transformer secondary matching the loudspeaker (or other load) impedance.

Low supply voltage. Another major advantage of transformer coupling is caused by the low voltage drop across the transformer winding. Because of this low voltage drop, it is possible to operate a transformer-coupled amplifier with a much lower supply voltage than with an *RC* amplifier. As a rule of thumb, the transformer-coupled amplifier can be operated at *one half the supply voltage* required for a comparable *RC* amplifier.

Typical uses. Transformers are often used in the audio-amplifier sections of transistorized radio receivers and portable hifi systems. In these applications, very little power is required, so the transformers can be made compact and lightweight. Since there is no loss in impedance match and low voltage is required, the transformer-coupled circuits are ideal for battery operation. Transformers are also used in high-power, hifi-stereo systems and television audio sections where the added weight is of little consequence.

2-7.1. Circuit Analysis

Figure 2-34 is the working schematic for a classic transformer-coupled audio amplifier. As shown, the circuit has a class A input or *driver* stage and a class B push–pull output stage. The class A stage provides both voltage and power amplification as needed to raise the low input signal to a level suitable for the class B power output stage.

The class A stage can be transformer coupled or *RC* coupled at the input, as needed. Transformer coupling is used at the input where a specific imped-ance-match problem must be considered in design. The class A input stage can be driven directly by the signal source, or can be used with a preamplifier

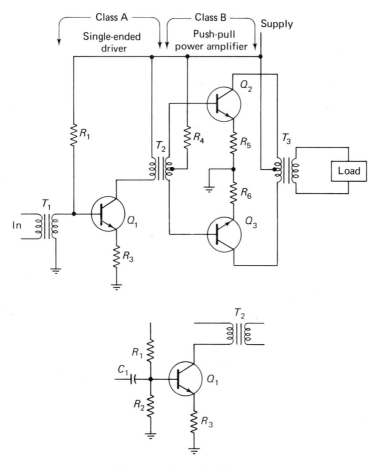

Alternate driver with *RC* input

Fig. 2-34   Transformer-coupled audio amplifier

for very low level signals. When required, a high-gain voltage amplifier (such as described in previous sections of this chapter, and in Chapter 4) is used as a *preamplifier*.

The push–pull output stage may be operated as a class B amplifier. That is, the transistors are cut off at the Q point and draw collector current only in the presence of an input signal. Class B is the most efficient operating mode for audio amplifiers, since it draws the least amount of current (and no current where there is no signal). However, class B operation can result in *crossover distortion*, as discussed in Sec. 1-8.3.

The effects of crossover distortion can be seen by comparing the input and output waveforms on Fig. 2-35a. In true class B operation, the transistor *remains cut off* at very low-signal inputs (because transistors have low current

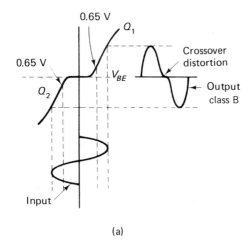

(a)

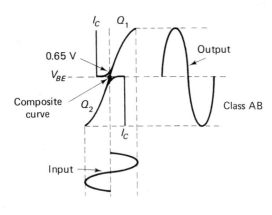

(b)

Fig. 2-35 Effects of crossover distortion, and how it is eliminated by forward bias of the base–emitter junction

gain at cutoff) and turns on abruptly with a large signal. As shown on Fig. 2-35a, there is no conduction when the base–emitter voltage V_{BE} is below about 0.65 V (for a silicon transistor). During the instantaneous pause when one transistor stops conducting and the other starts conducting, the output waveform is distorted.

Note that distortion of the signal is not the only bad effect of this crossover condition. The instantaneous cutoff of collector current can set up large voltage transients equal to several times the size of the supply voltage. This can cause the transistor to break down.

Crossover distortion can be minimized by operating the output stage as class AB (or somewhere between B and AB). That is, the transistors are forward biased just enough for a small amount of collector current to flow at the Q point. Therefore, some collector current is flowing at the lowest signal levels, and there is no abrupt change in current gain. The effects of this are shown in Fig. 2-35b. Note that the combined collector currents result in a *compositive curve* that is essentially linear at the crossover point. This produces an output that is a faithful reproduction of the input, at least as far as the crossover point is concerned. Of course, class AB is less efficient than class B, since more current must be used.

Some designers use an alternative method to minimize crossover distortion. This technique involves putting diodes in series with the collector or emitter leads of the push–pull transistors. Because the voltage must reach a certain value (typically 0.65 V for silicon diodes) before the diode will conduct, the collector-current curve is rounded (not sharp) at the crossover point.

Amplifier efficiency. The efficiency of an amplifier is determined by the ratio of collector input-to-output power. That is, an amplifier with 70 per cent efficiency will produce 7-W output for a 10-W input (with power input being considered as collector source voltage multiplied by total collector current).

Typically, class B amplifiers can be considered as 70 to 80 per cent efficient. Class A amplifiers are typically in the 35 to 40 per cent efficiency range, with class AB amplifiers showing 50 to 60 per cent efficiency.

In all cases, any amplifier circuit that produces an increase in collector current at the Q point will produce correspondingly lower efficiency. This results in a tradeoff between efficiency and distortion.

The efficiency produced by a class of operation also affects the heat-sink requirements. Any design that produces more collector current at the Q point requires a greater heat-sink capability. As a rule of thumb, a *class A amplifier requires double the heat-sink capability of a class B*, all other factors being equal.

Distortion of class B versus class AB. In practical terms, push–pull amplifiers are usually designed for true class B operation (transistors at or near cutoff at the Q point), and are then tested (in experimental form) for distortion. If the distortion is severe, the base–emitter forward bias is increased so that some current flows at the Q point. Then the amplifier is again tested for distortion. The amount of forward bias is adjusted until the desired distortion is reached, or until there is a compromise between distortion and power output. This approach is usually more realistic than trying to design an amplifier for a given class of operation.

Output power. The true output power of an audio amplifier can be determined by measuring the voltage across the load, and then solving the equation

$$\text{power output} = \frac{\text{voltage}^2}{\text{load impedance}}$$

The output of a transformer-coupled amplifier is also related to the collector current (produced by the signal) and the primary impedance of the transformer. In a push–pull amplifier, the relationship is

$$\text{power output} = \frac{\text{current}^2 \times \text{primary impedance}}{8}$$

In the single-ended amplifier, the relationship is

$$\text{power output} = \frac{\text{current}_2 \times \text{primary impedance}}{2}$$

These relationships provide a basis for design of transformer-coupled amplifiers.

Transformer characteristics. Audio transformers are listed by primary impedance, secondary impedance, and power-output capability. From a practical standpoint, it is not always possible to find an off-the-shelf transformer with exact primary/secondary relationships at a given power rating. Most manufacturers will produce transformers with exact impedance relationships on a special-order basis. However, this is usually not practical, except for special applications. Instead, the transformer is generally selected for exact secondary impedance and the nearest value of primary impedance within the given power rating.

The following rules can be applied to selection of the transformers shown in Fig. 2-34.

Push–pull output transformer T_3. The rated power capability of transformer T_3 should be about 1.1 times the desired power. The secondary impedance should match the load impedance (loudspeaker or other) into which the amplifier must operate. The primary impedance should be determined by maximum collector current passing through the primary windings and the total collector voltage swing. The relationship is

$$\text{primary impedance} = \frac{4(\text{supply voltage} - \text{min voltage})}{\text{max current}}$$

Minimum voltage can be determined by reference to the collector voltage–current curves for the transistors. As shown in Fig. 2-36, the minimum voltage point should be selected so that it is just to the right of the curved portion of the characteristics (where the curves start to straighten out). If curves are not available for the transistors, an arbitrary 2 V can be used for the minimum voltage. This will be satisfactory for most power transistors.

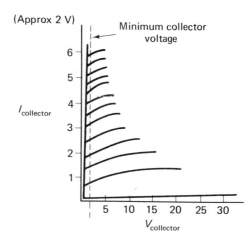

Fig. 2-36   Locating minimum collector voltage point on typical power transistor curves

Maximum current can be determined by the total collector voltage swing and the desired power output. The relationship is

$$\text{max current} = \frac{2.1 \times \text{power output}}{(\text{supply voltage} - \text{min voltage})}$$

Push–pull input transformer T_2.   Note that transformer T_2 is the input transformer for the push–pull output stage, and the output transformer for the single-ended driver. The secondary impedance should be chosen to match the signal input impedance of the push–pull stage, whereas the primary of T_2 is chosen to match the driver output.

The rated power capability of T_2 should be equal to the input power of the driver stage (which is generally about three times the output to the push–pull stage).

The total secondary impedance of T_2 should be four times the signal impedance of Q_2 and Q_3. This input impedance is found by dividing signal voltage by signal current.

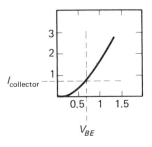 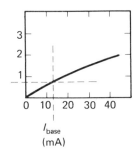

Fig. 2-37   Curves showing base voltage and current versus collector current for typical power transistor

The signal voltage and current are dependent upon the desired amount of collector signal current. In turn, the collector signal current is dependent upon the desired power output (from Q_2 to Q_3) and the primary impedance of T_3 (total primary impedance). The relationship is

$$\text{collector signal current} = \sqrt{\frac{8 \times \text{power output}}{\text{total primary impedance}}}$$

With the required collector signal current established, the input signal voltage and current for Q_2 and Q_3 can be found by reference to the *transfer characteristic curves*. Transfer curves are usually provided on power transistor datasheets. Typical transfer characteristics for power transistors are shown in Fig. 2-37. These characteristics illustrate the required base–emitter voltage (or signal voltage) and base current (or signal current) for a given collector current.

An additional factor must be considered in establishing signal voltage. The base–emitter voltages shown in Fig. 2-37 can be considered as the signal voltages only when the emitters of Q_2 and Q_3 are connected directly to ground. If emitter resistors (R_5 and R_6) are used to provide feedback stabilization, the voltage developed across the emitter resistors must be added to the base–emitter voltage to find a true signal voltage.

The values of R_5 and R_6 are chosen to provide a voltage drop *approximately equal* to the base–emitter voltage when the collector signal current is passing through the emitter resistors. Since the base–emitter voltage of a silicon power transistor is typically less than 1 V, the values of R_5 and R_6 can arbitrarily be set to provide between a 0.5- and 1-V drop with normal collector signal current.

The *primary impedance of T_2* should match the output of Q_1. The relationship is

$$\text{primary impedance} = \frac{\text{supply voltage} - \text{min voltage}}{\text{collector current at } Q \text{ point}}$$

However, for practical design it is generally easier to select the transformer on the basis of secondary impedance and power rating, and then adjust the Q_1 output to match a primary impedance available with an off-the-shelf transformer. Note that a large primary impedance requires a high supply voltage for a given collector current. It may be necessary to trade off between supply voltage, available primary impedances (for off-the-shelf transformers with correct secondary impedance and power rating), and collector current.

Once the primary impedance of T_2 is established, the collector signal current for Q_1 can be determined. The relationship is

$$\text{collector signal current} = \frac{2 \times \text{power output}}{\text{primary impedance}}$$

The power output from Q_1 must equal the required power input for Q_2 and Q_3. This power is determined by

$$\text{power input} = \text{signal voltage} \times \text{signal current}$$

Input transformer T_1. As shown in Fig. 2-34, transformer T_1 can be omitted if the RC circuit is used. Generally, the transformer-coupled circuit is used if it is of particular importance to match the impedance of the signal source to the amplifier. When used, the primary impedance of T_1 should equal that of the signal source. The secondary of T_1 should match the signal input of Q_1. The signal input impedance is found by

$$\text{signal input impedance} = \frac{\text{signal voltage}}{\text{signal current}}$$

The signal (input) voltage and current are dependent upon the desired amount of collector signal current for Q_1 (previously established in calculating the secondary impedance of T_2). With the required collector signal current established, the input signal voltage and current for Q_1 can be found by reference to the transfer characteristic curves.

If emitter resistor R_3 is used to provide feedback stabilization, the voltage developed across the emitter resistor must be added to the base–emitter voltage to find a true signal voltage. The value of R_3 can be selected to provide between 0.5- and 1-V drop with normal collector signal current (as discussed for emitter resistors R_5 and R_6).

If transfer characteristic curves are not available for Q_1 (as is sometimes the case for low-power transistors), it is still possible to find the approximate signal voltage and current required to produce the necessary output. Signal voltage can be approximated by assuming that the base–emitter drop is 0.5 V. This must be added to any drop across emitter resistor R_3 (arbitrarily chosen to be between 0.5 and 1 V). Signal current can be approximated by dividing the desired collector signal current by beta of Q_1. Of course, since beta is variable, the input signal current can only be a rough estimate.

Supply voltage. The source voltage required for Q_1 is approximately double that required for Q_2 and Q_3. However, since all three transistors are operated from the same supply, the source voltage is chosen to match the requirements of Q_1. As a guideline,

3 to 9 V is used for power outputs up to 2 W

6 to 15 V for power up to 20 W

15 to 50 V for power up to 50 W

These rules apply to the output of Q_1. Transistors Q_2 and Q_3 require approximately one half the voltage. For example, if Q_1 is to deliver 20 W (by itself), 6 to 15 V is required. If Q_2 and Q_3 are to deliver the same 20 W, the voltage required will run between 3 and 7.5 V.

In any amplifier circuit, a higher supply voltage permits lower currents (and vice versa), all other factors being equal.

Transistor selection. Both frequency limit and power dissipation must be considered in selecting transistors. In any audio system the transistors should have an f_{ae} (Sec. 2-3.1.1) higher than 20 kHz, and preferably higher than 100 kHz. The power dissipation of Q_1 should be at least three times the required output of Q_1. The power dissipation of Q_2 and Q_3 should be between 1.3 and 1.5 times the required output of the amplifier circuit (at T_3). If any power dissipation exceeds about 1 W, heat sinks will be required. Refer to Sec. 2-3.1.3.

Base bias resistances. Resistors R_1, R_2, and R_4 serve to drop the supply voltage to a level required at the transistor bases.

The drop across R_4 should (in theory) bias Q_2 and Q_3 at cutoff. That is, no base current should flow except in the presence of a signal. This is not true in a practical circuit, since the complete absence of base current will result in no drop across R_4. Also, such a bias condition will usually result in crossover distortion. As a starting point for Q_2 and Q_3 bias values, assume that a small residual base current is going to flow in each transistor. The value of this residual current can be assumed (arbitrarily) as 0.1 times the normal signal current. The combined residual currents of Q_2 and Q_3 flow through R_4. The value of R_4 is calculated by

$$R_4 = \frac{\text{supply voltage}}{2 \times \text{residual current}}$$

If transformer T_1 is used, resistor R_2 is omitted. Under these conditions, the drop across R_1 should bias Q_1 at the operating point. The base current of Q_1 flows through R_1 and produces the drop. The base current used to calculate the Q_1 input impedance can be used as a starting point for the R_1 value. The relationship is

$$R_1 = \frac{\text{supply voltage}}{\text{base current}}$$

If transformer T_1 is not used, resistor R_2 is added. The value of R_2 determines the *approximate* input impedance of the amplifier, and is generally so selected. Typically, R_2 is greater than 500 Ω, and less than 20 kΩ. When R_2 is used, the voltage drop from base to ground will also appear across R_2, resulting in some current flow through R_2. This current flow must be added to the base current in calculating the value of R_1.

Coupling capacitor. When transformer T_1 is used, capacitor C_1 is omitted. When used, C_1 forms a high-pass filter with R_2, producing some loss at the low-frequency end of the audio range. Using a 1-dB loss as the low-frequency cutoff point, the value of C_1 can be found by

$$\text{capacitance (F)} = \frac{1}{3.2 \text{ frequency (Hz)} \times \text{resistance } (\Omega)}$$

2-7.2. Single-Ended Class B Audio Amplifier

It is possible to have a class B audio amplifier with two transistors, but without an output transformer. Such an arrangement is generally referred to as single-ended class B, and is shown in Fig. 2-38. An interstage transformer T_1 is required. However, transformer T_1 need not carry the heavy current load found in the output transformer of a typical class B amplifier (Fig. 2-34).

In the circuit of Fig. 2-38, Q_1 is biased as class A, with Q_2 and Q_3 biased class B (or with some current flowing to eliminate crossover distortion). The polarities of the T_1 windings are arranged so that Q_2 and Q_3 conduct on alternate half-cycles. The series connection of Q_2 and Q_3 causes the signal current to flow through the load (loudspeaker) in *opposite directions* on alternate half-cycles. In this way, the loudspeaker is driven as if from a push–pull amplifier. One problem with the circuit of Fig. 2-38 is that a short in the load will produce heavy current, without reducing the collector voltage. This can result in the destruction of the transistors, as is the case with transformer-output class B circuit. The destructive condition is known as *secondary breakdown* or *second breakdown*, and results from a sudden channeling of collector current into a localized area of the transistor. Secondary breakdown can be prevented (or minimized) by limiting the collector current–voltage product. Likewise, there are some short-circuit protection circuits, such as described in Chapter 4 for the direct-coupled audio amplifiers.

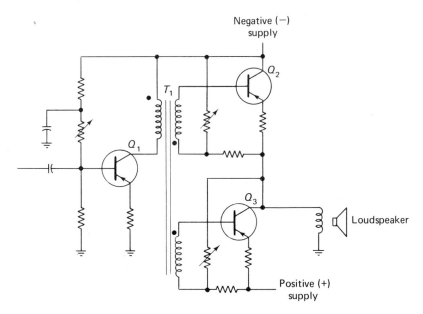

Fig. 2-38   Class B audio amplifier without an output transformer (single ended)

2-8. AUDIO-AMPLIFIER OPERATING AND ADJUSTMENT CONTROLS

The most common operating controls for audio circuits used with music or voice reproduction equipment (hi-fi, stereo, public address, etc.) are the *volume* or *loudness* control, the *treble* control, and the *bass* control. The other most common audio control is the *gain* control (found on such circuits as operational amplifiers, power control amplifiers, etc.).

The gain and volume controls are often confused, since they both affect output of the amplifier circuit. A true gain control sets the *gain of one stage* in the amplifier, thus setting the overall gain of the complete amplifier. A true volume control sets the *level of the signal* passing through the amplifier, without affecting the gain of any or all stages. A gain control is usually incorporated as part of a stage, whereas a volume control is usually found between stages, or at the input to the first stage.

In addition to volume, bass, and treble controls, most stereo amplifier systems have some form of *balance* control (so that both channels of audio can be balanced). Also, most stereo-hifi systems have a form of *playback equalization* (for tape and phonograph playback). In the following paragraphs we shall concentrate on analysis of the basic operating and adjustment controls to see how they affect the related audio-amplifier circuit.

2-8.1. Volume-Control Circuit Analysis

As shown in Fig. 2-39, the basic volume control is a variable resistor or potentiometer connected as a voltage divider. The voltage output (or signal level) is dependent upon the volume-control setting.

If the audio circuit is to be used with voice or music, the volume control is usually of the *audio taper* type where the voltage output is not linear through-out the setting range. (The resistance element is not uniform.) This produces a nonlinear voltage output to compensate for the human ear's nonlinear response to sound intensity. (The human ear has difficulty in hearing low-frequency sounds at low levels, and responds mainly to the high-frequency components.) Generally, the audio taper controls used with transistors are of the type where large changes are produced at the high-loss end. Audio taper potentiometers are also used with bass and treble controls, as is dis-cussed in later paragraphs.

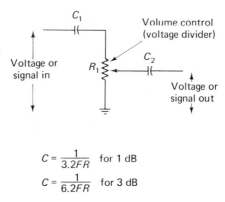

$$C = \frac{1}{3.2FR} \quad \text{for 1 dB}$$

$$C = \frac{1}{6.2FR} \quad \text{for 3 dB}$$

Fig. 2-39 Basic volume-control
circuit and equations

C in farads, F in hertz, R in ohms

If the audio circuit is not used with voice or music, the volume control is usually of the linear type (unless there is some special circuit requirement). With such a control, the actual voltage or signal is directly proportional to the control setting.

No matter what type of volume control is used, it should be isolated from the circuit elements. If a volume control is part of the circuit (such as the collector or base resistance), any change in volume setting can result in a change of impedance, gain, or bias. The simplest method for isolating a volume control is to use coupling capacitors, as shown in Fig. 2-39. However, the capacitors create a low-frequency response problem. As in the case of coupling capacitors described in previous sections, capacitor C_1 forms a high-pass RC filter with volume potentiometer R_1. Coupling capacitor C_2 forms another high-pass filter with the input resistance of the following stage.

The volume control should be located at a low-signal-level point in the amplifier circuit. The most common location for a volume control is at the amplifier input stage, or between the first and second stages.

When a volume control is located at the amplifier input, the control's resistance forms the input impedance (approximately). Volume controls are available in standard resistance values. Select the standard resistance value nearest the desired impedance.

When a volume control is located between stages, the resistance value should be selected to match the output impedance of the previous stage. Use the nearest standard resistance value to produce the least loss.

Very little current is required for a volume control that is isolated as shown in Fig. 2-39. Therefore, the power rating (in watts) required is quite low. Usually, 1 or 2 W are more than enough for any volume control used in transistor audio-amplifier circuits. Wirewound potentiometers should not be used for any audio application. The inductance produced by wirewound potentiometers can reduce the frequency response of the circuit.

Figure 2-39 shows the equations for low-frequency cutoff versus *RC* value relationships of typical audio volume controls. Note that these are the same as for high-pass filters discussed in previous sections.

Volume control using attenuators and pads. In some applications the volume of audio signals is controlled by various attenuators and pads (such as T, L, O, and H pads). These attenuators and pads are made up of several inter-related resistances, all mechanically coupled to a common control shaft. Such attenuators and pads are commercially available, and no detailed analysis will be given here. Further information on attenuators and pads is available in the author's *Handbook of Electronic Charts, Graphs, and Tables* (Prentice-Hall, Inc., Englewood Cliffs, N.J., 1970).

2-8.2. Gain-Control Circuit Analysis

As shown in Fig. 2-40, the basic gain control is a variable resistance or potentiometer, serving as one resistance element in the amplifier circuit. Any of the three resistors (base, emitter, or collector) could be used as the gain control, since stage gain is related to each resistance value (all other factors remaining constant). However, the emitter resistance is the most logical choice for a gain control. If the collector resistance is variable, the output impedance of the stage will change as the gain setting is changed. A variable base resistance will produce a variable input impedance. A variable emitter resistance (or source resistance in the case of a FET) has minimum effect on input or output impedance of the stage, but directly affects both current and voltage gain.

With all other factors remaining constant, a decrease in emitter resistance raises both current gain and voltage gain. An increase in emitter resistance lowers stage gain.

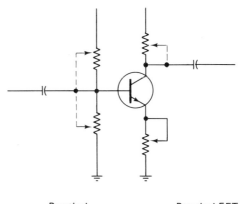

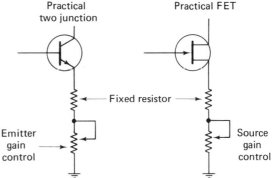

Fig. 2-40 Basic gain control for two-junction and FET stages

The resistance value of an emitter (or source) gain control should be chosen on the same basis as the emitter (or source) resistor, except that the desired value should be the approximate midpoint of the control range. For example, if a 500-Ω fixed resistor is normally used (or if 500 Ω is the calculated value for proper stage gain, bias stability, etc.), the variable gain control should be 1000 Ω.

In practical applications it is usually desirable to connect an emitter gain control in series with a fixed resistance. If the gain control is set to the minimum resistance value of 0 Ω, there will still be some emitter resistance to provide gain stabilization and prevent thermal runaway. As a guideline, the series resistance should be no less than one twentieth of the collector resistor value. This will provide a maximum stage gain of 20.

If the gain control must provide for reduction of the stage voltage gain from some nominal point down to unity, the maximum value of the control should equal the collector resistance.

If reduction to unity current gain is desired, the maximum value of the control should equal the input (base) resistance.

An audio taper potentiometer should not be used as a gain control, unless there is some special circuit requirement. However, the potentiometer used should be of the noninductive composition type. The wattage rating of an emitter (or source) gain control should be the same as for an emitter (or source) resistor. Usually, 1 or 2 W are sufficient for any stage of a voltage amplifier. Use of a gain control in a power amplifier should be avoided. If a gain control must be used in power amplifiers, the control should be incorporated in the input stage where emitter current is minimum.

2-8.3. Tone- and Balance-Control Circuit Analysis

Tone (treble and bass) controls are found in most hi-fi amplifier systems. Balance controls are used only in stereo amplifiers to balance the gain of both channels.

A treble control provides a means of adjusting the high-frequency response of an audio amplifier. Such adjustment may be necessary because of variation in response of the human ear, or to correct the frequency response of a particular recording.

A bass control provides a means of adjusting the low-frequency response of an audio amplifier. Such adjustment may be necessary because of variations in response of the human ear. As discussed, the human ear does not respond as well to low-frequency sounds at low levels as it does to high-frequency sounds at the same level. Also, coupling capacitors present high reactance to low-frequency signals. Both of these conditions require that the low-frequency signals be boosted (in relation to high-frequency signals).

There are many circuit arrangements for tone controls. Some involve the use of adjustable feedback (mainly in treble controls). Other circuits involve bypassing coupling capacitors with adjustable reactances (mainly in bass controls). However, the most common tone controls are RC filters using audio taper potentiometers as the adjustable R portion of the filter.

A typical tone-control network is shown in Fig. 2-41. This network is used in a stereo preamplifier. Although each stereo channel has its own tone controls, only one channel is shown. The tone control network of the second channel (not shown) is connected by means of the balance control R_{19}. In theory, the arm of R_{19} should be set to the exact midpoint. However, in a practical amplifier the gain of each channel is not exactly the same. Thus, R_{19} must be offset from the midpoint to balance both channels.

The bass and treble tone controls, R_{12} and R_{18}, are standard audio taper potentiometers. At 50 per cent rotation, the resistance is split, 90 per cent on one side of the wiper and 10 per cent on the other side. The relationship between wiper position and resistance is shown in Fig. 2-42.

In the bass-control circuit, when the control is in the center position the frequency response is flat from about 50 Hz to 20 kHz. This is shown in Fig. 2-43. The reactance of C_{11} is made equal to the 45-kΩ portion of R_{12} at 50 to 60 Hz, and the reactance of C_{12} is made equal to the 5-kΩ portion of R_{12}

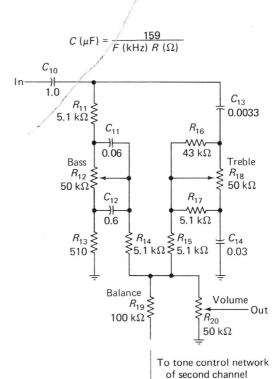

$$C\ (\mu F) = \frac{159}{F\ (kHz)\ R\ (\Omega)}$$

Fig. 2-41 Typical tone-control network for one channel of a stereo amplifier system

To tone control network of second channel

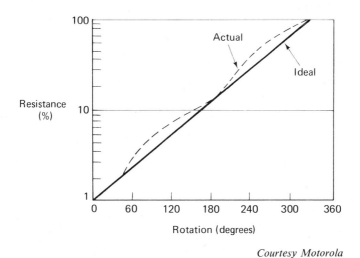

Courtesy Motorola

Fig. 2-42 Relationship of position (rotation) and resistance in tone controls

117

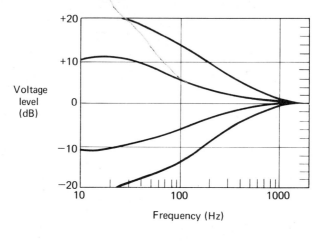

(a) Bass

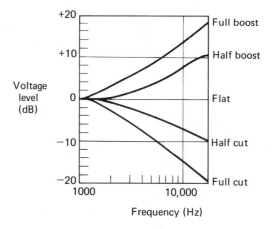

(b) Treble

Courtesy Motorola

Fig. 2-43   Normalized tone-control frequency-response curves

at 50 to 60 Hz. As frequency increases from 50 Hz, C_{12} couples more signal to the output, while C_{12} shunts more signal to ground through R_{13}. The net effect is a flat response from about 50 Hz to 20 kHz with a 20-dB insertion loss.

When the wiper of R_{12} is in the boost position, C_{12} with a reactance one-tenth the resistance of R_{12} at 50 to 60 Hz effectively shunts R_{11} out of the circuit, making R_{11} and C_{12} the dominant frequency response shaping components. Ideally, the full bass boost position will supply an output voltage (at about 50 Hz) that is 20 dB greater than the center position (flat response).

The full boost position represents zero attenuation in the tone control of the bass frequencies. The amplitude of the output will decrease at a 6 dB per octave rate to the frequency where the reactance of C_{12} is negligible. The output amplitude is then determined by the ratio of R_{11} to R_{13}.

When the wiper of R_{12} is in the "full cut" position, the output amplitude at about 50 Hz is determined by the ratio of C_{11} reactance to R_{13} resistance, and is about 40 dB below the input voltage. As frequency is increased, the reactance of C_{11} decreases until it is equal to the resistance R_{13}, again making the output amplitude dependent upon the ratio of R_{11} to R_{13}.

When R_{12} is in an intermediate position, the frequency at which rolloff begins (± 3 dB from the flat response curve) will vary, but the slope of the rolloff will change only slightly. Figure 2-43a shows the response of the bass control. The boost-cut axis uses the flat response position as the reference point or 0 dB, although, in fact, the point is 20 dB below the input signal (due to the approximate 20-dB insertion loss of the tone-control network).

The treble-control response curve is shown in Fig. 2-43b. At frequencies below about 2.1 kHz, the reactances of C_{13} and C_{14} become small when compared to the parallel divider combination of the control R_{18} and fixed resistance R_{16} and R_{17}. The resistive divider then provides the 10-to-1 voltage division to maintain the 20-dB insertion loss for the high frequencies. The net result is a 20-dB loss that is flat from about 20 Hz to 20 kHz.

The reactance of C_{13} should be about one half the resistance values of R_{18} at a frequency of 2.1 kHz (or about 25 kΩ). Thus, as shown by the equation of Fig. 2-41, $159/25$ k$\Omega \times 2.1 = 0.003$ μF. The value of C_{14} should be about 10 times the value of C_{13}, or $10 \times 0.003 = 0.03$ μF, to maintain the 10-to-1 voltage division.

The resistance of R_{16} is approximately one tenth the control (R_{18}) resistance, with the R_{17} resistance approximately 80 per cent of R_{18}. Resistors R_{14} and R_{15} are isolation resistors made equal to 10 per cent of the resistance of the respective control potentiometers R_{12} and R_{18}.

2-8.4. Playback-Equalization Network Circuit Analysis

There are many playback network circuits found in modern audio amplifiers. Most involve the use of frequency-selective feedback between stages, or from the output to the input of an amplifier. A typical feedback network consists of resistances and capacitances that form a feedback circuit. At any given frequency, the amount of feedback (and thus the frequency response) is set by selection of the appropriate RC combinations. As frequency increases, the capacitor reactance decreases. This results in a change of feedback (and frequency response).

The basic playback-equalization network using feedback is shown in Fig. 2-44. The voltage gain of two stages with feedback is approximately equal to the impedance of the feedback circuit divided by the source impedance. In this case, the source impedance is the emitter resistance R_E value. The feedback impedance Z_F is the vector sum of the potentiometer R_F resistance value and the capacitor C_F reactance value. The voltage gain of the two stages can be set to any desired level for any given frequency by means of the feedback circuit.

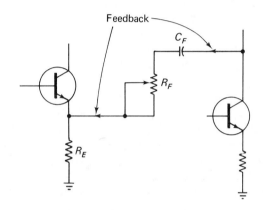

$R_F \approx R_E$ x maximum desired voltage gain

$$C_F \approx \frac{1}{125 \times R_E \times \text{high-frequency limit (Hz)}}$$

Fig. 2-44   Basic playback-equalization network

A more sophisticated playback-equalization network is shown in Fig. 2-45. This network is used in the same stereo preamplifier as discussed in Sec. 2-8.3, and forms a feedback circuit around one section of the preamplifier. In this case, the preamplifier is a differential IC, such as is described in Chapter 5.

The closed-loop (with feedback) voltage gain of the preamplifier section is set by the ratio of the feedback network to resistor R_2. The feedback (or playback-equalization) network for phonograph use is composed of C_3, C_4, R_3, and R_4, whereas the tape network is composed of R_5, C_6, and C_7.

RIAA playback equalization is used for the phonograph network. The standard RIAA equalization curve is shown in Fig. 2-46. The recording curve is the inverse of the playback curve, so that addition of the two gives a net flat frequency-versus-amplitude response. In phonograph recording, the high frequencies are emphasized to reduce effects of noise and low inertia of the cutting stylus. The low frequencies are attenuated to prevent large

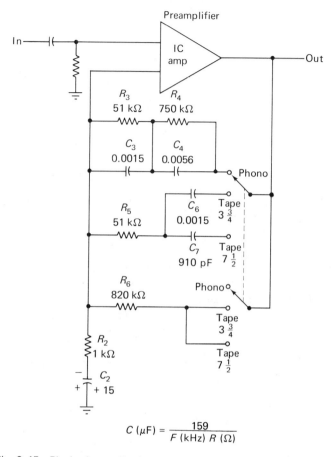

$$C\ (\mu F) = \frac{159}{F\ (kHz)\ R\ (\Omega)}$$

Fig. 2-45 Playback-equalization network for phonograph (RIAA) and tape (NAB) recordings

excursion of the cutting stylus. It is the job of the frequency-selective feedback network to accomplish the addition of the recording and playback responses.

It is impossible to have the playback network be the exact inverse of the recording compensation, since each recording system is slightly different. However, there are optional guidelines that can be applied. A typical audio range is from 20 Hz to 20 kHz. Thus, there is a rolloff at both the low and high ends. At the low end, the rolloff should start at some point between 10 and 20 Hz. This can be accomplished by making the 10-Hz point about 3 dB down from the 20-Hz point. As frequency increases from 20 Hz, there must be an almost linear (hopefully) rolloff. Ideally, the voltage gain at 20 Hz should be 100 times the gain at 20 kHz, and 10 times the gain at 1 kHz. This will produce the approximate RIAA curve of Fig. 2-46.

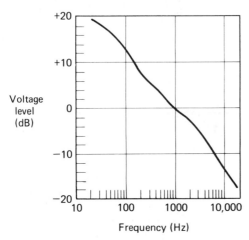

Courtesy Motorola

Fig. 2-46   RIAA playback-
equalization curve

In the circuit of Fig. 2-45, the linear rolloff is accomplished by dividing the playback network into three parts. The R_2C_2 section sets the 10-Hz point at 3 dB down from the 20-Hz point, the R_4C_4 section covers frequencies up to about 1 kHz, and the R_3C_3 network covers higher frequencies.

The value of R_3 should be 1000 times the desired voltage gain at 1 kHz, while the value of R_4 should be 15 kΩ times that of R_3. The value of R_2 is also based on the value of R_3, and is selected to provide the desired 1-kHz voltage gain (of 50 in this case). That is, the R_3/R_2 ratio sets the 1-kHz voltage gain.

The preamplifier of Fig. 2-45 produces an arbitrary minimum voltage gain of 5 at the highest frequencies (20 to 24 kHz). Most of the gain for the complete system is provided by another broadband amplifier (not shown). Using a minimum gain of 5 at the highest frequency, the gain at 20 Hz must be 100 times that amount, or 500. Likewise, the gain at 1 kHz must be 50. Using these desired gains, the value of R_3 is 1000 × 50 = 50 kΩ (use a 51-kΩ standard). With R_3 at 50 kΩ, the value of R_4 is 15 × 50 = 750 kΩ; and the value of R_2 is 50 kΩ/50 = 1 kΩ.

At low frequencies, the predominant impedance of the compensation feedback network is that of R_4. As frequency increases from about 50 Hz, the reactance of capacitor C_4 in parallel with R_4 begins to decrease the impedance of the C_4R_4 section. The reactance of C_4 is made to equal R_4 at about 35 to 40 Hz. At about 1 kHz, the net impedance of C_4R_4 is low compared to R_3, and R_3 sets the midband gain. As frequency increases to about 2 kHz, the parallel impedance of capacitor C_3 begins to shunt R_3, decreasing the impedance of the C_3R_3 section. The reactance of C_3 is made equal to R_3 at about 2.1 kHz.

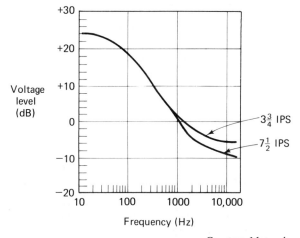

Courtesy Motorola

Fig. 2-47   NAB playback-equalization curve

Note that the equation shown on Fig. 2-45, based on frequency break-point and corresponding resistor, is used to find the values of C_2, C_3, and C_4.

NAB playback equalization is used for the tape network. The standard NAB equalization curve is shown in Fig. 2-47. Again, the recording curve is the inverse of the playback curve so that addition of the two gives a flat response. Likewise, the high frequencies are emphasized and the lows are attenuated. However, unlike phonograph playback, tape playback tends to flatten out after about 3 to 4 kHz. Also, a different response is required for different tape speeds. The playback response curves for both $3\frac{3}{4}$ and $7\frac{1}{2}$ IPS (inches per second) are shown in Fig. 2-47. Up to about 1 kHz, the curves are almost identical. Because there is only one frequency breakpoint (where the curve must start to flatten) for each tape speed, a simple *RC* series compensation network is all that is required (instead of the multisection network used for phonograph playback).

The breakpoint for $3\frac{3}{4}$ IPS occurs at about 1.85 kHz. The midband frequency gain is still 50, so the value of R_2 remains at 1 kΩ, and R_5 is made equal to R_3, or 51 kΩ. The reactance of C_6 is made equal to 51 kΩ (R_5) at 1.85 kHz (the nearest standard value is 0.0015 μF).

The breakpoint for $7\frac{1}{2}$ IPS is at about 3.2 kHz so that C_7 must have a reactance of 51 kΩ at this frequency. A C_7 capacitor value of 910 pF is the nearest standard.

Because C_6 and C_7 block the direct current path for the IC preamplifier feedback input, resistor R_6 is added when the phono-tape switch is in either tape position. The use of R_6 prevents realization of a full 20-dB bass boost because of the shunting action across the tape compensation network. However, the network does provide about 15 dB of boost, which is generally satisfactory.

It should be noted that the accuracy of both the RIAA and NAB compensation will be only as good as the components used. In a practical amplifier it is usually recommended that 5 per cent (or better) tolerance resistors and capacitors be used. Likewise, it may be necessary to "trim" the values to get an "exact" performance curve (required for truly good hi-fi performance).

2-9. TUNED AUDIO AMPLIFIERS (ACTIVE FILTERS)

In addition to passive audio filters (either *LC* or *RC*), it is possible to use amplifiers to form *active filters*. There are prime advantages in using these active filters. First, it is possible to obtain the equivalent of of an inductive reactance, without actually using a heavy and bulky inductance required for a typical *LC* filter. (*LC* filters are generally not practical in the AF range). Second, the use of an active filter eliminates the signal loss normally associated with passive filters (either *RC* or *LC*).

There are two common approaches to active filters. One approach involves the use of operational amplifiers with feedback networks. Such active filters are described in Chapter 6. The other approach is to use low-gain amplifier stages with simple *RC* feedback networks. This latter approach is described in the following paragraphs.

2-9.1. Active Low-Pass (High-Cut) Filter

Figure 2-48 shows the basic circuit of an active low-pass filter, together with the corresponding characteristic curves for several sets of component values. Note that these values are approximate and will usually require trimming to achieve an exact curve. The typical voltage gain is slightly less than 1 (unity) for transistors with a minimum beta of 20.

The amount of gain, as well as the shape of the curves, is set by the amount of feedback in relation to signal (which, in turn, is set by component values). Note that the feedback is positive, and thus adds to the signal. However, the feedback amplitude (across the entire frequency range) is just below the point necessary for oscillation. The circuit of Fig. 2-48 is an emitter follower, which typically has no voltage gain.

The circuit of Fig. 2-48 requires a bias of approximately -10 V (one half the -20-V supply) at the input. This can be obtained from a previous stage. If no such stage exists, the bias can be obtained by the addition of the 20-kΩ

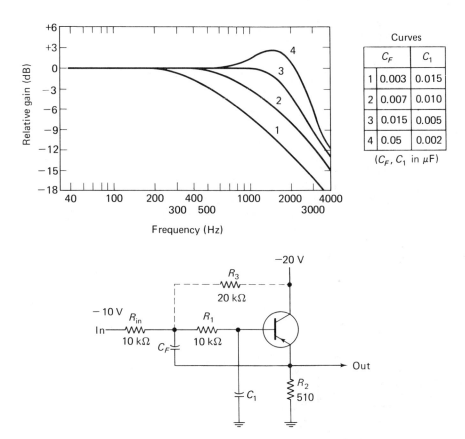

Fig. 2-48   Active low-pass (high-cut) filter and corresponding response curves

resistor (shown in phantom as R_3) and by changing the value of R_{IN} to 20 kΩ. Such an arrangement will introduce a loss of about 6 dB. Therefore, it is better to operate the circuit of Fig. 2-48 by direct coupling from the output of a previous stage.

2-9.2. Active High-Pass (Low-Cut) Filter

Figure 2-49 shows the basic circuit of an active high-pass filter, together with characteristic curves. The circuit of Fig. 2-49 is the inverse of the Fig. 2-48 circuit. That is, the Fig. 2-49 circuit uses capacitors in series with the base, with feedback obtained through R_F rather than C_F. The gain and shape of the curves is set by the amount of feedback (determined by circuit values).

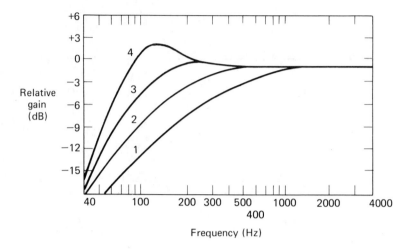

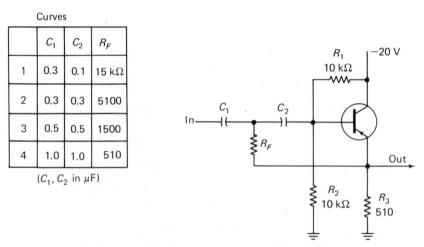

Fig. 2-49 Active high-pass (low-cut) filter and corresponding response curves

2-9.3. Active Bandpass Filter

The circuits of Fig. 2-48 and 2-49 can be cascaded to provide a bandpass filter. Any of the curves can be used. However, curve 3 is the most satisfactory, since it has the sharpest break at cutoff. Curves 1 and 2 have considerable slope with no sharp break. Curve 4 produces some peaking at the breakpoints.

If the circuits are cascaded, the low-pass filter (Fig. 2-48) should follow the high-pass filter (Fig. 2-49). This provides the necessary bias at the input of the low-pass filter (-10 V from the emitter of the high-pass filter).

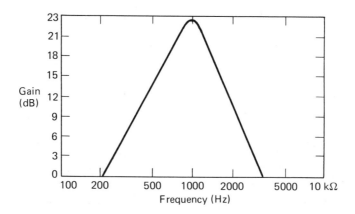

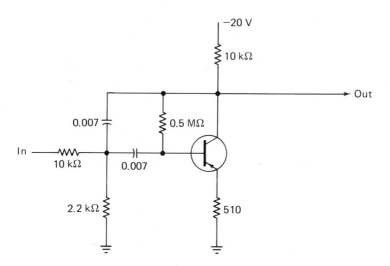

Fig. 2-50   Active peaking filter and corresponding response curve

2-9.4. Active Peaking Filter

The circuits of Figs. 2-48 and 2-49 can be cascaded to provide a peaking filter with the proper selection of components. However, a single-stage tuned amplifier will produce the same results. Such a circuit is shown in Fig. 2-50.

As shown by the characteristic curve, the center or peak frequency is approximately 1 kHz. If desired, the center frequency can be changed by as much as 3 decades when both capacitors are changed by a common factor. However, in a practical circuit the input resistance values will require some trimming.

Note that the characteristic curve shows a gain of approximately 24 dB. This is a no-load voltage gain at the peak or center frequency. If the circuit is loaded, as it must be in any practical application, the gain will be reduced. (Ideally, the circuit should work into an approximate 10-kΩ load.) Although there is a reduction in gain for a load, the shape of the output peak should remain substantially the same.

2-10. TRANSFORMERLESS PHASE INVERTER

From the discussion of transformer-coupled, push–pull audio circuits (Sec. 2-7), we may gather two facts concerning the signal input to the bases of the power transistors. First, the signal voltage applied to each base must be approximately equal in amplitude. Second, since the base of one transistor swings toward the positive as the base of the other swings negative, these signal voltages must be 180° out of phase with each other. This is accomplished by the center-tapped secondary winding of the input transformer.

The same result can be accomplished with a stage of resistance-coupled amplification, called a *phase inverter*. (See Fig. 2-51.) In a common-emitter amplifier with a resistive load, the collector and emitter are 180° out of phase with each other. If the input signal voltages for a push–pull stage are obtained from these two points (collector and emitter), we have the necessary 180° out-of-phase relationship. Also, since approximately the same collector current flows through R_C and R_E, if their resistances are equal, the voltage drops will be equal. That is, point A becomes as much negative as point B becomes positive.

In the circuit of Fig. 2-51, resistors R_1 and R_2 form the voltage divider that forward biases the emitter–base junction of Q_1. The collector resistor R_C and the emitter resistor R_E are equal in value, as are the coupling capacitors

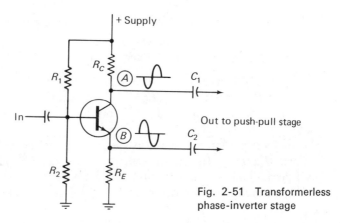

Fig. 2-51   Transformerless phase-inverter stage

C_1 and C_2. Note that R_E is unbypassed. This provides inverse-current feedback that reduces distortion and stabilizes gain.

The main advantage for using a phase-inverter stage is to eliminate the need for the input transformer. This results in a smaller, lighter, and less expensive amplifier, and eliminates the transformer's magnetic field (which can produce distortion of the signal, unless the transformer is properly shielded).

2-11. INTEGRATED-CIRCUIT AUDIO AMPLIFIERS

There are a great variety of audio amplifiers available in *IC* form, or in *hybrid* form. Such hybrid circuits consist of resistors, capacitors, diodes, and transistors, all contained in a single, hermetically sealed package. Hybrid circuits are similar to integrated circuits (*ICs*) except *ICs* usually are complete functioning circuits.

Most *IC* and hybrid audio amplifiers involve some form of *direct coupling* and/or *differential amplifiers*. For that reason, these audio circuits are discussed in their related chapters, 4 and 5.

3. RADIO-FREQUENCY AMPLIFIERS

When electrical signals reach frequencies of about 15 kHz and higher, they take on the properties of radio-frequency or RF signals. That is, the signals generate electromagnetic radio waves, which are radiated (transmitted) from the conductor. Amplifiers designed to amplify signals of such frequencies are known as RF amplifiers. Useful radio frequencies may be as high as several thousands of megahertz (MHz) or several gigahertz (GHz).

It is not practical to design any amplifier circuit that will cover the entire frequency range, or to use all radio frequencies for all purposes. Instead, the RF spectrum is broken into various *bands*, each used for a specific purpose. In turn, amplifier circuits are generally designed for use in one particular band. Figure 3-1 shows the most common assignment of radio frequency bands. Both commercial and military bands are shown. Note that the commercial RF bands run from about 3 kHz to 300 GHz, whereas the military band assignment runs from 225 MHz to 56 GHz.

Radio waves with frequencies greater than about 1 GHz are known as *microwaves*. The amplifier circuits used with microwaves are quite different from those used at lower frequencies. Because of their specialized nature, microwave amplifiers and related circuits are not discussed in this book. Instead, we shall concentrate on RF amplifiers operating at frequencies up to and including the UHF band.

3-1. TYPES OF RADIO-FREQUENCY AMPLIFIERS

Although there is an infinite variety of amplifier circuits, RF amplifiers may be divided into two general types: *narrowband amplifiers* with bandwidths up to several hundred kilohertz), and *wideband amplifiers*

Commercial Bands

Very low frequency (VLF) 3-30 kHz
Low frequency (LF) 30-300 kHz
Medium frequency (MF) 300 kHz-3 MHz
High frequency (HF) 3-30 MHz
Very high frequency (VHF) 30-300 MHz
Ultrahigh frequency (UHF) 300 MHz-3 GHz
Superhigh frequency (SHF) 3-30 GHz
Extrahigh frequency (EHF) 30-300 GHz

Military Bands

P-band 225-390 MHz
L-band 390-1550 MHz
S-band 1.55-5.2 GHz
X-band 5.2-10.9 GHz
K-band 10.9-36 GHz
Q-band 36-46 GHz
V-band 46-56 GHz

United States Broadcast Bands

Amplitude modulated (AM) 535-1605 kHz
Frequency modulated (FM) 88-108 MHz
VHF television 54-216 MHz
UHF television 470-890 MHz

Fig. 3-1   Assignment of radio-frequency bands in the United States

(with bandwidths in the order of megahertz). The reason for this division or classification merits some discussion.

As shown in Fig. 3-1, the amplitude-modulated (AM) broadcast band for the United States is from 535 to 1605 kHz. The frequencies of transmitting stations within this band are spaced from 10 to 15 kHz apart to prevent interference with each other. In the frequency-modulated (FM) broadcast band, the transmitting stations are spaced 200 kHz apart. In the television broadcast bands, the stations are approximately 6 MHz apart.

Within a specific band, each transmitting station is assigned a specific frequency at which it is to operate. However, each station transmits not only at this frequency but at a relatively narrow band of frequencies lying at either side of this assigned frequency. Such a band of frequencies is required if the signal is to convey intelligence. For example, an AM broadcast band station that is assigned a certain frequency will transmit a signal whose frequencies encompass a band extending 5 to 7.5 kHz on either side of the assigned frequency.

An RF amplifier used in an AM broadcast radio receiver is adjusted to cover not the entire AM broadcast band simultaneously, but rather a portion of the band about 15 kHz wide, corresponding to the spread of a single station. Under these conditions, the *bandwidth* of the RF amplifier is said to be 15 kHz. The amplifier is adjusted (or *tuned*) to one station at a time. In the FM broadcast band, where each station is spaced 200 kHz apart, the bandwidth of the RF amplifier is about 150 kHz. Both AM and FM broadcast band RF amplifiers are essentially *narrowband amplifiers*. In the television bands, where the stations are 6 MHz apart, the RF amplifiers are of the *wideband type*, since the transmitted TV signal is approximately 4.5 MHz wide.

As can be seen from this discussion, the RF amplifier serves two purposes. One purpose is as a *band-pass filter*, which passes signals from the desired station and rejects all others. The other purpose is to amplify these signals to a suitable voltage (or power) level.

3-2. BASIC NARROWBAND RADIO-FREQUENCY AMPLIFIER THEORY

The circuit of Fig. 3-2 is a typical narrowband RF amplifier, such as those found in broadcast- and communications-type radio receivers. The circuit is a single stage of *tuned radio frequency* (TRF) *voltage amplification*. Input to the transistor is by means of a *tuned RF transformer*, and output is obtained by a similar device. Transformer T_1 is the *input* transformer and

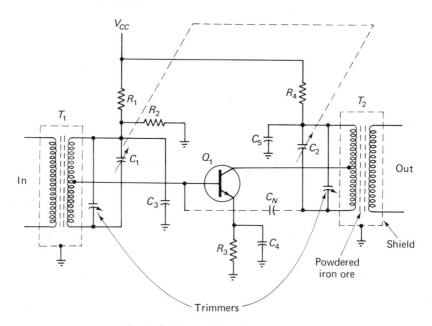

Fig. 3-2   Tuned RF voltage amplifier

T_2 is the *output* transformer. The secondary winding of T_1 is tuned to resonance at the frequency of the incoming signal by means of variable capacitor C_1. The primary of T_2 is tuned to the same resonant frequency by means of variable capacitor C_2.

At the resonant frequency, the secondary of T_1 and capacitor C_1 form a parallel-resonant circuit, as do the primary of T_2 and capacitor C_2. A parallel-resonant circuit offers a very high impedance to a current at the resonant frequency, but a low impedance to currents at other frequencies. (Resonant circuits used in RF amplifiers are discussed in later sections of this chapter.) Thus, if C_1 is adjusted to tune the secondary winding of T_1 to resonance at the frequency of the desired signal, a relatively large voltage will appear across this resonant circuit (and the transistor base) for signals of this frequency. For all other signals, the voltage will be low.

Likewise, if C_2 is adjusted to tune the primary winding of T_2 to resonance at the frequency of the desired signal, this resonant circuit will show a large impedance for signals of this frequency, and a very low impedance for other signals. This resonant circuit is the collector load. As discussed in Chapter 2, voltage amplification of a stage is set by collector load impedance (all other factors remaining equal). These conditions provide high voltage amplification *at the desired frequency*. At all other frequencies, where the collector load impedance is low, the amplification will also be low.

The shunting effect of the transistor input and output capacitances is also minimized by the capacitances of the tuned resonant circuits. For example, the typically small input capacitance of the transistor is in parallel with the relatively large capacitance of variable capacitor C_1, and thus has but a small additive effect.

Laminated iron-core transformers cannot be used in RF amplifiers. For one reason, such transformers do not have the required bandwidth, as discussed in Chapter 2. Furthermore, the effect of the stray capacitance of the many turns in an iron-core transformer, although fairly small at audio frequencies, is large enough at radio frequencies to reduce the amplification to almost zero. Besides, the eddy-current losses that would occur in the laminated iron-core transformers at radio frequencies would be tremendous.

Accordingly, RF amplifiers frequently use *air-core transformers*. The difficulty with the air-core type is that there is very little magnetic linkage between the primary and secondary windings. Modern RF transformers usually use a core of *powdered iron*, which increases the inductance and linkage without excessive eddy-current losses.

Bias network. Resistors R_1 and R_2 form a voltage divider across the power supply (V_{CC}) to forward bias the emitter–base junction of the *PNP* transistor. Resistor R_2 and capacitor C_3 form a decoupling network to prevent the RF signal from entering the power supply (through which the signal may be fed to the output circuit, or another stage). Resistor R_3 is the emitter stabilization resistor and C_4 is its bypass capacitor. Resistor R_4 and capacitor C_5 form the decoupling network for the collector circuit.

Impedance match. Note that the base of the transistor is connected to a *tap* on the secondary winding of T_1. Since the input impedance of the transistor is relatively small, only a portion of the secondary winding is used to obtain a proper impedance match between the two. However, the entire secondary winding is tuned by C_1 to form the parallel-resonant circuit at the signal frequency. For similar reasons the collector is connected to a tap on the primary winding of T_2.

Feedback problems. One difficulty often found in RF amplifiers is the prevention of feedback from the output of a stage to its own input, or to another stage. There are two types of undesired feedback: *radiated feedback* and *feedback through the transistor*. (Of course, there is feedback that is deliberately introduced to stabilize gain, temperature response, etc., as discussed in Chapters 1 and 2.) The danger of undesired feedback is greater for RF amplifiers than for AF amplifiers.

To eliminate radiated feedback, the amplifier must be properly designed to separate the base and collector leads. Also, nonmagnetic *shielding*, usually of aluminum or copper, may be used to isolate the base and collector circuits from each other, as well as each complete stage from the others. These shields are well grounded to the chassis or printed circuit board. In extreme cases, the wires that connect the base and collector circuits may be encased in flexible copper-braid tubing, which, in turn, is grounded.

At lower frequencies, instead of shielding the entire stage of amplification, it is usually sufficient to shield the RF transformer by enclosing it in a grounded metal can, usually of aluminum. This shielding is indicated by the dashed-line boxes around the transformers in Fig. 3-2. At higher frequencies, complete stage shielding is used, with feed-through-type capacitors in the shielding to provide for connection to and from the circuit.

In any form of shielding, eddy currents are generated in the shields by the magnetic fields around the windings of the transformers. The power for these currents comes from the enclosed components. Thus, eddy currents represent a loss of electrical energy at the expense of the signal. For this reason, shielding is used only where required, and the shield cans are not mounted too close to the RF transformers.

Fortunately, most modern transistors are so constructed that there is little danger of *feedback through* the transistor at moderately high frequencies. However, at higher frequencies, internal feedback can produce undesired conditions in an *RF* amplifier. One feedback problem is known as the *Miller effect*. As shown in Fig. 3-3, there is a capacitance between base and emitter of a two-junction transistor (or between gate and source of a FET). This forms the input capacitance of the circuit. There is also a capacitance between the base and collector (or gate and drain). This capacitance feeds back some

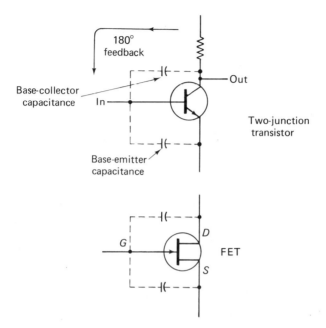

Fig. 3-3   Input and feedback capacitances in two-junction transistors and FETs

of the collector signal to the base. The collector signal is amplified, and is 180° out of phase with the base signal (in a common-emitter amplifier). The collector signal feedback opposes the base signal and tends to distort the input signal. Likewise, the collector–base capacitance is, in effect, in series with the base–emitter capacitance, and thus changes the input capacitance.

These conditions make for a constantly changing amplitude-modulated relationship of signals in an amplifier. For example, if the input signal amplitude changes, the amount of feedback changes, changing the input capacitance. In turn, the change in input capacitance changes the match between the transistor and the input tuned circuit, changing the amplitude. Likewise, if the input signal frequency changes, the feedback changes (since the collector–base capacitive reactance changes), and there is a corresponding change in amplification.

The Miller effect is not necessarily a problem in all solid-state RF amplifiers. The FET RF amplifier is usually more susceptible to Miller effect than two-junction transistors. However, when the Miller effect becomes severe with any RF amplifier, it can be eliminated or minimized to a realistic level by *neutralization*.

Neutralization is a method for reducing the amount of unwanted feedback, either from radiation or internal feedback. With neutralization, a portion of the voltage from the output circuit of the stage of amplification is fed back to the input circuit in such a way as to cancel out the base voltage caused by the unwanted feedback. Neutralization is accomplished by impressing a voltage on the base that is equal in magnitude, but opposite in phase, to the undesired feedback. Thus, the two voltages will "buck" each other out.

The two ends of the primary winding of the output transformer (such as T_2 in Fig. 3-2) are of opposite phase. If this opposite-phase voltage is fed to the base through the neutralizing capacitor (C_N of Fig. 3-2), the two voltages will cancel out. As a guideline, the neutralizing capacitor should equal the collector–base capacitance (typically a few picofarads).

Another method for reducing the amount of unwanted feedback, without neutralization, is to use the *common-base* amplifier configuration. (Refer to Chapter 1.) A common-base RF amplifier circuit is shown in Fig. 3-4. The input transformer T_1 is tuned to resonance by variable capacitor C_1. The output transformer T_2 is tuned to the same resonant frequency by C_2. Resistor R_1 is the emitter resistor and C_3 is its bypass capacitor. Resistor R_2 and capacitor C_4 form a decoupling filter.

The base is grounded. The input signal is applied to the emitter. The output is obtained between the collector and base, which is common to the input and output circuits. The grounded base acts as a shield between the input and output circuits, thus reducing feedback.

Tuning methods. The windings of the RF transformer may be tuned by the ordinary air-type variable capacitor. Thus, by adjustment of this capacitor,

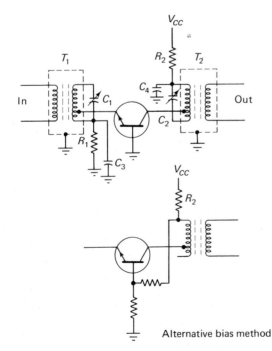

Fig. 3-4   Common-base
Alternative bias method   (grounded-base) RF amplifier

the transformer may be tuned to resonance at different frequencies. Of course, the transformer may be tuned by using a variable inductance and a fixed capacitance. Accordingly, many RF transformers have a powdered-iron core that can be moved in or out to vary the inductance of the windings. The fixed capacitor generally is of the mica or ceramic type.

Both the input and output transformers must be tuned simultaneously to resonance at the frequency of the incoming signal. Also, if several stages of RF amplification are coupled together, the tuned RF transformers may be used to couple these stages. Here, too, all the transformers must be tuned simultaneously to resonance. Although the use of several tuned circuits increases the overall *selectivity* of the amplifier, the need for manipulating a number of variable capacitors can be a problem. To eliminate the problem, the capacitor shafts usually are connected so that they tune simultaneously when one dial is manipulated. This is called *ganging*. Most commonly, all the rotors of the variable capacitors are mounted on one shaft. Thus, we have a *two-gang capacitor*, a *three-gang capacitor*, and so forth, depending upon the number of sections ganged together. The dashed lines connecting the variable capacitors in Fig. 3-2 indicate that these capacitors are ganged together. Where the transformers are tuned by movable powdered-iron cores, these cores may be similarly ganged.

Since all the tuned circuits must tune to the same frequency, and since it is virtually impossible to construct two amplifiers that are exactly alike, small semivariable capacitors, called *trimmers*, are usually connected across the larger variable capacitors. As the trimmer is in parallel with the large variable capacitor, the trimmer varies the overall capacitance of the tuned circuit slightly, thus compensating for small differences between the circuits.

3-2.1. Intermediate-Frequency Amplifier

Most radio and TV receivers operate upon the *superheterodyne* principle whereby the frequency of the received radio signal is first converted to a lower, predetermined frequency called the *intermediate frequency*. The amplifier is fixed to operate at this frequency, rather than being tunable over the entire band. The amplifier, now called an *intermediate-frequency* (IF) *amplifier*, is shown in Fig. 3-5. This IF amplifier is similar to the RF amplifier of Fig. 3-2, except that the IF amplifier transformers are tuned to the pre-determined frequency by means of small fixed capacitors instead of variable capacitors.

Since it is not necessary to tune the IF transformers over the entire spread of the band, it is practical to tune both the primary and secondary windings of each transformer to the intermediate frequency. Thus, by adding tuned circuits the selectivity of the receiver is increased.

To compensate for small variations between the IF transformers, each winding has a movable powdered-iron core that can be moved in or out, thus varying the inductance slightly. In some of the earlier transformers the

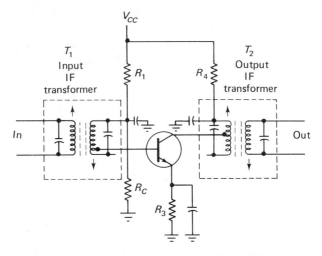

Fig. 3-5   Intermediate frequency (IF) amplifier

inductances were fixed, and the windings were tuned by semivariable trimmers.

As is the case in AF amplifiers, several stages of RF or IF amplifiers may
be connected in cascade to produce a greater overall amplification of the
signal. In modern solid-state receivers, a single stage of RF amplification is
used. The same is true of IF amplification, except in communications and
other high-quality receivers. Such receivers often use two (or possibly more)
stages of IF amplification.

Figure 3-6 is the circuit of a two-stage IF amplifier. Note that only the
primary windings of the transformers are tuned. Since the impedance of the

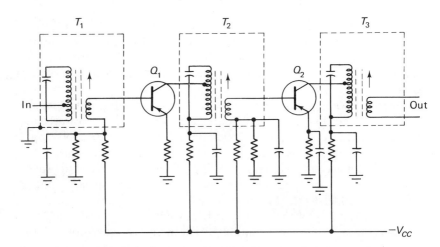

Fig. 3-6   Two-stage IF amplifier

primary is reflected to the secondary, the effect is the same regardless of which winding is tuned. Nevertheless, because there are fewer tuned circuits, the overall selectivity of the amplifier is somewhat reduced. All other factors being equal, the more tuned circuits in any amplifier, the greater the selectivity, and vice versa.

The collector of each transistor is connected to a tap on the primary winding of its output transformer to match the impedance of the winding to the relatively low output impedance of the transistor. Similarly, the secondary of each input transformer has fewer turns than the primary winding so that it may match the relatively low input impedance of the transistor.

3-2.2. Radio-Frequency Power Amplifier

Most radio transmitters use some form of power amplifier to raise the low-amplitude signal developed by the oscillator to a high-amplitude signal suitable for transmission. For example, most oscillators develop signals of less than 1 W, whereas a solid-state transmitter may require 100 W (or more) output.

Figure 3-7 shows two basic RF power amplifier circuits. In the circuit of Fig. 3-7a, the collector load is a parallel-resonant circuit (called a *tank* circuit) consisting of variable capacitor C_1 and inductor L_1, tuned to resonance at the desired frequency. The output, which is an amplified version of the input voltage, is from L_2, which together with L_1 forms an output transformer.

The circuit of Fig. 3-7a has certain advantages and disadvantages. The winding of L_2 can be made to match the impedance of the load (by selecting the proper number of turns and by positioning L_2 in relation to L_1). While that may prove an advantage in some cases, it also makes for an interstage coupling network that is subject to mismatch and detuning by physical movement of shock. Another disadvantage of the Fig. 3-7a circuit is that all the current must pass through the tank circuit coil. Also, for best transfer of power the impedance of L_1 should match that of the transistor output. Since two-junction transistor output impedances are generally low, the value of L_1 must be low, often resulting in an impractical size for L_1. The circuit of Fig. 3-7a is a carry-over from vacuum-tube circuits and, as such, is not often found in modern two-junction transistor amplifiers. However, the circuit is found in FET RF amplifiers (which are generally low power and higher impedance).

The circuit of Fig. 3-7b, or one of its many variations, is commonly found in solid-state radio transmitters using two-junction transistors. The collector load is a resonant circuit formed by the network of L_1, C_1, and C_2. Note that C_1 is marked "Loading adjust," whereas C_2 is marked "Resonant tuning adjust." As is discussed in later paragraphs, these networks provide the *dual function* of frequency selection (equivalent to the tank circuit) and

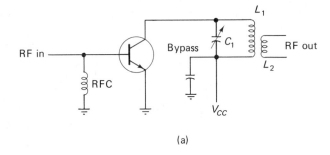

(a)

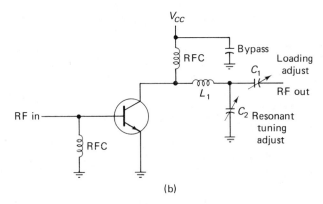

(b)

Fig. 3-7   Radio-frequency power amplifiers

impedance matching between transistor and load. To properly match impedances, both the resistive (so-called real part) and reactive (so-called imaginary
part) components of the impedance must be considered.

Both Fig. 3-7 circuits are operated class B, which is typical for RF amplifiers, Class B operation is obtained by connecting the emitter directly to
ground and applying no bias to the base–emitter junction. Since any two-
junction transistor requires some forward bias to produce current flow, the
transistor remains cut off except in the presence of a signal.

3-2.3. Radio-Frequency Multiplier

The circuits of Fig. 3-7 can be used as a frequency multiplier.
That is, the collector circuit is tuned to a higher whole-number multiple
(harmonic) of the input frequency. Many radio transmitters use some form
of multiplier to raise the low-frequency signal developed by the oscillator
to a high-frequency signal. For example, most crystals used in oscillators
have a fundamental frequency of less than 10 MHz, whereas a solid-state
transmitter may produce an output in the UHF range.

Although the circuits of power multipliers and power amplifiers are essentially the same, the efficiency is different. That is, an amplifier operating at the same frequency as the input signal will have a higher efficiency than an identical circuit operating at a multiple of the input frequency.

3-2.4. Radio-Frequency Amplifier–Multiplier Combinations

The circuits of Fig. 3-7 can be cascaded to provide increased power amplification and/or frequency multiplication. Typically, no more than three stages are so cascaded. The stages can be mixed. That is, one or two stages can provide frequency multiplication, with the remaining one or two stages providing power amplification. Such arrangements are discussed in later paragraphs of this chapter.

3-3. BASIC WIDEBAND RADIO-FREQUENCY AMPLIFIER THEORY

Except for pure sinewaves, all signals are found to contain not only the fundamental frequency, but harmonic frequencies as well. These harmonics are whole-number multiples of the fundamental frequency. Pulse signals have an especially high harmonic content. For example, the pulses used in television contain frequencies ranging from about 30 Hz to 4 MHz. An ordinary RF amplifier with a bandwidth of several hundred kilohertz is unable to amplify *uniformly* signals that have such a broad range of frequencies. For this reason it is necessary to use special *broadband* or *wideband* amplifiers for such applications. These amplifiers are usually known as *video amplifiers* in television equipment, or as *pulse amplifiers* in radar and similar equipment.

The resistance-coupled (*RC*) amplifiers described in Chapter 2 are, in effect, wideband amplifiers. That is, such circuits amplify uniformly at all frequencies of the entire audio range, dropping off only at the low- and high-frequency ends. If we can extend the uniform amplifying action at both ends of the frequency range, we have a wideband amplifier, capable of passing (and amplifying) RF signals (including pulses).

The basic *RC* amplifier circuit is shown in Fig. 3-8. Capacitor C_{OUT} represents the output capacitance of Q_1. Capacitor C_D represents the *distributed capacitance* of the various components and their connecting wires. Capacitor C_{IN} represents the input capacitance of Q_2.

Coupling capacitor C_C and base resistor R_b form a voltage divider across the input of Q_2. At low frequencies, the impedance of C_C is large, and relatively little of the signal voltage is applied to the base of Q_2. Accordingly, the low-frequency response of the amplifier is lowered.

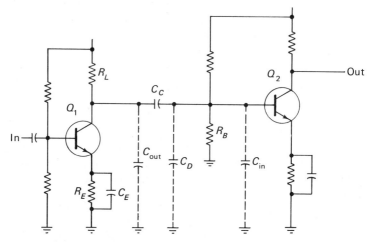

Fig. 3-8   Capacitances C_{OUT}, C_D, and C_{IN} in RC amplifiers

Capacitances C_{OUT}, C_D, and C_{IN} acting in parallel shunt the load resistor R_L of Q_1. This lowers the effective resistance of R_L and the high-frequency response of the amplifier. As discussed in Chapters 1 and 2, a lower value of R_L lowers the gain, all other factors being equal.

3-3.1. Increasing Wideband Response

There are several methods for improving the low- and high-frequency response of wideband amplifiers (or RC amplifiers designed for wideband use). In all cases, transistors with small input and output capacitances should be used. Likewise, the components must be carefully placed so that their leads and distributed capacitances are kept at a minimum.

The emitter-bypass capacitor C_E affects the low-frequency gain of the amplifier. As discussed in Chapter 2, the impedance of C_E is higher at lower frequencies. Thus, the amplifier gain is lower at lower frequencies. Accordingly, the capacitance of C_E must be great enough to offer a low impedance (with respect to R_E) at the lowest frequency to be amplified.

The value of collector resistor R_L also affects the frequency response and gain of the amplifier. The graph of Fig. 3-9 shows the effects produced by various values of R_L. Note that a large-value collector resistor produces a high gain at the middle frequencies and a steep drop in gain at the high and low frequencies. A small-value collector resistor produces a much smaller overall gain, but the proportional drop in gain at the high and low frequencies is also much less than for the larger-value collector resistors. With the small-value collector resistor, the amplifier gain is uniform over a much wider range of frequencies. In effect, the amplifier sacrifices gain for bandwidth.

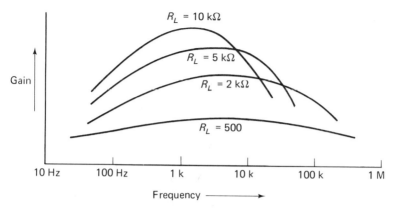

Fig. 3-9   Graph showing the effects of collector load resistance values at different frequencies

Because of these conditions wideband amplifiers use low-value collector resistances and transistors with high gain (large betas).

There are several other methods of compensating for the drop in gain at the high and low ends of the frequency band. At low frequencies, the effects of the transistor input–output capacitances and the distributed capacitances are negligible, but the impedance of the coupling capacitor becomes increasingly important. One way to compensate for the effects of the coupling capacitor is shown in Fig. 3-10, which is the video amplifier of a typical TV receiver.

Here, the load resistance for Q_1 is made up of two parts, R_5 and R_6, connected in series. Capacitor C_3 is the bypass capacitor for R_6. At the higher frequencies the collector load is effectively R_5, since the small impedance of C_3 at these frequencies permits C_3 to completely bypass R_6 (in effect removing R_6 from the circuit).

At low frequencies the impedance of C_3 becomes high and the bypassing effect is greatly reduced. The collector load resistance then becomes $R_5 + R_6$. This greater resistance produces a greater output voltage, thus compensating for the low-frequency drop produced by C_3.

Also, since the drop in high-frequency response is due to the shunting effect of the transistor capacitances (and distributed capacitances) upon the load resistor, a small inductor L_1 (called a *shunt peaking coil*) is inserted in series with the load resistance. At low frequencies L_1 offers very little impedance and the collector load is, essentially, the resistance of $R_5 + R_6$. At high frequencies the impedance of L_1 is high, and the collector load is the sum of the $R_5 + R_6$ resistances and the reactance of L_1. Thus, the gain of the amplifier is increased.

Inductor L_1 sets up a resonant circuit with the distributed capacitances of the circuit (and the capacitances of the transistor). The value of L_1 is so selec-

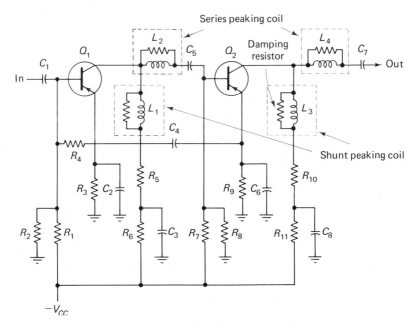

Fig. 3-10   Video amplifier for a typical TV receiver

ted that the circuit is resonant at a frequency where the high-frequency response of the amplifier begins to drop. In this way an additional boost is given to the gain, and the high-frequency end of the response curve is flattened.

Another similar compensating circuit is illustrated in Fig. 3-10. Here, *series peaking coil* L_2 is connected in series with the coupling capacitor C_5. At high frequencies L_1 forms a low-impedance series-resonant circuit with the capacitances, causing a larger voltage to appear at the base of Q_2. As before, at low frequencies, L_1 has virtually no effect on the circuit.

Video and pulse amplifiers usually use both shunt and series peaking coils for high-frequency compensation. As a general guideline, the values of these coils are such that resonance is obtained at the highest desired frequency. That is, the capacitances are calculated (or measured), and a corresponding value of inductance is chosen for resonance at the high-frequency end.

Note that the coils of Fig. 3-10 are shown with resistances connected in parallel. As is discussed in later paragraphs of this chapter, when resistances are connected across coils, the resonant point of the coils is flattened or broadened. Such resistances are often known as *damping* resistances.

Another method for overcoming the effects of the drop in gain at the low and high ends of the frequency band involves the use of *inverse feedback*. As discussed in Chapters 1 and 2, inverse feedback tends to oppose any change in signal level. Thus, when the signal level tends to drop at either end of the frequency range, inverse feedback will oppose this change. In the circuit of Fig. 3-10, inverse feedback is provided by C_4 and R_4.

3-4. RESONANT CIRCUITS FOR RADIO-FREQUENCY AMPLIFIERS

Radio-frequency amplifier design is based on the use of resonant circuits (tank circuits) consisting of a capacitor and a coil (inductance) connected in series or parallel, as shown in Fig. 3-11. At the resonant frequency the inductive and capacitive reactances are equal, and the circuit acts as a high impedance (if it is a parallel circuit) or a low impedance (if it is a series circuit). In either case, any combination of capacitance and inductance has some resonant frequency.

Either (or both) the capacitance or inductance can be variable to permit tuning of the resonant circuit over a given frequency range. When the inductance is variable, tuning is usually done by means of a metal slug (usually powdered iron) inside the coil. The metal slug is screwdriver adjusted to change the inductance (and thus the inductive reactance) as required.

Typical RF circuits used in receivers (AM, FM, communications, etc.) often include two resonant circuits in the form of a transformer (RF or IF transformer, etc.). Either the capacitance or inductance can be variable. Since such transformers are available commercially, no detailed analysis will be given here. However, measurement of RF (and IF) transformer resonant values, as they affect amplifiers, is discussed in Chapter 7.

In the case of RF transmitters, it is sometimes necessary to design the coil portion of the resonant circuit. This is because coils of a given inductance and physical size may not be available from commercial sources. Therefore, the following analysis of resonant circuits is given.

3-4.1. Basic Design Considerations for Resonant Circuits

The two most important considerations for RF resonant circuits are *resonant frequency* and the *Q* (or *quality* factor).

Resonant frequency. Figure 3-11 contains equations which show the relationship between capacitance, inductance, reactance, and frequency as they relate to resonant circuits. Note that there are two sets of equations. One set includes reactance (inductive and capacitive). The other set omits reactance. The reason for two sets of equations is that some design approaches require the reactance to be calculated for resonant networks. Solid-state RF transmitter circuits are a classic example of this.

Quality factor and selectivity. A resonant circuit has a *Q*, or quality, factor. This *Q* factor is directly related to the selectivity of the circuit, and is dependent upon the ratio of reactance to resistance. If a resonant circuit has pure

Capitive Reactance

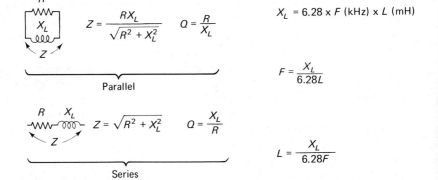

Capacitive Reactance

Parallel

$$Z = \frac{RX_C}{\sqrt{R^2 + X_C^2}} \qquad Q = \frac{R}{X_C}$$

Series

$$Z = \sqrt{R^2 + X_C^2} \qquad Q = \frac{X_C}{R}$$

$$X_C = \frac{159}{F\,(\text{kHz}) \times C\,(\mu F)}$$

$$F\,(\text{kHz}) = \frac{159}{(X_C) \times C\,(\mu F)}$$

$$C\,(\mu F) = \frac{159}{F\,(\text{kHz}) \times X_C}$$

Inductive Reactance

Parallel

$$Z = \frac{RX_L}{\sqrt{R^2 + X_L^2}} \qquad Q = \frac{R}{X_L}$$

Series

$$Z = \sqrt{R^2 + X_L^2} \qquad Q = \frac{X_L}{R}$$

$$X_L = 6.28 \times F\,(\text{kHz}) \times L\,(\text{mH})$$

$$F = \frac{X_L}{6.28 L}$$

$$L = \frac{X_L}{6.28 F}$$

Impedance and Resonance

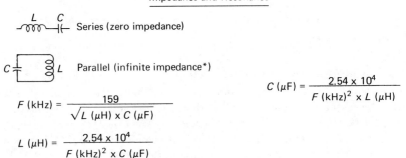

Series (zero impedance)

Parallel (infinite impedance*)

$$F\,(\text{kHz}) = \frac{159}{\sqrt{L\,(\mu H) \times C\,(\mu F)}}$$

$$L\,(\mu H) = \frac{2.54 \times 10^4}{F\,(\text{kHz})^2 \times C\,(\mu F)}$$

$$C\,(\mu F) = \frac{2.54 \times 10^4}{F\,(\text{kHz})^2 \times L\,(\mu H)}$$

*When circuit Q is 10 or higher

Fig. 3-11   Resonant circuit equations

reactance, the Q is high (theoretically infinite). However, this is not practical. For example, any coil will have some dc resistance, as will the leads of a capacitor. Also, as frequency increases, the ac resistance presented by the leads will increase due to skin effect. The sum total of these resistances is usually lumped together and is considered as a resistor in series or parallel with the resonant circuit. The total resistance is usually termed the *effective* resistance, and is not to be confused with the reactance.

The resonant circuit Q is dependent upon the individual Q factors of the inductance and capacitance used in the circuit. For example, if both the inductance and capacitance have a high Q, the circuit will have a high Q, provided that a minimum of resistance is produced when the inductance and capacitance are connected to form a resonant circuit.

Usually, resonant circuit Q is measured at points on either side of the resonant frequency where the signal amplitude is down 0.707 of the peak resonant value, as shown in Fig. 3-12. (Resonant circuit Q measurement

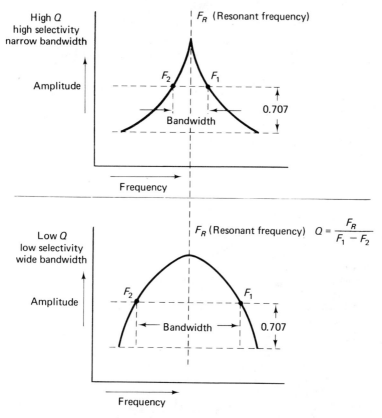

Fig. 3-12   Relationship of Q to bandwidth and selectivity

techniques are described in Chapter 7.) As shown, a resonant circuit with a high Q produces a sharp resonance curve (narrow bandwidth), whereas a low Q produces a broad resonance curve (wide bandwidth). For example, a high Q resonant circuit will provide good harmonic rejection (tend to pass only the fundamental frequency) and efficiency in comparison with a low Q circuit, all other factors being equal. Thus, the *selectivity* of a resonant circuit is related directly to Q.

Note that a very high Q (or high selectivity) is not always desired. In some applications, it is necessary to add resistance to a resonant circuit to broaden the response (increase the bandwidth, decrease the selectivity). An example of this is the damping resistor used across peaking coils in a video amplifier, as described in Sec. 3-3.1.

Note that if a given bandwidth must be maintained, but the resonant frequency increased, the Q must also increase. For example, if the resonant frequency is 10 MHz with a bandwidth of 2 MHz, the required circuit Q is 5. If the resonant frequency is increased to 50 MHz, with the same 2-MHz bandwidth, the required Q is 25. Also, note that Q must be decreased for increases in bandwidth, if the same resonant frequency is to be maintained.

3-4.2. Basic Design Considerations for Radio-Frequency Coils

The equations necessary to calculate the self-inductance of a single-layer air-core coil are given in Fig. 3-13. Note that such coils are most efficient (that is, maximum inductance for minimum physical size) when the

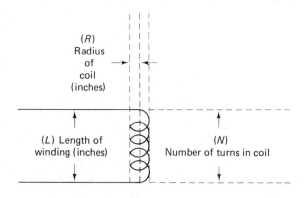

(R)
Radius
of
coil
(inches)

(L) Length of
winding (inches)

(N)
Number of turns in coil

Inductance (μH) $\approx \dfrac{(RN)^2}{9R + 10L}$

When $L = 0.8 \times R$ or $R/L = 1.25$

$N \approx \sqrt{\dfrac{17 \times \text{inductance } (\mu\text{H})}{R}}$

inductance (μH) $\approx \dfrac{(RN)^2}{R17}$

Fig. 3-13 Calculations for self-inductance of single-layer air-core coil

ratio of coil radius to coil length is 1.25, that is, when the length is 0.8 of the radius.

The equations of Fig. 3-13 are approximations only, and do not take into account such factors as uneven sizes of turns, spacing between turns, and the like. From a practical standpoint, use the equations to find the nearest number of turns (for a given inductance), and then spread or compress the turns as necessary to obtain a precise value of inductance (as measured on an inductance bridge).

3-5. RADIO-FREQUENCY VOLTAGE-AMPLIFIER CIRCUIT ANALYSIS

Radio-frequency voltage amplifiers are used primarily in receivers and receiver-type circuits. An IF (intermediate frequency) amplifier, or IF limiter amplifier, is an example of an RF voltage amplifier. The input or first stage of a receiver may include a separate RF voltage amplifier (such as with some communications receivers). However, most solid-state receivers combine the RF voltage-amplifier function with that of the local oscillator. Such circuits are discussed in Sec. 3-6.

Figure 3-14 shows the working schematic of a typical RF voltage amplifier. Such a circuit could be used as an IF amplifier, IF limiter, or separate RF amplifier with few modifications. Both the input and output are tuned to the desired operating frequency by means of the resonant circuits. In this case, the resonant circuits are composed of transformers with a capacitor across the primary. The capacitors could be variable, but are usually fixed. The resonant circuit is tuned by an adjusting slug between the windings.

3-5.1. *Circuit Analysis*

The considerations for transformer-coupled RF amplifiers are similar to those of audio amplifiers, as described in Chapter 2. However, the rules of thumb for trial values are somewhat different.

Transistor characteristics. The considerations for selection of transistors described in Chapters 1 and 2 apply to RF amplifiers. Of particular importance are interpreting datasheets and determining parameters at different frequencies. The temperature-related problems described in Chapters 1 and 2 generally do not apply, since RF voltage amplifiers usually operate at very low power levels.

The main concern is that the transistor will provide the required gain at the frequency of interest. In general, the transistor should provide 1.5 times the required gain at the operating frequency. This will compensate for mismatch, variation in gain due to differences in transistors, and so forth.

Transformer characteristics. It is often practical to design an RF amplifier around the characteristics of an existing commercial transformer (interstage

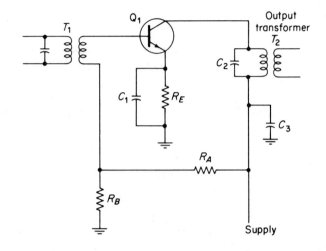

Voltage gain $\approx$ beta $\times (\frac{1}{N})$

where $N = \sqrt{\dfrac{Z_P}{Z_S}}$ $N^2 = Z_P/Z_S$

Voltage drop across	At operating frequency
$R_E \approx$ emitter-base voltage 0.5 for silicon 0.2 for germanium	$X_{C1} \lessgtr Z_{in}$ of Q_1 $X_{C2} \approx X_{primary}$ of T_2 $X_{C3} \lessgtr 100\ \Omega$

$R_B \approx 10\ R_E$

Voltage drop across R_A = supply – drop across R_B

Supply $\approx$ 3 to 4 times desired output voltage

Fig. 3-14   Basic RF voltage amplifier

IF transformer, IF detector transformer, RF/IF transformer, etc.). Such transformers are usually rated as to primary and secondary impedance (rather than turns ratio), and possibly current capacity. However, the typical low currents involved present no design problems.

Some commercial transformers are provided with a built-in fixed capacitor across the primary (and/or secondary in some cases). When a capacitor is used, the transformers are rated as to the resonant frequency range or mid-point (455 kHz for an AM broadcast intermediate frequency; 500 to 1600 kHz for an RF input transformer, often of the ferrite "loopstick" type; 10.7 MHz for an FM broadcast intermediate frequency; etc.). In other transformers, a fixed or variable capacitor must be connected across the transformer windings. For example, a "loopstick" requires a variable capacitor of the given range to provide full tuning across the AM broadcast band. When it is necessary to calculate the capacitance value for a given inductance or reactance, use the equations of Fig. 3-11.

Stage gain. Voltage gain of a fully bypassed RF amplifier is approximately equal to transistor gain (beta) and the turns ratio of the output transformer, as shown in Fig. 3-14. When the turns ratio is considered as primary/secondary, the stage gain equals transistor gain times the *inverse* of the turns ratio. For example, if the transistor gain is 10, and the transformer has a turns ratio of 10 (primary) to 1 (secondary), the net voltage gain is 1 (or unity).

The required stage gain depends upon the circuit application. As guidelines, a communications-receiver RF amplifier requires a voltage gain between 10 and 20, an AM broadcast IF stage requires a gain of between 30 and 40, and an IF amplifier for FM requires a gain between 40 and 50; a television IF amplifier (broadband) requires a gain of between 15 and 20.

Supply voltage. The value of supply voltage for an RF voltage amplifier is not critical. Of course, the supply voltage cannot exceed the transistor collector voltage limits. The supply voltage should be between three and four times the desired output voltage for the stage. In most receiver RF circuits, the desired output is between 1 and 2 V, so the supply could be between 3 and 9 V. A higher supply voltage can be used, provided the transistor characteristics are not exceeded.

Emitter resistance. When the emitter resistance R_E is bypassed, the resistance value should be chosen on the basis of direct current, rather than signal. The value of R_E should provide a voltage drop equal to the emitter–base voltage differential when the normal (dc operating point) collector current is flowing. A typical silicon emitter–base differential is about 0.5 V (0.2 V for germanium transistors). The drop across R_E will serve to stabilize the gain of Q_1, as discussed in Chapters 1 and 2.

To get a perfect impedance match between transistor and output transformer (a practical impossibility), the total impedance presented by R_E and the transistor should match the transformer primary impedance. As a guideline, assume that the impedance represented by the full collector voltage and current (V_C/I_C) is the total transistor and R_E impedance. This will establish the desired collector current I_C and a corresponding voltage drop across R_E. For example, assume that the supply voltage is 10 V, and the primary impedance is 10 kΩ. Ignoring the small dc drop across the primary, the collector is at 10 V. The desired collector current to match impedance is

$$I = \frac{E}{R} = \frac{10}{10,000} = 0.001 \text{ A}$$

or 1 mA. With 1 mA flowing through R_E (at the operating point) and a desired 0.5-V drop, the value of R_E is

$$R_E = \frac{E}{I} = \frac{0.5}{0.001} = 500 \,\Omega$$

Bypass capacitors. The value of the emitter-bypass capacitor C_1 should be such that the reactance, at the lowest operating frequency, is less than the input impedance of the transistor. This will effectively remove the emitter resistor from the circuit, so far as signal is concerned. The input impedance of a typical two-junction transistor for RF applications is on the order of a few ohms and is given on the datasheet. If the input impedance is not known, use a capacitor value that will produce a reactance of less than 10 Ω at the operating frequency.

The value of supply line bypass capacitor C_3 is between 0.001 and 0.01 μF in a typical RF voltage amplifier. As a first trial value, use that value for C_3 that will produce a reactance of less than 100 Ω at the operating frequency.

Bias resistance network. The values of the bias resistance network should be chosen to place transistor Q_1 at the desired operating point. For example, if the desired collector current is 1 mA and Q_1 has a nominal gain of 10, the base current must be 0.1 mA. Likewise, if the emitter–base voltage differential is assumed to be 0.5 V, with another 0.5-V drop across R_E, the base should be at 1 V under no-signal conditions. Any combination of R_B and R_A that produces these relationships would be satisfactory. As a first trial value, make R_B 10 times the value of R_E. Then calculate a corresponding value for R_A, using the equations of Fig. 3-13.

When the circuit of Fig. 3-13 is to incorporate an AVC–AGC (automatic volume control–automatic gain control) function, the bias network is also used as the AVC–AGC line. Thus, the bias network values must be calculated on that basis. A discussion of AVC–AGC circuits is covered in Sec. 3-7.

3-6. FREQUENCY MIXERS AND CONVERTERS

Figure 3-15 shows the working schematic of a typical frequency mixer and converter. Such a circuit is a combination of an RF voltage amplifier and an RF oscillator. The individual outputs of the two sections are combined to produce an intermediate frequency (IF) output. Usually, the RF oscillator operates at a frequency above the RF amplifier, with the difference in frequency being the intermediate frequency. The resonant circuit of T_1 is tuned to the incoming RF signal, T_2 is tuned to the oscillator frequency (RF + IF), and T_3 is tuned to the intermediate frequency (IF). The resonant circuits of T_1 and T_2 are usually tuned by means of variable capacitors ganged together so that both the oscillator and RF amplifier will remain at the same frequency relationship over the entire tuning range. For example,

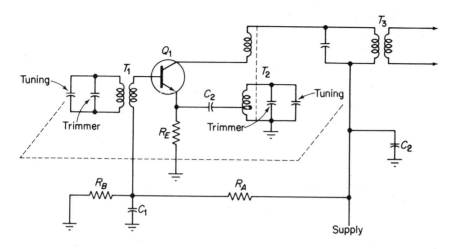

$$R_E \approx \text{impedance of tap on } T_2$$

Drop across $R_B \approx 0.2 \times$ supply voltage

Drop across $R_A \approx 4 \times R_B$

$R_A + R_B \approx 15{-}20 \times R_E$
at operating frequency X_{C1} and $X_{C2} \approx 50 \, \Omega$ (or less)

$$\text{Power output of } T_2 \approx 0.5 \times \frac{\text{collector voltage}^2}{Z \text{ of } T_2 \text{ collector winding}}$$

$$\text{Power output of } T_3 \approx 0.125 \times \frac{\text{collector voltage}^2}{Z \text{ of } T_2 \text{ collector winding}}$$

Fig. 3-15   Basic RF mixer and converter (RF amplifier and local oscillator)

if T_1 tunes from 550 to 1600 kHz and T_3 is at a fixed IF of 455 kHz, T_2 must tune from 1005 kHz to 2055 kHz. Usually, trimmer capacitors are placed in parallel with the variable capacitors to permit adjustment over the tuning range.

3-6.1. Circuit Analysis

The RF portion of the circuit can be designed on the same basis as the RF voltage amplifier described in Sec. 3-5. However, design of the oscillator section usually sets the operating characteristics for the remainder of the circuit. For example, the oscillator may require a power transistor, typically in the order of 0.5-W maximum output. Assuming that the oscillator is 50 per cent efficient, an output of 0.5 W requires an input of 1 W. Many transistors are capable of this power dissipation without heat sinks. However, the transistor power dissipation characteristics should be checked, as described in Chapters 1 and 2.

Transformer characteristics. As in the case of RF voltage amplifiers, converters are (or can be) designed around the characteristics of existing commercial transformers. The input impedance of T_1 should match the antenna (or other source) impedance. The output impedance of T_1 should match the input impedance of the stage. As a guideline, the stage input impedance can be considered as the emitter resistance R_E value (since R_E is unbypassed). However, a mismatch in T_1 is usually not critical. The output impedance of T_3 should match the input impedance of the following stage (an IF amplifier). The input impedance of T_3 should match the collector winding impedance of T_2. The impedance at the emitter tap of T_2 should match the oscillator input impedance (approximately equal to the value of R_E).

Although both signal currents (intermediate frequency and oscillator) are present in the collector circuit, the IF signal power is approximately 25 per cent of the oscillator signal power. Also, T_2 and T_3 are resonant at different frequencies. Thus, the impedance presented by T_3 has little effect on the oscillator signal in the collector circuit.

Some commercial transformers specify the fixed capacitance necessary to provide the desired resonant frequency (or variable capacitance limits for a given tuning range). If it is necessary to calculate the capacitance values for a given inductance or reactance, use the equations of Fig. 3-11.

Emitter resistance. Ideally, the emitter resistance R_E should be chosen to match the output impedance of T_1 and the emitter-tap impedance of T_2. When the design must be adapted to existing transformers and there is a mismatch between transformers, use a value of R_E that matches T_2 (the oscillator resonant circuit).

Bias relationships. One problem in the design of any oscillator circuit is that the circuit should be operated as a class A amplifier for starting, and switch to class C operation as the oscillations build up. That is, the emitter base should be forward biased initially, and then be reverse biased, except on peaks of the oscillation. (Ideally, the transistor conducts during approximately 140° of each 360° cycle.) Thus, a variable bias is required.

A variable bias is obtained when the capacitors are charged and discharged through the emitter resistance and bias resistance. However, if the capacitors are too small in value, the oscillator may not start, or the output waveform will be distorted. If the capacitors are too large, the change in charge (to produce a variable bias) will be slow, causing the oscillator to operate intermittently.

A fixed forward bias should be placed on the base by means of R_A and R_B. As a first trial value, the drop across R_B should be 0.2 times the supply voltage at the collector. Therefore, R_A should be approximately four times the value of R_B. Any combination of R_B and R_A that would produce the required bias could be used. However, the total series resistance of $R_A + R_B$ should be 15 to 20 times that of R_E. This will minimize excess current drain by the bias network.

Bypass and coupling capacitors. The values of the bypass and coupling capacitors are typically between 0.001 and 0.1 μF. For a first trial, choose a capacitance value that will produce a reactance of 50 Ω at the operating frequency. As discussed, the final value of the capacitors can be critical (to produce continuous oscillations with good waveforms). Thus, the only final test of correct capacitor values is the display of the output waveform on an oscilloscope.

Power output. Power output of the oscillator is *approximately* equal to

$$P_{\text{OUT}} = 0.5 \times \frac{\text{collector voltage}^2}{\text{impedance of } T_2 \text{ collector winding}}$$

Power output of T_3 to the following IF stage is

$$P_{T_3} = 0.125 \times \frac{\text{collector voltage}^2}{\text{impedance of } T_2 \text{ collector winding}}$$

Transistor characteristics. The considerations for selection of transistors described in Chapters 1 and 2 apply to converters and mixers. The main concern is that the transistor will provide the required power at the frequency of interest.

3-7. AVC–AGC CIRCUITS FOR AMPLIFIERS

Most receivers have some form of AVC–AGC (automatic volume control–automatic gain control) circuit. The terms AVC and AGC are used interchangeably. AGC is a more accurate term since the circuits involved control the gain of an IF or RF stage (or several stages simultaneously), rather than volume of an audio signal in an AF stage. However, in a broadcast receiver the net result is an automatic control of volume. Either way, the circuit functions to provide a constant output despite variations in signal strength. An increased signal will reduce stage gain, and vice versa.

Figure 3-16 shows the working schematic of two AGC systems that are common to broadcast and communications receivers. Diode CR_1 acts as a variable shunt resistance across the input of the IF stages. Diode CR_2 functions as the detector and AGC bias source.

Under no-signal conditions, or in the presence of a weak signal, diode CR_1 is reverse biased and has no effect on the circuit. In the presence of a very large signal, CR_1 is forward biased and acts as a shunt resistance to reduce gain.

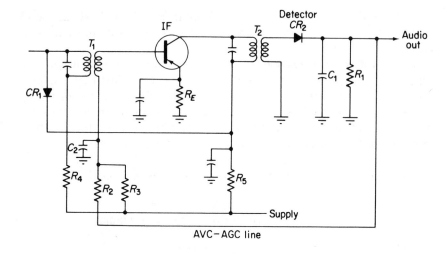

CR₁ is reverse biased with no signal
$C_2 \approx 10\,\mu F$
Drop across $R_1 \approx 0.5 - 1.0$ V
Drop across $R_1 + R_2 \approx 1.0 - 2.0$ V
Drop across $R_3 =$ supply $- (R_1 + R_2)$
$R_1 + R_2 \approx 10 \times R_E$

Fig. 3-16   Basic AVC–AGC circuit

The output of CR_2 is developed across R_1 and applied to the audio stages. Resistor R_1 also forms part of the bias network for the IF stage transistor. The combined fixed bias (from the network) and variable bias (from the detector) is applied to the IF stage base–emitter circuit. The detector bias varies with signal strength, and is of a polarity that opposes variations in signal. That is, if the signal increases, the detector bias will be more positive (or less negative) for the base of a *PNP* transistor, and vice versa for an *NPN* transistor.

3-7.1. *Circuit Analysis*

Both AGC systems (shunt diode and variable bias) are often found in the same receiver. The variable-bias system handles normal variations in signal. The shunt diode handles large signal variations.

Shunt diode. The shunt diode CR_1 should have a maximum reverse (peak inverse) voltage rating equal to the supply voltage. In most cases, the diode will never have a reverse voltage greater than 1 or 2 V. However, if the diode is capable of handling the full supply voltage, there will be no danger of breakdown. The forward-current capability of CR_1 should be such that the diode can pass the current if there is a full voltage drop across the collector resistors. The values of R_4 and R_5 must be such that CR_1 is reverse biased under no-signal conditions (with the IF stages at the Q point).

Bias network. The values of the bias network (R_1, R_2, and R_3) should be chosen to provide the desired fixed bias for the IF stages, as described in Sec. 3-5. The drop across R_1 and R_2 is the bias value applied to the base of the IF stage. The drop across R_1 is combined with the pulsating detector signal output. Typically, the drop across R_1 is on the order of 0.5 to 1 V. The drop across R_1 and R_2 is between 1 and 2 V. The value of C_2 is quite large in relation to other bypass capacitors, and is typically 10 μF, or larger.

3-7.2. Television-Amplifier AGC Circuits

Most solid-state TV receivers use a *keyed*, saturation-type AGC circuit. The RF tuner and IF stage transistors connected to the AGC line are forward biased at all times. On strong signals the AGC circuits *increase* the forward bias, driving the transistors into saturation, thus reducing gain. Under no-signal conditions the forward bias remains fixed.

Although the AGC bias is a dc voltage, it is partially developed (or controlled) by bursts of IF signals. A portion of the IF signal is taken from the IF amplifiers and is pulsed, or keyed, at the horizontal sweep frequency rate (15,750 Hz). The resultant keyed bursts of signal control the amount of dc voltage produced on the AGC line.

Figure 3-17 is the schematic diagram of a typical solid-state AGC circuit for RF and IF amplifiers. Transistor Q_1 is an IF amplifier with its collector tuned to the IF center frequency of 42 MHz by transformer T_1. No dc voltage

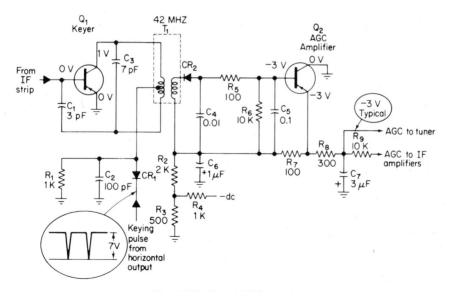

Fig. 3-17 Keyed AGC circuits

as such is supplied to Q_1. The keying pulses from the horizontal flyback transformer (at 15,750 Hz) are applied to the collector through diode CR_1. This produces an average collector voltage of about 1 V. When Q_1 is keyed on, the bursts of IF signals pass through T_1 and are rectified by CR_2. A corresponding dc voltage is developed across C_4, and acts as a bias for AGC amplifier Q_2. Transistor Q_2 is connected as an emitter follower, with the AGC line being returned to the emitter. Variations in IF signal strength cause corresponding variations in Q_2 bias, Q_2 emitter voltage, and the AGC line voltage.

3-8. FET RADIO-FREQUENCY AMPLIFIERS

Field-effect transistors are quite useful as RF *voltage* amplifiers. However, because of their high impedance and generally low power ratings, FETs are not particularly useful as power amplifiers. In this section, we discuss the basic procedures for design of FET RF voltage amplifiers. We shall also introduce the reader to the concept of considering a transistor as a *linear active two-port network* (or LAN) with *admittance parameters* (known as *y* parameters). This concept is quite common in design of RF amplifiers, both two junction and FET. In later sections of this chapter, we discuss variations of the concept and how they apply specifically to two-junction transistors.

It is difficult, at best, to provide a simple, step-by-step procedure for designing FET RF amplifiers to meet all possible circuit conditions. In practice, the procedure often results in considerable trial and error. There are several reasons for this problem of FET RF amplifier design.

First, not all the FET characteristics are always available in datasheet form. For example, input and output admittance may be given for some low frequency, but not at the frequency of interest.

Often, manufacturers do not agree on terminology. A classic example of this is in *y* parameters, where one manufacturer uses letter subscripts (y_{fs}) and another uses number subscripts (Y_{21}). Of course, this can be solved by conversions, as described in later paragraphs of this section.

In some cases, manufacturers will give the required information on datasheets, but not in the required form. For example, instead of showing input admittance in mhos, the input capacitance is given in farads. The input admittance is found when the input capacitance is multiplied by $6.28F$ (where F is the frequency of interest). This is based on the assumption that the input admittance is primarily capacitive, and thus dependent upon frequency. The assumption is not always true for the frequency of interest. It may be necessary to make actual tests of the FET, using complex admittance measuring equipment.

The input and output tuning circuits of a RF amplifier must perform two functions. Obviously, the circuits must tune the amplifier to the desired frequency. The circuits must also match the input and output impedances of the FET to the impedances of the source and load. Otherwise, there will be considerable loss of signal. (Note that although impedances are involved admittances are used instead, since admittances greatly simplify the measurement and calculations.)

Finally, as is the case with any amplifier, there is some feedback between output and input of a FET RF amplifier. If the admittance factors are just right, the feedback can be of sufficient amplitude and of proper phase to cause distortion and/or oscillation. The amplifier is considered as *unstable* when this occurs. The condition is always undesirable, and can be corrected by feedback (neutralization) or by changes in the input–output tuning networks. Although the neutralization and tuning circuits are relatively simple, the equations for determining stability (or instability) and impedance matching are long and complex. Generally, such equations are best solved by *computer-aided design* methods.

In an effort to cut through this maze of information and complex equations, we discuss all the steps involved in FET RF design in the following paragraphs. Armed with this information, the reader should be able to interpret datasheet or test information, and use the information to design tuning networks that will provide stable RF amplification at the frequencies of interest.

With each step, we discuss the various *alternative* procedures and types of information available. Specific design examples are given at the end of this section. These examples summarize the information contained in the various steps. On the assumption that all readers may not be familiar with two-port networks, we shall start with a summary of the *y*-parameter system.

3-8.1. y Parameters

Impedance (Z) is a combination of resistance (the real part) and reactance (the imaginary part). Admittance (y) is the reciprocal of impedance, and is composed of conductance (the real part) and susceptance (the imaginary part). A y parameter is an expression for admittance in the form

$$y_{is} = g_{is} + jb_{is}$$

where g_{is} is the real (conductive) part of common-source input admittance, and b_{is} is the imaginary (susceptive) part of common-source input admittance.

Since a FET can be treated as a linear active two-port network in small-signal applications, all the standard y-parameter stability criteria and design parameters are directly applicable. The y parameters are very useful in

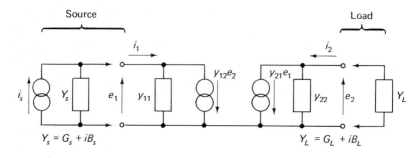

Y = admittance
G = conductance (real part of Y)
B = susceptance (imaginary part of Y)

Fig. 3-18   FET y-equivalent circuit with source and load

comparing different device types, in choosing a particular configuration (common source, common gate, neutralized, unneutralized, cascode, etc.), and in the final design of the RF amplifier.

The y-parameter circuit is shown in Fig. 3-18. The following is a summary of the four y parameters of primary interest.

Note that RF designers have traditionally used the nomenclature y_{11}, y_{12}, y_{21}, and y_{22} for all active devices—two-junction transistors, integrated circuits, and other devices. Some manufacturers still do. Other manufacturers use descriptive letter subscripts y_{is}, y_{rs}, y_{fs} and y_{os} for the same parameters. Both systems are given in the following summary. (Note that the letter s refers to common-source operation.)

Input admittance, with Y_L = infinity (short circuit), is expressed as

$$y_{is} = y_{11} = g_{11} + jb_{11} = y_i = \frac{i_1}{e_1}\bigg|_{e_2=0}$$

Note that all datasheets do not necessarily show y_{is} or y_{11} at any frequency. However, input capacitance c_{iss} is generally listed. If one assumes that the input admittance is entirely (or mostly) capacitive (jb_{11}), then the input impedance can be found when c_{iss} is multiplied by $6.28F$ (F = frequency in hertz). For example, if the frequency is 100 MHz, and the c_{iss} is 7 pF, the input impedance is $6.28 \times (100 \times 10^6) \times (7 \times 10^{-12}) \approx 4.4$ mmhos. The assumption is accurate only if the real part of y_{is} (or g_{is}) is negligible.

Figure 3-19 shows input admittance curves for a typical FET. Note that the imaginary part (jb_{is}) is the more significant factor across the entire frequency range.

Forward transadmittance, with Y_L = infinity (short circuit), is expressed as

$$y_{fs} = y_{21} = g_{21} + jb_{21} = y_f = \frac{i_2}{e_i}\bigg|_{e_2=0}$$

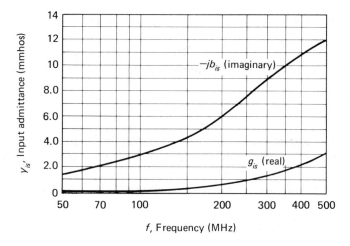

Fig. 3-19   Input admittance, $y_{is} = g_{is} + jb_{is}$, of MFE3007

Some datasheets show y_{fs} at a low frequency, typically 1 kHz, and also show a factor representing the real part of y_{fs} at a higher frequency, typically 100 MHz. On such datasheets the real part factor is listed as $R_e(y_{fs})$, or a similar term. The minimum values listed for $R_e(y_{fs})$ are quite similar to the minimum values of y_{fs}, and are generally realistic. However, the maximum values of $R_e(y_{fs})$, when listed, are not realistic.

A more accurate and complete method of showing y_{fs} is by means of curves. Figure 3-20 shows the y_{fs} curves, both real and imaginary, for a

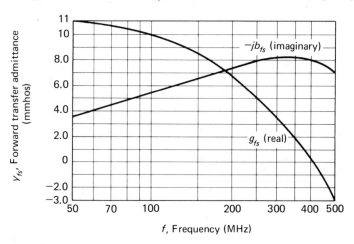

Fig. 3-20   Forward transfer admittance, $y_{fs} = g_{fs} + jb_{fs}$, of MFE3007

typical FET. Note that the real and imaginary parts actually cross over at about 180 MHz. Also, the real part becomes a negative quantity at about 400 MHz.

Output admittance, with y_S = infinity (short circuit), is expressed as

$$y_{os} = y_{22} = g_{22} + jb_{22} = y_o = \left. \frac{i_2}{e_2} \right|_{e_i = 0}$$

Generally, a datasheet will show y_{os} at only one frequency (typically 1 kHz). A more accurate and complete method of showing y_{os} is by means of curves such as illustrated in Fig. 3-21.

Reverse transadmittance, with Y_S = infinity (short circuit), is expressed as

$$y_{rs} = y_{12} = g_{12} + jb_{12} = y_r \left. \frac{i_1}{e_2} \right|_{e_i = 0}$$

Many datasheets omit y_{rs} or y_{12} completely. However, most datasheets do give reverse transfer capacitance C_{rss}. If one assumes that the reverse transadmittance (or reverse transfer admittance as it is sometimes called) is entirely (or mostly) capacitive (jb_{12}), then the reverse transadmittance can be found when C_{rss} is multiplied by 6.28F. This assumption is generally accurate in the case of y_{rs} or y_{12}, as shown in Fig. 3-22. Note that the real part of y_{rs} (or g_{rs}) is zero across the entire frequency range. When the term $R_e y_{12}$ or $R_e y_{rs}$ appears, it can be considered as zero for all practical purposes.

Measurement of y parameter. As can be seen thus far, y-parameter information is not always available or in a convenient form. In practical amplifier

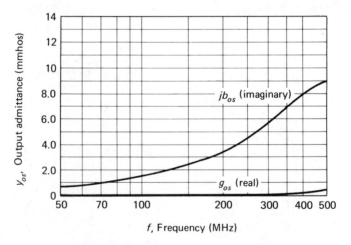

Fig. 3-21   Output admittance, $y_{os} = g_{os} + jb_{os}$, of MFE3007

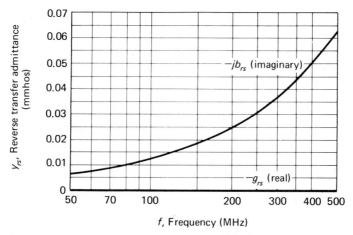

Courtesy Motorola

Fig. 3-22   Reverse transfer admittance, $y_{rs} = g_{rs} + jb_{rs}$, of MFE3007

work it may be necessary to measure the y parameters using laboratory equipment.

Note all y parameters are based on ratios of input–output current to input–output voltage. For example, y_{fs} is the ratio of output current to input voltage.

y_{fs} and y_{os} can be measured using signal generators, voltmeters, and simple circuits. Likewise, y_{is} and y_{rs} can be found by measuring c_{iss} and c_{rss} (using a simple capacitance meter) and then calculating the y_{is} and y_{rs} based on frequency of interest. Measurement procedures are described in the author's *Practical Semiconductor Databook for Electronic Engineers and Technicians* (Prentice-Hall, Inc., Englewood Cliffs, N.J., 1970).

More accurate results are obtained if precision laboratory equipment is used. All four y parameters can be measured on a General Radio Transfer Function and Immittance Bridge. A possible exception is y_{rs}, which is typically very small in relation to the other parameters. In the case of y_{rs} it is often more practical to measure c_{rss} and multiply by 6.28F.

The main concern in measuring y parameters, from a practical standpoint, is that the measurements are made under conditions simulating those of the final amplifier circuit. For example, the supply or drain-source voltage, bias (if any), and operating frequency should be identical (or close) to the final amplifier circuit. Otherwise, the tests can be misleading.

3-8.2. Stability Factors

There are two factors used to determine the potential stability (or instability) of FETs in RF amplifiers. (Note that these same factors are used with other devices such as two-junction transistor RF amplifiers, IC amplifiers, etc.)

One factor is known as the Linvill C factor; the other is the Stern k factor. Both factors are calculated from equations requiring y-parameter information (to be taken from datasheets, or by actual measurement at the frequency of interest.)

The Linvill C factor assumes that the FET (or other device) is not connected to a load. The Stern k factor includes the effect of a given load, and is thus more practical.

The Linvill C factor is calculated from

$$C = \frac{y_{12}y_{21}}{2g_{11}g_{22} - R_e(y_{12}y_{21})}$$

If C is less than 1, the device (FET) is *unconditionally stable*. That is, using a conventional (unmodified) circuit, no combination of load and source admittance can be found that will cause oscillations. If C is greater than 1, the FET is potentially unstable. That is, certain combinations of load and source admittance could cause oscillation.

The Stern k factor is calculated from

$$k = \frac{2(g_{11} + G_S)(g_{22} + G_L)}{y_{12}y_{21} + R_e(_{12}y_{21})}$$

where G_S and G_L are source and load conductance, respectively. ($G_S = 1/$source resistance; $G_L = 1/$load resistance.)

If k is greater than 1, the amplifier circuit is stable (this is the opposite from Linvill). If k is less than 1, the amplifier is unstable. In a practical amplifier, it is recommended that a k factor of 3 or 5 be used, rather than 1, to provide a margin of safety. This will accommodate parameter and component variations (particularly with regards to bandpass response of the amplifier).

Note that both equations are fairly complex and require considerable time for their solution (unless computer-aided design techniques are used). Some manufacturers provide alternative solutions to the stability and load matching problem, usually in the form of a datasheet graph. Such a graph is shown in Fig. 3-23, which is a Linvill C factor chart for a typical FET. Note that the FET is unconditionally stable at frequencies above 250 MHz, but potentially unstable at frequencies below 250 MHz. At frequencies below about 50 MHz, the FET becomes highly unstable.

3-8.3. Solutions to Stability Problems

There are two basic solutions to the problem of unstable RF amplifiers. First, the amplifier can be *neutralized*. That is, part of the output

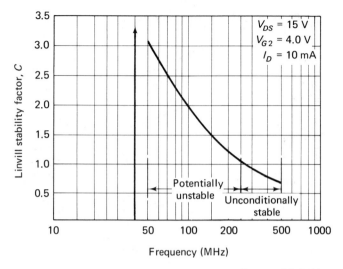

Courtesy Motorola

Fig. 3-23   Linvill stability factor, *C*, for the MFE3007 between 50 and 500 MHz

can be fed back to the input so as to cancel oscillation. This solution requires extra components, and creates a problem when frequency is changed. The other solution is to introduce some *mismatch* into either the source or load tuning networks. This solution requires no extra components, but does produce a reduction in gain.

A comparison of the two methods is shown in Fig. 3-24. The upper gain curve represents the unilateralized (or neutralized) method of operation. That is, the higher curve represents maximum amplifier gain without regard to stability, or with neutralization to produce stability. The lower gain curve represents the circuit power gain, when the amplifier is mismatched as necessary to produce a Stern *k* factor of 3, but neutralization is not used.

Assume that the frequency of interest is 100 MHz. If neutralization is used (top gain curve), the power gain is about 38 dB. If the amplifier is matched to a load and source where the Stern *k* factor is 3 (probably some mismatch), the lower curve applies, and the power gain is about 29 dB.

The upper curve of Fig. 3-24 is found by the *general power gain equation*:

$$G_P = \frac{\text{power delivered to load}}{\text{power delivered to input}}$$

$$= \frac{(Y_{21})^2 G_L}{(Y_L + y_{22})^2 \, R_e\!\left(y_{11} - \dfrac{y_{12}\,y_{21}}{y_{22} + Y_L}\right)}$$

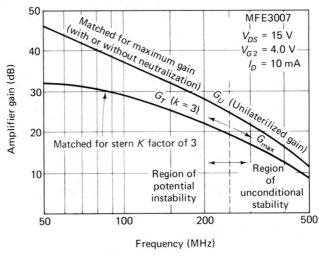

Courtesy Motorola

Fig. 3-24   Amplifier gain characteristic in common-source configuration

The general power gain equation applies to circuits with no external feedback, and to circuits that have external feedback (neutralization), provided the *composite y* parameters of both FET and feedback network are substituted for the FET y parameters in the equation.

The lower gain curve of Fig. 3-24 is found by the *transducer gain expression*:

$$G_T = \frac{\text{power delivered to load}}{\text{max power available from source}}$$

$$= \frac{4G_s G_L (y_{21})^2}{[(y_{11} + Y_s)(y_{22} + Y_L) - y_{12}y_{21}]^2}$$

The transducer gain expression includes input mismatch. The lower curve of Fig. 3-24 assumes that the mismatch is such that a Stern k factor of 3 is produced. That is, the circuit tuning networks are adjusted for admittances that produce a Stern k factor of 3. The transducer gain expression considers the input and output networks as part of the source and load.

With either gain expression, the input and output admittances of the FET are modified by the load and source admittance.

The input admittance of the FET is given by

$$Y_{\text{IN}} = y_{is}(\text{or } y_{11}) = \frac{y_{12}y_{21}}{y_{22} + Y_L}$$

The output admittance of the FET is given by

$$Y_{OUT} = y_{os} \text{ (or } y_{22}) - \frac{y_{12}y_{21}}{y_{11} + Y_S}$$

At low frequencies, the second term in the input and output admittance equations is not particularly significant. At VHF, the second term makes a very significant contribution to the input and output admittances.

The imaginary parts of Y_S and Y_L (B_S and B_L, respectively) must be known before values can be calculated for power gain, transducer gain, input admittance, and output admittance. Except for some very special cases, exact solutions for B_S and B_L consist of time-consuming complex algebraic manipulations.

To find fairly good simplifying approximations for the equations, let $B_S \approx -b_{11}$ and $B_L \approx -b_{22}$ so that

General power gain expression:

$$G_P \approx \frac{(y_{21})^2 G_L}{(G_L + g_{22})^2 \, R_e\left(y_{11} - \dfrac{y_{12}y_{21}}{g_{22} + G_L}\right)}$$

Transducer gain expression:

$$G_T \approx \frac{4G_S G_L (y_{21})^2}{[(g_{11} + G_S)(g_{22} + G_L) - y_{12}y_{21}]^2}$$

Input admittance:

$$Y_{IN} \approx y_{11} - \frac{y_{12}y_{21}}{g_{22} + G_L}$$

Output admittance:

$$Y_{OUT} \approx y_{22} \frac{y_{12}y_{21}}{g_{11} + G_S}$$

3-8.4. Stern Solution

A stable design with a potentially unstable FET is possible without external feedback (neutralization) by proper choice of source and load admittances. This can be seen by inspection of the Stern k factor equation. G_S and G_L can be made large enough to yield a stable circuit, regardless of the degree of potential instability. Using this approach, a circuit stability factor (typically $k = 3$) is selected, and the Stern k factor equation is used to arrive at values of G_S and G_L that will provide the desired k. Of course, the *actual G* of the source and load cannot be changed. Instead, input and output tuning circuits are designed as if the actual G values were changed. This results in a mismatch and a reduction in power gain, but does produce the desired degree of stability.

To get a particular circuit stability factor, the amplifier designer may choose any of the following combinations of matching or mismatching of G_S and G_L to the FET input and output conductances, respectively:

G_S matched and G_L mismatched
G_L matched and G_S mismatched
Both G_S and G_L mismatched

Often a decision on which combination to use will be dictated by other performance requirements or practical considerations.

Once G_S and G_L have been chosen, the remainder of the design may be completed using the relationships that apply to the amplifier *without feedback*. Power gain and input–output admittances may be computed using the appropriate equations (Sec. 3-8.3).

Simplified Stern approach. Although the above procedure may be adequate in many cases, a more systematic method of source and load admittance determination is desirable for designs that demand maximum power gain per degree of circuit stability. Stern has analyzed this problem and developed equations for computing the optimum G_S, G_L, B_S, and B_L for a particular circuit stability factor (Stern k factor).

Unfortunately, these equations are very complex and quite tedious if they must be done frequently. The complex Stern solution is best solved by computer. As a matter of interest, a program has been written in BASIC to provide essential information for FETs used as RF amplifiers. A second program has been written to include the effects of a specific source and load. This second program permits the designer to experiment with theoretical breadboard circuits in a matter of seconds. Other programs perform parameter conversions and the network synthesis for FET RF amplifier design.

When a Stern solution must be obtained without the aid of a computer, it is best to use one of the many shortcuts that have been developed over the years. The following shortcut is by far the simplest and most widely accepted, yet provides an accuracy close to that of the computer solutions.

1. Let $B_S \approx -b_{11}$ and $B_L \approx -b_{22}$. This method permits the designer to closely approximate the exact Stern solution for Y_S and Y_L, while avoiding that portion of the computations which is the most complex and time consuming. Furthermore, the circuit can be designed with tuning adjustments for varying B_S and B_L, thereby creating the possibility of experimentally achieving the true B_S and B_L for maximum gain as accurately as if all the Stern equations had been solved.

2. Mismatch G_S to g_{11} and G_L to g_{22} by an *equal ratio*. That is, find a ratio that produces the desired Stern k factor, then mismatch G_S to g_{11} (and G_L to g_{22}).

For example, if the ratio is 4 to 1, make G_S 4 times the value of g_{11} (and G_L 4 times the value of g_{22}).

If the mismatch ratio, R, is defined as

$$R = \frac{G_L}{g_{22}} = \frac{G_S}{g_{11}}$$

then R may be computed for any particular circuit stability (k) factor using the equation

$$R = \left(\sqrt{k \left[\frac{(y_{21}y_{12}) + R_e(y_{12}y_{21})}{2g_{11}g_{22}} \right]} \right) - 1$$

This shortcut method may be advantageous if source and load admittances and power gains for several different values of k are desired. Once the R for a particular k has been determined, the R for any other k may be quickly found from the equation

$$\frac{(1 + R_1)^2}{(1 + R_2)^2} = \frac{k_1}{k_2}$$

where R_1 and R_2 are values of R corresponding to k_1 and k_2, respectively.

Stern solution with datasheet graphs. It is obvious that the Stern solution, even the shortcut method, is somewhat complex. For this readon, some manufacturers have produced datasheet graphs that show optimum source and load admittances for a particular FET over a wide range of frequencies.

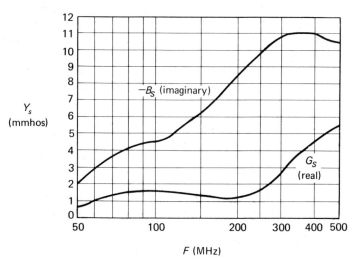

Courtesy Motorola

Fig. 3-25 Optimum source admittance, $Y_S = G_S + jB_S$

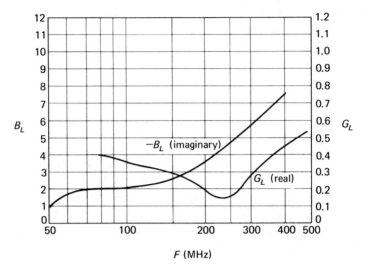

Courtesy Motorola

Fig. 3-26   Optimum load admittance, $Y_L = G_L + jB_L$

Examples of these graphs are shown in Figs. 3-25 and 3-26. Figure 3-25 shows both the real (G_S) and imaginary (B_S) values that will produce maximum gain, but with a stability (Stern k) factor of 3 at frequencies from 50 to 500 MHz. Figure 3-26 shows corresponding information for G_L and B_L.

To use these illustrations, simply select the desired frequency, and note where the corresponding G and B curves cross the frequency line. For example, assuming a frequency of 100 MHz, $Y_L = 0.35 - j2.1$ mmhos, and $Y_S = 1.30 - j4.4$ mmhos.

If the tuning circuits are designed to match these admittances, rather than the actual admittance of the source and load, the circuit will be stable. Of course, the gain will be reduced. Use the transducer gain expression (G_T) of Sec. 3-8.3 to find the resultant power gain.

3-8.5. Systematic FET Radio-Frequency Amplifier Design Procedure

A review of the two-port (y parameter) network design method may be helpful at this point.

1. Determine the potential instability of the FET (or other device). This involves extracting the y parameters of the FET from the datasheet, or determining y parameters from actual test, and then plugging the y parameters into the Linvill C and/or Stern k equations to find potential stability or instability.

Use the Linvill C factor where source and load impedances are not involved (or known). Use the Stern k factor when load and source impedances are known. As a practical matter, it is usually more convenient to go directly to the Stern k factor, since this serves as a starting point if the circuit must be modified to produce stability.

2. If the FET (or other device) is not *unconditionally stable*, decide on a course of action to ensure circuit stability. Usually, this involves going to neutralization, or to some form of mismatching input–output tuning circuits. Mismatching is, by far, the most popular course of action. If the FET is unconditionally stable, without neutralization or mismatch, the design can proceed without fear of oscillation. Under these circumstances, the usual object is to get maximum gain by matching the tuning circuits to the actual source and load.

3. Determine source and load admittances based on gain and stability considerations (together with practical circuit limitations). If the FET is potentially unstable at the frequency of interest with actual source and load impedances, use another source and load that will guarantee a certain degree of amplifier stability.

This involves the Stern solution described in Sec. 3-8.4. If optimum source and load impedances are given by manufacturer's datasheets, use these as a first choice. As a second choice, use computer-aided design techniques to get a Stern solution for the desired stability and gain. If neither of these two are available, use the shortcut Stern technique.

Note that it is a good idea to check circuit stability (Stern k) factor even when an unconditionally stable FET has been found by the Linvill C factor. A FET may be stable without a load, or with certain loads, but not stable with some specific load.

Once the optimum source and load admittances have been selected, verify that the required gain will be available. In practical terms, it is possible to mismatch almost any RF amplifier sufficiently to produce a stable circuit. However, the resultant power gain may be below that required. In that case, a different FET (or other device) must be used (or a lower gain accepted).

4. Design appropriate networks (input–output tuning circuits) to provide the desired (or selected) source and load admittances. First, the networks must be resonant at the desired frequency. (That is, inductive and capacitive reactance must be equal at the selected frequency.) Second, the network must match the FET to the load and source. Sometimes it will be difficult to achieve a desired source and load due to tuning range limitations, excess network losses, component limitations, and the like. In such cases the source and load admittances will be a compromise or tradeoff between desired performance and practical limitations. Generally, this tradeoff involves a sacrifice of gain to achieve stability.

The remainder of this section is devoted to the step-by-step design procedures for two FET RF amplifier circuits. Although FETs are involved, the same basic procedures can be used for two-junction transistors. Additional examples and special techniques for two-junction RF amplifier design are described in Sec. 3-9.

3-8.6. Design Example of 100-MHz Common-Source FET Radio-Frequency Amplifier

Assume that the circuit of Fig. 3-27 is to be operated at 100 MHz. The source and load impedance are both 50 Ω. The characteristics of the FET are given in Figs. 3-19 through 3-26. The problem is to find the optimum values of C_1 through C_4, as well as L_1 and L_2.

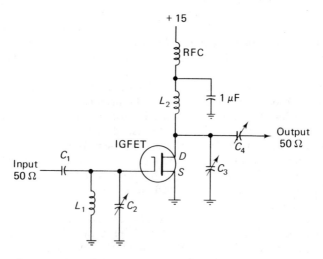

Courtesy Motorola

Fig. 3-27   Common-source FET RF amplifier

From Fig. 3-23, the Linvill stability factor C is seen to be 2.0. A Linvill C factor greater than 1 indicates that the device is potentially unstable. Thus, mismatching or neutralization is necessary to prevent oscillation. Mismatching is used in this example. Figure 3-24 shows that for a circuit stability (Stern k) factor of 3.0 the transducer gain is about 29 dB. The load and source admittances for the required mismatch and gain (found in Figs. 3-25 and 3-26) are

$$Y_L = 0.35 - j2.1 \text{ mmhos}$$
$$Y_s = 1.30 - j4.4 \text{ mmhos}$$

At 100 MHz, y parameters for the FET are

$$y_{is} = y_{11} = 0.15 + j3.0 \text{ (Fig. 3-19)}$$
$$y_{fs} = Y_{21} = 10 - j5.5 \text{ (Fig. 3-20)}$$
$$y_{os} = y_{22} = 0.04 + j1.7 \text{ (Fig. 3-21)}$$
$$y_{rs} = y_{12} = 0 - j0.012 \text{ (Fig. 3-22)}$$

Assuming the load and source impedances are both 50 Ω, networks can be designed to match the FET to the load and source. The calculations are more easily performed with impedances rather than admittances. The procedure will first be discussed for the output matching.

The 50-Ω load impedance must be transformed to the optimum load for the FET ($Y_L = 0.35 - j2.1$). This transformation can be performed by the

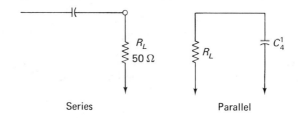

(a) Output Impedance Transformation

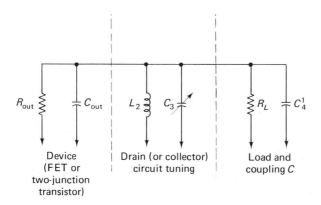

(b) Equivalent Output Circuit

Courtesy Motorola

Fig. 3-28   Impedance transformation and equivalent output circuit of common-source RF amplifier

network shown in Fig. 3-28a. In effect, R_L is in series with C_4. The 50 Ω must be transformed to

$$R_L = \frac{1}{G_L} = \frac{1}{0.35 \times 10^{-3}} \approx 2.86 k\Omega$$

The series capacitive reactance required for this matching can be found by

$$X_{C_4} = X_{series} = R_S \sqrt{\frac{R_P}{R_S} - 1}$$

where R_P is the parallel resistance and R_S is the series resistance.

$$X_{C_4} = 50 \sqrt{\frac{2.86 \times 10^3}{50} - 1} \approx 372 \, \Omega$$

The capacitance that provides this reactance at 100 MHz is

$$C_4 = \frac{1}{6.28 \, FX_{C_4}} = \frac{1}{6.28(10^8)(372)} \approx 4.3 \, pF$$

The parallel equivalent of this capacitance is needed for determining the bandwidth and resonance later in the design:

$$X'_{C_4} = X_{parallel} = X_S \left[1 + \left(\frac{R_S}{X_S} \right)^2 \right]$$

$$= 372 \left[1 + \left(\frac{50}{372} \right)^2 \right] \approx 378 \, \Omega$$

and the equivalent parallel capacitance is therefore

$$C'_4 \approx 4.2 \, pF$$

An equivalent circuit for the output tank, after transformation of the load, is shown in Fig. 3-28b. Since the resistance across the output circuit is fixed by the parallel combination of R_{OUT} and R_L (after transformation), the desired bandwidth of the output tank will be determined by C_3.

It should be noted that the output admittance (Y_{OUT}) of the FET will not equal y_{os} under most conditions. Only when the input is terminated in a short circuit, or the feedback admittance is zero, does Y_{OUT} equal y_{os}. When y_{rs} is not zero, and the input is terminated with a practical source admittance, the true output admittance is found from

$$Y_{OUT} = y_{os} - \frac{y_{fs} y_{rs}}{y_{is} + Y_S}$$

$$= 0.04 + j1.7 - \frac{(10 - j5.5)(0 - j0.012)}{(0.15 + j3.0) + (1.3 - j4.4)}$$

$$= -0.066 + j2.05 \text{ mmhos}$$

Therefore,

$$R_{OUT} = \frac{1}{G_{OUT}} = \frac{1}{-0.066 \times 10^{-3}} = -15.2 \text{ k}\Omega$$

$$C_{OUT} = \frac{B_{OUT}}{6.28F} = \frac{2.05 \times 10^{-3}}{6.28(10)^8} \approx 3.2 \text{ pF}$$

The negative output impedance indicates the instability of the unloaded amplifier.

Now the total impedance across the output tank can be calculated:

$$R_T = \frac{1}{G_{OUT} + G_L}$$

$$= \frac{1}{(-0.066 \times 10^{-3}) + (0.35 \times 10^{-3})} \approx 3.52 \text{ k}\Omega$$

Since the output impedance is several times higher than the input imped-ance of the FET, amplifier bandwidth is primarily dependent upon output loaded Q. For a bandwidth of 5 MHz (3-dB points),

$$C_T = \frac{1}{6.28 R_T (\text{BW})}$$

$$= \frac{1}{6.28(3.52 \times 10^3)(5 \times 10^6)} \approx 9 \text{ pF}$$

Hence,

$$C_3 = C_T - C_{OUT} - C_4' = 9.0 - 3.2 - 4.2 = 1.6 \text{ pF}$$

The output inductance that resonates with C_T at 100 MHz is 280 nH. This completes the output circuit design.

Input calculations performed in a similar manner yield the following results:

$$Y_S = 1.30 - j4.4 \text{ mmhos}$$
$$X_{C_1} = 190 \ \Omega; \quad \text{therefore,} \quad C_1 = 8.4 \text{ pF}$$
$$X_{C_1}' = 203 \ \Omega; \quad \text{therefore,} \quad C_1' = 7.8 \text{ pF}$$

$$Y_{IN} = y_{is} = \frac{y_{fs} y_{rs}}{y_{os} + Y_L} = -0.25 + j4.52 \text{ mmhos}$$

Therefore,

$$R_{IN} = 14 \text{ k}\Omega, \quad C_{IN} = 7.2 \text{ pF}, \quad R_T = 950 \ \Omega$$

The bandwidth of the input tuned circuit is chosen to be 10 MHz. Hence,

$$C_T = 17 \text{ pF}; \quad \text{therefore,} \quad L_1 = 150 \text{ nH}$$
$$C_2 = 17 - 7.2 - 7.8 = 2 \text{ pF}$$

This completes design of the tuned circuits. It is important that the circuit be well bypassed to ground at the signal frequency, since only a small impedance to ground may cause instability or loss in gain. The bypass capacitor should be such that the reactance is about 1 to 2 Ω at the operating frequency. A 1-μF capacitor will provide less than 1-Ω reactance at 100 MHz.

3-8.7. Design Example of 87.5- to 108.5-MHz Common-Gate FET Radio-Frequency Amplifier

Assume that the circuit of Fig. 3-29 is to be tuned across the range from 87.5 to 108.5 (broadcast FM) MHz. An equivalent circuit of the RF amplifier to be designed is given in Fig. 3-30. Coil turns are defined and high-frequency conductances are shown in Fig. 3-30.

The following design values are assumed:

1. $G_S^* = 3.33$ mmho (300-Ω antenna).
2. $G_{IN}^* = G_S^*$ (for minimum antenna VSWR).
3. $G_L^* = 4.0$ mmho (typical input conductance of a two-junction transistor mixer).
4. Selectivity at 21.4 MHz above center frequency = 45 dB (needed for image rejection = 45 dB).
5. Tuning range = 87.5 to 108.5 MHz.

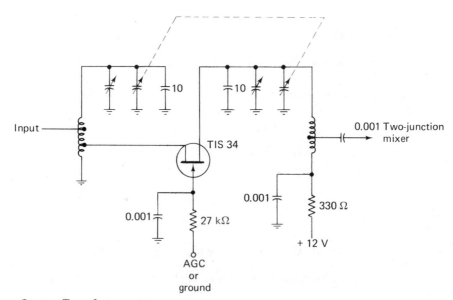

Courtesy Texas Instruments

Fig. 3-29   Common-gate FET RF amplifier

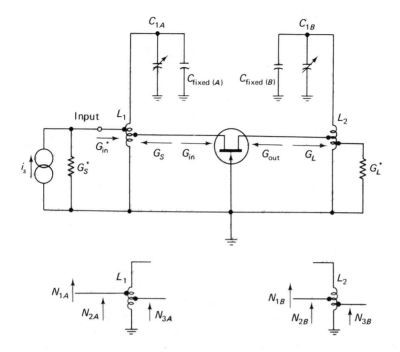

G_S^* = stage driving source conductance

G_{in}^* = stage input conductance

G_S = FET driving source conductance

G_{in} = FET input conductance with a finite load

G_{out} = FET output conductance with a finite driving source conductance

G_L = FET load conductance

G_L^* = stage load conductance

F_1 = lowest signal frequency

F_2 = highest signal frequency

$$C_{\text{fixed }(A)} = \frac{\Delta C_{1A}}{\left(\frac{F_2}{F_1}\right)^2 - 1} - C_{1A}\,(\text{min})$$

$$L_1 = \frac{1}{(6.28F_2)^2\left[\dfrac{\Delta C_{1A}}{\left(\frac{F_2}{F_1}\right)^2 - 1}\right]}$$

$$C_{\text{fixed }(B)} = \frac{\Delta C_{1B}}{\left(\frac{F_2}{F_1}\right)^2 - 1} - C_{1B}\,(\text{min})$$

$$L_2 = -\frac{1}{(6.28F_2)^2\left[\dfrac{\Delta C_{1B}}{\left(\frac{F_2}{F_1}\right)^2 - 1}\right]}$$

Courtesy Texas Instruments

Fig. 3-30 Alternating-current equivalent circuit of common-gate FET RF amplifier

6. Tune with standard variable capacitor: $C_{1A} = C_{1B} = 6 - 21$ pF ($\Delta C = 15$ pF).
7. Coils must have practical tapping ratios, wire sizes, and shape factors.
8. Gain control will be reverse AGC.
9. Observing the above restrictions, achieve high power gain and low noise figure, using an N-channel silicon TIS34 JFET.

Given a ΔC for the tuning capacitor, and a desired tuning range, the input and output coil inductances (L_1 and L_2) and fixed capacitances [$C_{\text{fixed}(A)}$ and $C_{\text{fixed}(B)}$] can be calculated using the equations of Fig. 3-30.

For this particular example $C_{1A(min)} = C_{1B(min)}$, $\Delta C_{1A} = \Delta C_{1B}$, $F_1 = 87.5$ MHz, and $F_2 = 108.5$ MHz, so that

$$L_1 = L_2 = 0.077\ \mu\text{H}$$

$$C_{\text{fixed}(A)} = C_{\text{fixed}(B)} = 22\text{ pF}$$

These values for the capacitances $C_{\text{fixed}(A)}$ and $C_{\text{fixed}(B)}$ include trimmer capacitances, stray capacitances, and reflected FET input and output capacitances, in addition to any actual fixed capacitors placed across the input or output tanks. With a typical trimmer setting of about 8 pF, and allowing a total of about 4 pF for stray capacitance and FET reflected capacitance, a 10-pF capacitor can be placed across the output tank.

The remainder of the design is carried out at the signal frequency $F_0 = 100$ MHz, where

$$X_C = X_L = 6.28 F_0 L = 48.4\ \Omega$$

Based on the discussions of biasing in Chapters 1 and 2, the values of the drain and gate resistance shown in Fig. 3-30 result in an operating point at I_{DSS}.

Typical *common-source* y parameters for the TIS34 at I_{DSS} are

$$y_{is} = y_{11} = 0.01 + j3.00\text{ mmho}$$

$$y_{fs} = y_{21} = 4.80 - j1.20\text{ mmho}$$

$$y_{rs} = y_{12} = 0 - j0.95\text{ mmho}$$

$$y_{os} = y_{22} = 0.05 + j1.05\text{ mmho}$$

Since the circuit of Fig. 3-30 is *common gate*, the common-source parameters must be converted to common gate as follows:

$$y_{ig} = y_{is} + y_{fs} + y_{rs} + y_{os} = 4.95 + j1.9\text{ mmho}$$

$$y_{fg} = -(y_{fs} + y_{os}) = -4.85 + j0.15\text{ mmho}$$

$$y_{rg} = -(y_{rs} + y_{os}) = -0.05 - j0.1 \text{ mmho}$$
$$y_{og} = y_{os} = 0.05 + j1.05 \text{ mmho}$$

Using these common-gate y parameters and the Linvill C factor, the stability is

$$C\frac{y_{12}y_{21}}{2g_{11}g_{22} - R_e(y_{12}y_{21})} = \frac{0.545}{2(4.95)(0.05) - 0.258} = 2.3$$

Since C is greater than 1, the FET is potentially unstable in the common-gate configuration.

Mismatching is to be used for the required stability. A Stern k factor of 4 is chosen. Using the shortcut Stern solution (mismatching G_S and G_L to g_{11} and g_{22} by an equal ratio R), the value of R is

$$R = \left(\sqrt{\frac{4(0.545 + 0.258)}{2(4.95)(0.05)}}\right) - 1 \approx 1.56$$

Using a value of 1.56 for R, and the equation,

$$1.56 = \frac{G_S}{g_{11}} = \frac{G_L}{g_{22}}$$

then

$$G_S = (1.56)(4.95)(10^{-3}) = 7.72 \text{ mmho}$$
$$\left(R_S = \frac{1}{G_S} = 130 \ \Omega\right)$$
$$G_L = (1.56)(0.05)(10^{-3}) = 0.078 \text{ mmho}$$
$$\left(R_L = \frac{1}{G_L} = 12.8 \text{ k}\Omega\right)$$

If it were practical to realize such a small G_L (or a high R_L), the power gain and transducer gain (from the equations of Sec. 3-8.3) would be $G_P \approx 15.8$ dB, and $G_T \approx 14.3$ dB.

Unfortunately, G_L is too small to be conveniently realized. These calculations are informative, however, in that they suggest, for the *common-gate configuration*, that the FET should work into as high an impedance load as is practical. This says that the drain terminal of the FET will go *to the top of the output tank* $\left(\frac{N_{1B}}{N_{2B}} = 1.0\right)$, that the actual value of R_L^* will be tapped as low on the output tank as is practical, and that the highest practical value for the unloaded coil Q should be chosen.

Given as a guide for the selection of Q_U, Figs. 3-31, 3-32, and 3-33 represent data taken at 100 MHz on band-wound coils using a Boonton Radio Co. 190-A high-frequency Q meter.

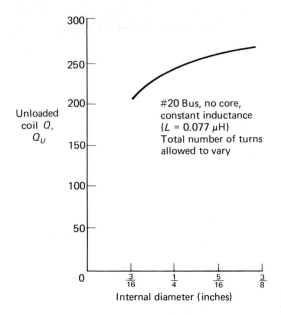

#20 Bus, no core,
constant inductance
($L = 0.077 \ \mu H$)
Total number of turns
allowed to vary

Fig. 3-31 Unloaded coil Q as a function of coil internal diameter

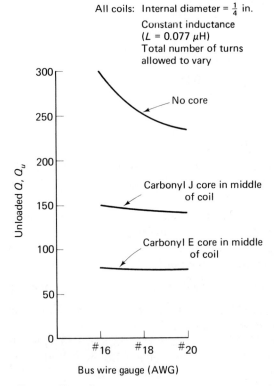

All coils: Internal diameter = $\frac{1}{4}$ in.
Constant inductance
($L = 0.077 \ \mu H$)
Total number of turns
allowed to vary

Fig. 3-32 Unloaded coil Q as a function of wire gauge

Figure 3-31 shows how Q_U is a function of the internal diameter of a coil, when no core is used and the wire gauge and inductance of the coil are held constant. Assuming no differences in the proximity of the coils to a metal chassis due to different coil sizes, the larger-diameter coils have the higher Q_U.

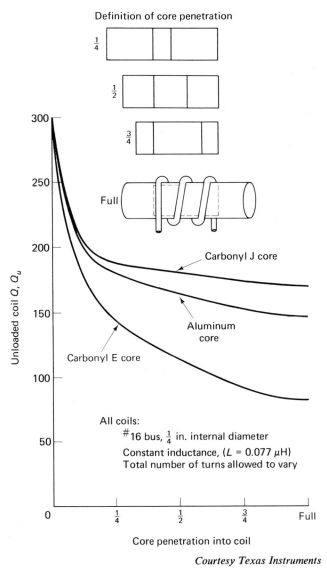

Courtesy Texas Instruments

Fig. 3-33   Unloaded coil Q as a function of core penetration into the coil

Figure 3-32 illustrates how Q_U varies with wire gauge for a constant inductance and a constant internal diameter. The Q_U of coils without cores are fairly sensitive to the wire gauge, but those using Carbonyl J and Carbonyl E cores are almost independent of the wire gauge.

In Figure 3-33, Q_U is given as a function of the penetration of different types of cores into the coil. Curves are given for Carbonyl J, Carbonyl E, and aluminum cores, with the wire gauge, internal diameter, and inductance held constant.

The circuit of Fig. 3-29 is taken from an FM receiver front end. Practical layout and shielding requirements usually require that the FM front end be placed in its own small metal chassis. Because the coils are close to metal with such an arrangement, the unloaded Q is reduced. Likewise, the presence of solder on the coils can reduce unloaded Q.

With these comments in mind, in selecting as high an impedance (for the output circuit) as practical, the following somewhat arbitrary limits are imposed:

$$Q_U = 200 \text{ min } (Q_U X_C = 200 \times 48.6 \approx 9.6 \text{ k}\Omega)$$

$$\frac{N_{1B}}{N_{3B}} = 5.0 \text{ max} \qquad \frac{N_{1B}}{N_{2B}} = 1.0$$

These restrictions give

$$G_L \text{ min} = \frac{1}{Q_U X_C(\text{max})} = \frac{G_L^*}{\left(\dfrac{N_{1B}}{N_{3B}} \text{ min}\right)^2}$$

$$G_L \text{ min} = 0.304 \text{ mmho} \quad \text{or} \quad R_L \text{ max} = 3.29 \text{ k}\Omega$$

Temporarily assuming that the output resistance of the common-gate FET will not appreciably load the output tank, the loaded Q of the output circuit will be

$$Q_L \approx \frac{R_L(\text{max})}{X_L} \approx 68$$

It will be seen that the FET output resistance is quite high, after the FET driving source conductance is calculated from the input coil design.

The loaded Q of the input circuit must be high enough to ensure that the frequency response of the RF stage will satisfy the selectivity requirements of 45-dB attenuation at 21.4 MHz *off resonance*.

The basic attenuation equation for tuned circuits is given in Fig. 3-34. Attenuation values calculated from this equation at 5.25 MHz and 21.4 MHz off resonance from a center frequency of 100 MHz are also given in Fig. 3-34 for several values of loaded Q. (The 5.35-MHz values are included

Q_L	Δf (MHz)	Attenuation (dB)
10	5.35	3.3
10	21.40	12.9
20	5.35	7.5
20	21.40	18.7
30	5.35	10.5
30	21.40	22.2
40	5.35	12.9
40	21.40	24.7
50	5.35	14.7
50	21.40	26.6
60	5.35	16.3
60	21.40	28.2
70	5.35	17.6
70	21.40	29.5
80	5.35	18.7
80	21.40	30.7
90	5.35	19.7
90	21.40	31.7
100	5.35	20.6
100	21.40	32.6
110	5.35	21.4
110	21.40	33.5

$$\text{Attenuation at } \Delta F \text{ Hz off resonance (dB)} = 20 \log_{10} \sqrt{1 + \left(\frac{Q_L \, 2\Delta F}{\begin{array}{c}\text{resonant} \\ \text{frequency} \\ (100 \text{ MHz})\end{array}} \right)^2}$$

Courtesy Texas Instruments

Fig. 3-34   Attenuation versus frequency for a single-tuned transformer; center frequency = 100 MHz

because they are useful in predicting $F_0 + \frac{1}{2}$ IF rejection. The IF amplifiers in a broadcast FM receiver are typically 10.7 MHz.)

The attenuation at ΔF Hz off resonance for *two* single-tuned transformers is found simply by summing the two appropriate attenuation values. The attenuations at the image frequency from Fig. 3-34 are also presented in graphic form in Fig. 3-35 to point out how increasing values of loaded Q rapidly reach a point of diminishing returns. As discussed, the output Q_L is somewhat less than 68, but still in the neighborhood of 68.

From Fig. 3-34, a loaded Q of 60 will provide 28.2-dB attenuation at the image frequency. Choosing an input loaded Q of 20 adds an additional 18.7-dB attenuation at the image frequency so that the selectivity requirement is satisfied.

The input conductance of the FET is found from

$$G_{\text{IN}} \approx g_{11} - R_e \left[\frac{y_{12} y_{21}}{g_{22} + G_L} \right] \approx 4.22 \text{ mmho}, \qquad R_{\text{IN}} = \frac{1}{G_{\text{IN}}} = 236 \, \Omega$$

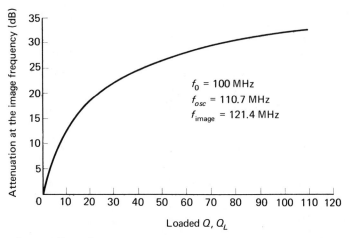

Fig. 3-35   Attenuation at the image frequency as a function of loaded
Q for a single-tuned transformer

To minimize input loss, let the input coil $Q_U = 200$. Next, the *tapping ratios* on the input coil need to be determined. In Fig. 3-36, all conductances are *referred to the top of the input tank*. Using the tapping ratio equations of Fig. 3-36, solve both equations simultaneously for N_{1A}/N_{2A} and N_{1A}/N_{3A}, with the following previously established values:

$$G_{IN}^* = 3.33 \text{ mmho (match antenna admittance)}$$

$$Q_L = 20 \text{ (for selectivity)}$$

$$X_C = 48.4 \text{ }\Omega$$

$$Q_U = 200 \text{ (minimize input loss)}$$

$$G_{IN} = 4.22 \text{ mmho}$$

$$G_S^* = 3.33 \text{ mmho (300-}\Omega \text{ antenna)}$$

This results in tapping ratios of $N_{1A}/N_{2A} = 2.5$, and $N_{1A}/N_{3A} = 3.2$.

The FET *driving source conductance* is found by plugging the same values into the appropriate equation of Fig. 3-36, which gives $G_S = 6.35$ mmho ($R_S = 158 \text{ }\Omega$).

Using this value of G_S and the previously established values results in a Stern k stability factor (Sec. 3-8.2) of

$$k = \frac{2(4.95 + 6.35)(0.05 + 0.304)}{0.545 + 0.258} = 10$$

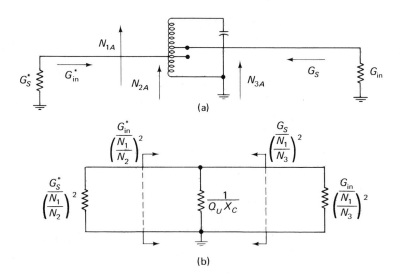

Tapping ratio

$$\frac{G_{in}^{*}}{\left(\dfrac{N_{1A}}{N_{2A}}\right)^{2}} = \frac{1}{Q_{U}X_{C}} + \frac{G_{in}}{\left(\dfrac{N_{1A}}{N_{3A}}\right)^{2}}$$

FET driving source conductance

$$\frac{G_{S}}{\left(\dfrac{N_{1A}}{N_{3A}}\right)^{2}} = \frac{1}{Q_{U}X_{C}} + \frac{G_{S}^{*}}{\left(\dfrac{N_{1A}}{N_{2A}}\right)^{2}}$$

Tapping ratio

$$\frac{1}{Q_{L}X_{C}} = \frac{1}{Q_{U}X_{C}} + \frac{G_{in}}{\left(\dfrac{N_{1A}}{N_{3A}}\right)^{2}} + \frac{G_{S}^{*}}{\left(\dfrac{N_{1A}}{N_{2A}}\right)^{2}}$$

Output power loss

$$10 \log \frac{\left[\dfrac{G_{L}}{\left(\dfrac{N_{1B}}{N_{2B}}\right)^{2}}\right]}{\left[\dfrac{G_{L}^{*}}{\left(\dfrac{N_{1B}}{N_{3B}}\right)^{2}}\right]}$$

Input power loss

$$10 \log \frac{\left[\left(\dfrac{G_{S}^{*}}{\left(\dfrac{N_{1A}}{N_{2A}}\right)^{2}}\right) + \dfrac{G_{in}}{\left(\dfrac{N_{1A}}{N_{3A}}\right)^{2}}\right]^{2}}{\left[\dfrac{G_{S}^{*}}{\left(\dfrac{N_{1A}}{N_{2A}}\right)^{2}}\right]\left[\dfrac{G_{in}}{\left(\dfrac{N_{1A}}{N_{3A}}\right)^{2}}\right]} + 20 \log \left[\frac{Q_{U}}{Q_{U} - Q_{L}}\right]$$

Output tank loaded Q

$$\frac{1}{Q_{L}X_{C}} = \frac{1}{Q_{U}X_{C}} + \frac{G_{out}}{\left(\dfrac{N_{1B}}{N_{2B}}\right)^{2}} + \frac{G_{L}^{*}}{\left(\dfrac{N_{1B}}{N_{3B}}\right)^{2}}$$

Courtesy Texas Instruments

Fig. 3-36   Input coil design

A k factor of 10 produces a highly stable RF amplifier stage.

Although the optimum driving source impedance for the TIS34 is approximately 500 Ω, the above value of 158 Ω should be satisfactory since the FET noise figure is excellent for a broad range of driving source impedances.

Then *power loss of the input single-tuned network* is calculated using the equation of Fig. 3-36. The first term of the equation represents mismatch between antenna admittance and the FET input admittance. The second term represents the *insertion loss* of the tuned circuit. Since the second term degrades noise figure, the necessity of having a high Q_U to Q_L ratio is apparent. For this particular example, *input power loss* is

$$10 \log \frac{(0.52 + 0.416)^2}{4(0.52)(0.416)} + 20 \log \frac{200}{200 - 20} = 1 \text{ dB}$$

The FET driving source conductance having been determined, the FET *output conductance* may be calculated as follows:

$$G_{\text{OUT}} = g_{22} - R_e \left[\frac{y_{12} y_{21}}{G_{11} + G_S} \right]$$

$$= 0.027 \text{ mmho} \qquad R_{\text{OUT}} = \frac{1}{G_{\text{OUT}}} = 37 \text{ k}\Omega$$

With G_{OUT} established, a more exact calculation for output tank loaded Q can now be made using the following equation:

$$\frac{1}{Q_L X_C} = \frac{1}{Q_U X_C} + \frac{G_{\text{OUT}}}{(N_{1B}/N_{2B})^2} + \frac{G_L^*}{(N_{1B}/N_{3B})^2}$$

The appropriate values for this equation are $X_C = 48.4 \, \Omega$, $Q_U = 200$, $G_{\text{OUT}} = 0.027 \text{ mmho}$, $G_L^* = 4.00 \text{ mmho}$, $N_{1B}/N_{2B} = 1$, $N_{1B}/N_{3B} = 5$, so that Q_L (output) is 62.5. This confirms the initial assumption that the FET output impedance would not appreciably load the output coil.

The output power loss, which is not included in the general power gain expression of Sec. 3-8.5, cannot be calculated in the same fashion as the input loss. The mismatch loss that is accounted for in the power gain expression is mismatch between the FET's output conductance and the *total* load seen by the FET.

The mismatch term in the *input power loss* equation of Fig. 3-36 is the mismatch only between the stage driving source conductance and the FET input conductance. Thus, if the general power gain expression is to be used in calculating the overall RF stage power gain, the additional output power loss that must be considered is not a simple function of Q_U and Q_L, but instead is the ratio of power delivered to the total FET load G_L to the power delivered to the actual load G_L^*, or

$$\text{output power loss} = \frac{\text{power delivered to } G_L}{\text{power into } G_L^*}$$

Using the *output power loss* equation of Fig. 3-36, the output power loss is

$$10 \log \frac{(0.304/1)}{(4.00/25)} = 2.8 \text{ dB}$$

Using the general power gain expression, $G_P \approx 13.5$ (or 11.3 dB). The overall RF stage power gain then is

$$PG = G_P - \text{input power loss} - \text{output loss}$$
$$PG = 11.3 - 1.0 - 2.8$$
$$PG = 7.5 \text{ dB}$$

The overall voltage gain of the *RF* stage in decibels is

$$VG \text{ (db)} = PG - 10 \log \frac{R_{IN}^*}{R_L^*} = 7.5 - 10 \log \frac{300}{200} = 5.7 \text{ dB}$$

Summarizing the coil design;
Input coil:

$$Q_U = 200, \quad Q_L = 20, \quad L = 0.07 \; \mu\text{H}, \quad \frac{N_{1A}}{N_{2A}} = 2.5, \quad \frac{N_{1A}}{N_{3A}} = 3.2$$

Output coil:

$$Q_U = 200, \quad Q_L = 62.6, \quad L = 0.077 \; \mu\text{H}, \quad \frac{N_{1B}}{N_{2B}} = 1, \quad \frac{N_{1B}}{N_{3B}} = 5$$

3-9. RADIO-FREQUENCY POWER AMPLIFIERS AND MULTIPLIERS

Figure 3-37 shows the working schematics of typical RF power amplifiers. The same basic circuits can be used as frequency multipliers. However, in a multiplier circuit, the output must be tuned to a multiple of the input. A multiplier may or may not provide amplification. Usually, most of the amplification is supplied by the final amplifier stage, which is not operated as a multiplier. That is, the input and output of the final stage are at the same frequency.

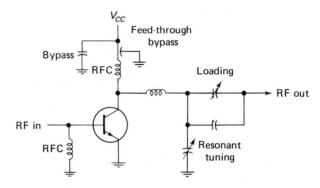

(a) Typical for Final Amplifier

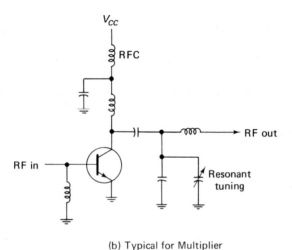

(b) Typical for Multiplier

Fig. 3-37   Typical RF power amplifier and multiplier circuits

A typical RF transmitter will have three stages: an oscillator to provide the basic signal frequency, an intermediate stage that provides amplification and/or frequency multiplication, and a final stage for power amplifications. In some cases there are three stages of amplification following the oscillator.

3-9.1. Circuit Analysis for Power Amplifiers and Multipliers

All the considerations in Chapters 1 and 2 apply to RF power amplifiers and multipliers. Of particular importance are interpreting data-sheets, determining parameters at different frequencies, and temperature-related problems. In addition to the basic considerations, the following problems must be considered in RF power amplifier design.

Tuning and adjustment controls. The circuit of Fig. 3-37a has two tuning controls (variable capacitors) in the output network. The network of the Fig. 3-37b circuit has only one adjustment control. The circuit of Fig. 3-37a is typical for power amplifiers, where the output must be tuned to the resonant frequency by one control and adjusted for proper impedance match by the other control (often called the loading control). In practice, both controls affect tuning and loading (impedance matching). The circuit of Fig. 3-37b is typical for multipliers or intermediate amplifiers where the main concern is tuning to the resonant frequency.

Minimum capacitance. Also note that variable capacitors are connected in parallel with fixed capacitors in both networks. This parallel arrangement serves two purposes. First, it provides a minimum fixed capacitance in case the variable capacitor is adjusted to its minimum value. In some cases, if a minimum capacitance is not included in the network, a severe mismatch can occur when the variable capacitor is at its minimum, resulting in damage to the transistor. Another reason for the parallel capacitor is to reduce required capacitance rating (an thus the physical size) of the variable capacitor.

Tuning range. The maximum capacitance range of the tuning network is dependent upon the required tuning range of the circuit. A wide frequency range requires a wide capacitance range. In a practical circuit, use a capacitor with a midrange capacitance equal to the desired capacitance at the center frequency. For example, if 25 pF is required to produce resonance at the center operating frequency, use a variable capacitor with a range of 1 to 50 pF. If such a capacitor is not readily available, use a fixed capacitor of 15 pF in parallel with a 15-pF variable capacitor. This provides a capacitance range of 16 to 30 pF, with a midrange of about 23 pF.

Class of operation. Normally, amplifiers and multipliers using the circuits of Fig. 3-37 are operated class C. The transistors remain cut off until a signal is applied, and are never conducting for more than 180° (half-cycle) of the 360° input signal cycle. In practice, the transistors conduct for about 140° of the input cycle, either on the positive half or negative half, depending upon the transistor type (*NPN* or *PNP*). No bias, as such, is required for this class of operation.

Emitter connection. In RF power amplifiers, the emitter is connected *directly* to ground. In those transistors where the emitter is connected to the case (typical in many RF power transistors), the case can be mounted on a chassis that is connected to the ground side of the supply voltage. A direct connection between emitter and ground is of particular importance in high-frequency applications. If the emitter is connected to ground through a resistance (or even a long lead wire), an inductive or capacitive reactance can develop at high frequencies, resulting in undesired changes in the network. Another reason for a direct connection between emitter and ground is to produce maximum gain. All other factors being equal, a decrease in emitter resistance (in relation to collector impedance) produces an increase in amplifier gain.

Power-supply connections. The transistor base is connected to ground through an RF choke (RFC). This provides a dc return for the base, as well as RF signal isolation between base and emitter or ground. The transistor collector is connected to the supply voltage through an RFC and (in some cases) through the coil portion of the resonant network. The RFC provides dc return, but RF signal isolation, between collector and power supply. When the collector is connected to the power supply through the resonant network, the coil must be capable of handling the full collector voltage. For this reason, final amplifier networks are often chosen so that collector current does not pass through the coil (Fig. 3-37a). The circuit of Fig. 3-37b is used for power applications where the current is low.

Radio-frequency-choke ratings. The ratings for RFCs are sometimes confusing. Some manufacturers list a full set of characteristics: inductance, dc resistance, ac resistance, Q, current capability, and nominal frequency range; ac resistance and Q are usually frequency dependent. A nominal frequency-range characteristic is helpful, but usually not critical.

All other factors being equal, the dc resistance should be at a minimum for any circuit carrying a large amount of current. For example, a large dc resistance in the collector of a final power amplifier can result in a large voltage drop between power supply and collector. Usually, the selection of a trial value for an RFC is based on a tradeoff between inductance and current capability. The minimum current capacity should be greater (by at least 10 per cent) than the maximum anticipated direct current. The inductance is dependent upon operating frequency. As a trial value, use an inductance that will produce a reactance between 1000 and 3000 Ω at the operating frequency.

Bypass capacitances. The power supply circuits must be bypassed. The feed-through bypass capacitors shown in Fig. 3-37 are used at higher frequencies where the RF circuits are physically shielded from the power supply and other circuits. The feed-through capacitor permits direct current to be applied through the shield, but prevents radio frequencies from passing outside the shield (radio frequencies are bypassed to the ground return). As a trial value, use a total bypass capacitance range of 0.001 to 0.1 μF.

From a practical standpoint, the best test for adequate bypass capacitance and RFC inductance is the absence of RF signals on the power-supply side of the dc voltage line. If RF signals are present on the power-supply side, the bypass capacitance and/or RFC inductance is not *adequate*. (A possible exception to this is when the RF signals are being picked up due to inadequate shielding.) If the shielding is good and RF signals are present in the power supply, increase the bypass capacitance value. Also increase RFC inductance. Of course, circuit performance must be checked with each increase in capacitance or inductance value. For example, too much bypass capacitance can cause undesired feedback and oscillation; too much RFC inductance can reduce amplifier output and efficiency. The procedures for the measurement of RF signals are described in Chapter 7.

Efficiency. A class C RF amplifier has a typical efficiency of about 65 to 70 per cent. That is, the RF power output is 65 to 70 percent of the dc input power. To find the required dc input power, divide the desired RF power output by 0.65 or 0.7. For example, if the desired RF output is 50 W, the dc input power is 50/0.7 or about 70 W. Ignore the slight voltage drop across the RFC and/or coil, and divide the input power by the power supply voltage to find the collector current. With a dc input of 70 W and a 28-V supply, the collector current is 2.7 A.

Transistor characteristics. Transistors must be capable of handling the full power-supply voltage at their collectors, and the transistor current and/or power rating must be greater than the maximum required values. Likewise, the transistor must be capable of producing the necessary power output at the operating frequency. These problems are discussed in Chapters 1 and 2.

The transistors must also provide the necessary *power gain at the operating frequency*. Likewise, the input power to an amplifier must match the desired output and gain. For example, assume that a 50-W RF amplifier is to be designed, and that the available transistors have a power gain of 10. The final amplifier must have an input signal of at least 5 W, with a dc input of about 7 W.

Multiplier efficiency. Typically, the efficiency of a second harmonic amplifier or multiplier (with the output tuned to twice the input frequency) is about 42 per cent. The efficiencies of third, fourth, and fifth harmonic amplifiers are 28, 21, and 18 per cent, respectively, Therefore, if an intermediate amplifier is to be operated at the second harmonic and produce 5-W RF output, the dc power input is approximately 12 W (5/0.42 = 12).

Another problem to be considered in frequency multiplication is that the power gain (as listed on the datasheet) may not remain the same as when amplifier input and output are at the same frequency. Some datasheets specify power gain at the basic frequency, and then derate the power gain for second harmonic operation. As a guideline, always use the *minimum power gain factor* when calculating power input and output values.

3-9.2. Resonant Network Analysis

In any RF amplifier, the tuning network must be resonant at the desired frequency. (Inductive and capacitive reactance must be equal at the selected frequency.) Also, the tuning network must match the transistor output impedance to the load.

Generally, an antenna load impedance is about 50 Ω whereas the output impedance of a typical two-junction transistor at radio frequencies is a few ohms. In the case where one amplifier feeds into another amplifier, the network must match the output impedance of one transistor to the input impedance of another transistor. Any mismatch can result in a loss of power between stages, or to the final load.

Transistor impedance (both input and output) has both resistive and reactive components, and therefore varies with frequency. To design a resonant network for the output of a transistor, it is necessary to know the output reactance (usually capacitive), the output resistance at the operating frequency, and the output power. Likewise, it is necessary to know the input resistance and reactance of a transistor at a given frequency and power when designing the resonant network of the stage feeding into the transistor.

Generally, the input resistance, the input capacitance, and the output capacitance of RF power transistors are shown by means of graphs similar to those of Fig. 3-38. The reactance can then be found using the correspond-

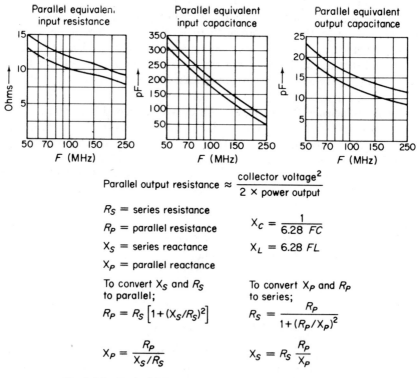

Parallel output resistance $\approx \dfrac{\text{collector voltage}^2}{2 \times \text{power output}}$

R_S = series resistance

R_P = parallel resistance $X_C = \dfrac{1}{6.28\ FC}$

X_S = series reactance $X_L = 6.28\ FL$

X_P = parallel reactance

To convert X_S and R_S To convert X_P and R_P
to parallel;   to series;

$R_P = R_S \left[1 + (X_S/R_S)^2 \right]$ $R_S = \dfrac{R_P}{1 + (R_P/X_P)^2}$

$X_P = \dfrac{R_P}{X_S/R_S}$ $X_S = R_S \dfrac{R_P}{X_P}$

Fig. 3-38 Typical RF power amplifier transistor characteristics

ing frequency and capacitance. For example, the output capacitance shown on the graph of Fig. 3-38 is about 15 pF at 80 MHz. This produces a capacitive reactance of about 130 Ω at 80 MHz. The reactance and resistance can then be combined as necessary.

Input and output transistor impedances are generally listed on datasheets in *parallel from*. That is, the datasheets assume that the resistance is in parallel with the capacitance. However, some tuning networks require that the impedance be calculated in series form. It is therefore necessary to convert between series and parallel impedance forms. The necessary equations are listed in Fig. 3-38.

The output resistance of RF power transistors is usually not shown on datasheets, but may be calculated using the equation of Fig. 3-38.

Typical resonant networks. Figures 3-39 through 3-43 show five typical resonant networks, together with the equations necessary to find component values. Note that these equations are best solved by computer-aided design techniques, since many are tedious and time consuming.

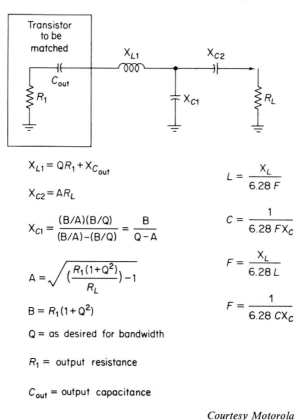

$$X_{L1} = QR_1 + X_{C_{out}}$$

$$X_{C2} = AR_L$$

$$X_{C1} = \frac{(B/A)(B/Q)}{(B/A)-(B/Q)} = \frac{B}{Q-A}$$

$$A = \sqrt{\left(\frac{R_1(1+Q^2)}{R_L}\right)-1}$$

$$B = R_1(1+Q^2)$$

Q = as desired for bandwidth

R_1 = output resistance

C_{out} = output capacitance

$$L = \frac{X_L}{6.28\,F}$$

$$C = \frac{1}{6.28\,FX_C}$$

$$F = \frac{X_L}{6.28\,L}$$

$$F = \frac{1}{6.28\,CX_C}$$

Courtesy Motorola

Fig. 3-39   Radio-frequency network where R_1 is less than R_L

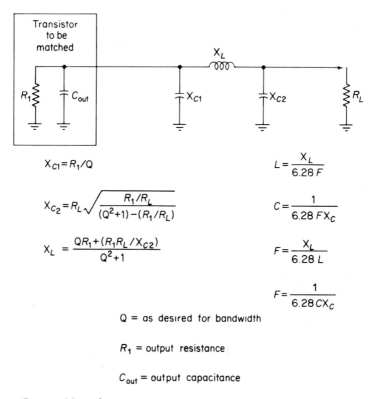

$$X_{C1} = R_1/Q$$

$$X_{C2} = R_L \sqrt{\frac{R_1/R_L}{(Q^2+1)-(R_1/R_L)}}$$

$$X_L = \frac{QR_1 + (R_1 R_L / X_{C2})}{Q^2+1}$$

$$L = \frac{X_L}{6.28\,F}$$

$$C = \frac{1}{6.28\,FX_C}$$

$$F = \frac{X_L}{6.28\,L}$$

$$F = \frac{1}{6.28\,CX_C}$$

Q = as desired for bandwidth

R_1 = output resistance

C_{out} = output capacitance

Courtesy Motorola

Fig. 3-40   Radio-frequency network where R_1 is approximately equal to R_L

The networks can be used as amplifiers and/or multipliers. Note that the network of Fig. 3-39 is similar to that of Fig. 3-37a, whereas Fig. 3-41 is similar to Fig. 3-37b.

Impedance matching. The resistor and capacitor shown in the box labeled "Transistor to be matched" represent the *complex output impedance* of a transistor. When the network is to be used with a final amplifier, the resistor labeled R_L is the antenna impedance, or other load. When the network is used with an intermediate amplifier, R_L represents the input impedance of the following transistor. It is therefore necessary to calculate the input impedance of the transistors being fed by the network, using the data and equations of Fig. 3-38.

The complex impedances are represented in series form in some cases and parallel form in others, depending on which form is the most convenient for network calculation. The resultant impedance of the network, when terminat-

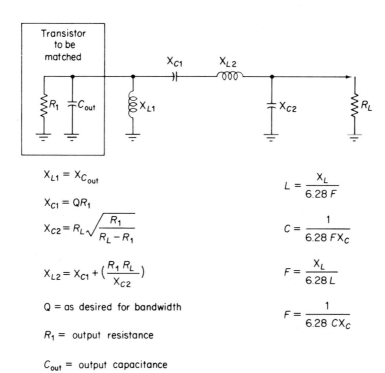

$$X_{L1} = X_{C_{out}}$$

$$X_{C1} = QR_1$$

$$X_{C2} = R_L \sqrt{\frac{R_1}{R_L - R_1}}$$

$$X_{L2} = X_{C1} + \left(\frac{R_1 R_L}{X_{C2}}\right)$$

Q = as desired for bandwidth

R_1 = output resistance

C_{out} = output capacitance

$$L = \frac{X_L}{6.28\, F}$$

$$C = \frac{1}{6.28\, F X_C}$$

$$F = \frac{X_L}{6.28\, L}$$

$$F = \frac{1}{6.28\, C X_C}$$

Courtesy Motorola

Fig. 3-41 Radio-frequency network where R_1 is very small in relation to R_L

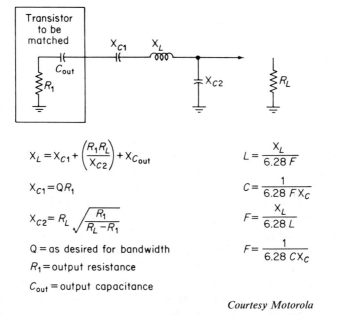

$$X_L = X_{C1} + \left(\frac{R_1 R_L}{X_{C2}}\right) + X_{C_{out}}$$

$$X_{C1} = QR_1$$

$$X_{C2} = R_L \sqrt{\frac{R_1}{R_L - R_1}}$$

Q = as desired for bandwidth

R_1 = output resistance

C_{out} = output capacitance

$$L = \frac{X_L}{6.28\, F}$$

$$C = \frac{1}{6.28\, F X_C}$$

$$F = \frac{X_L}{6.28\, L}$$

$$F = \frac{1}{6.28\, C X_C}$$

Courtesy Motorola

Fig. 3-42 Radio-frequency network where R_1 is very small in relation to R_L (alternative network)

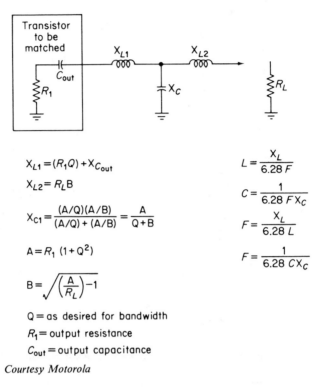

$$X_{L1} = (R_1 Q) + X_{Cout}$$

$$X_{L2} = R_L B$$

$$X_{C1} = \frac{(A/Q)(A/B)}{(A/Q) + (A/B)} = \frac{A}{Q+B}$$

$$A = R_1 (1 + Q^2)$$

$$B = \sqrt{\left(\frac{A}{R_L}\right) - 1}$$

$$L = \frac{X_L}{6.28\,F}$$

$$C = \frac{1}{6.28\,F X_C}$$

$$F = \frac{X_L}{6.28\,L}$$

$$F = \frac{1}{6.28\,C X_C}$$

Q = as desired for bandwidth

R_1 = output resistance

C_{out} = output capacitance

Courtesy Motorola

Fig. 3-43 Radio-frequency network where R_1 is very small, or very large, in relation to R_L

ed with a given load, must be equal to the *conjugate* of the impedance in the box. For example, assume that the transistor has a series output impedance of $7.33. - j3.87$. That is, the resistance (real part of impedance) is $7.33\ \Omega$, while the capacitance reactance (imaginary part of impedance) is 3.87.

For maximum power transfer from the transistor to the load, the load impedance must be the conjugate of the output impedance, or $7.33 + j3.87$. If the amplifier is designed to operate into a $50\text{-}\Omega$ external load, the network must transfer the $50 + j0$ external load to the $7.33 + j3.87$ transistor load. In addition to performing this transformation, the network provides harmonic rejection (unless a harmonic is needed in a multiplier stage), low loss, and provisions for adjustment of both loading and tuning.

Network characteristics. Each network has its advantages and limitations.

The network of Fig. 3-39 is applicable to most RF power amplifiers, and is especially useful when the series real part of the transistor output impedance, or R_1, is *less than* $50\ \Omega$. With a typical $50\text{-}\Omega$ load, the required reactance of C_1 rises to an impractical value when R_1 is close to $50\text{-}\Omega$.

The network of Fig. 3-40 (often known as a Pi network) is best suited where the parallel resistor R_1 is high (*near the value of R_L, typically $50\ \Omega$*). If the network of Fig. 3-40 is used with a low value of R_1 resistance, the inductance of L_1 must be very small, whereas C_1 and C_2 become very large (beyond practical limits).

The networks of Figs. 3-41 and 3-42 produce practical values of C and L, *especially where R_1 is very low*. The main limitation for the networks of Figs. 3-41 and 3-42 is that R_1 *must be* substantially lower than R_L. These networks, or variations thereof, are often used with intermediate stages where a low output impedance of one transistor must be matched to the input impedance of another transitor.

The network of Fig. 3-43 (often known as a Tee network) is best suited when R_1 *is much less or much greater than R_L*. That is, the Fig. 3-43 network is well suited to drastic mismatch situations (such as matching a transistor with an output impedance of 10 Ω to a an 300-Ω antenna).

3-10. USING DATASHEET GRAPHS TO DESIGN RADIO-FREQUENCY AMPLIFIER NETWORKS

High-frequency characteristics are especially important in the design of RF networks. Unfortunately, the high-frequency information provided on many datasheets in tabular form is not adequate for simplified design. To properly match impedances, both the resistive and reactive components must be considered. The reactive component (either inductive or capacitive) changes with frequency. In practical amplifier work it is necessary to know the reactance values over a wide range of frequencies, not at some specific frequency (unless you happen to be designing for that frequency only). The best way to show how resistance and reactance vary in relation to frequency for a particular transistor is by means of graphs or curves. Fortunately, manufacturers who are trying to sell their transistors for high-frequency power-amplifier use generally provide a set of curves showing the characteristics over the anticipated frequency range.

The following paragraphs describe the basic design steps for an RF power amplifier, using typical datasheet curves and the equations of Figs. 3-38 through 3-43.

The amplifier shown in Fig. 3-44 delivers 30-W output at 175 MHz, with a gain of about 29 dB and 50 per cent overall efficiency.

3-10.1. *Circuit Analysis*

A summary of the amplifier's performance is given in the table in Fig. 3-44. The circuit operates from a positive-grounded 28-V supply, and draws 2.1 A for a 30-W output. All stages are class C amplifiers with zero no-signal bias. Conventional single-tuned matching networks are used throughout.

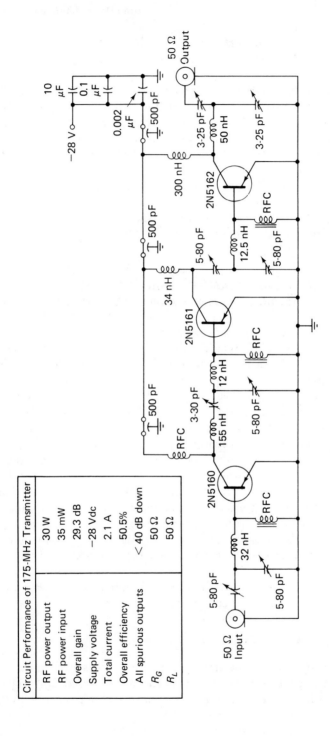

Fig. 3-44 A 175-MHz amplifier circuit

Courtesy Motorola

Circuit Performance of 175-MHz Transmitter	
RF power output	30 W
RF power input	35 mW
Overall gain	29.3 dB
Supply voltage	−28 Vdc
Total current	2.1 A
Overall efficiency	50.5%
All spurious outputs	< 40 dB down
R_G	50 Ω
R_L	50 Ω

The circuit uses the first commercially available *PNP* VHF high-power transistors, the Motorola 2N5160, 2N5161, and the 2N5162. The gain and power output of these transistors are shown in Figs. 3-45 through 3-47.

The 2N5161 and 2N5162 transistors, which are used in the driver and final stages, are furnished in TO-60 stud packages with the emitter internally connected to the case. This is important for class C operation, since a low im-

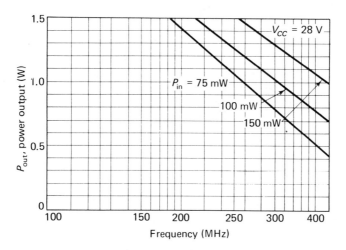

Courtesy Motorola

Fig. 3-45   Power output versus frequency of 2N5160

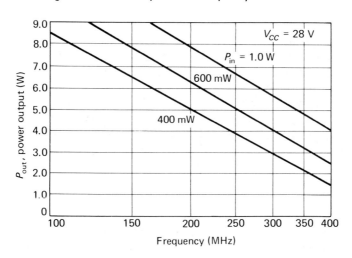

Courtesy Motorola

Fig. 3-46   Power output versus frequency of 2N5161

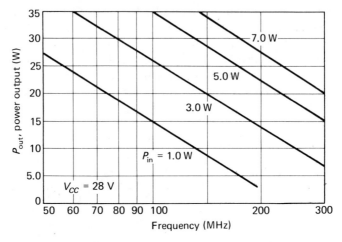

Fig. 3-47   Power output versus frequency of 2N5162

pedance emitter ground is necessary to obtain maximum gain from each stage. With the emitter connected internally to the stud, a good ground results from bolting the stud to a copper heat sink fastened to the conductive chassis ground. This mounting also aids in conducting the heat away from the transistor, helping to maintain the junction operating temperature at a safe value below the specified maximum of 200°C (typical for silicon power transistors).

The other transistor in the circuit is a 2N5160 used as a predriver. The 2N5160 is supplied with a TO-39 case.

Basic design procedure. Figures 3-48 through 3-56 contain information about large-signal complex impedances for the three transistors. This is all the data required for design, except the collector load resistance (parallel output resistance), which can be computed using the equation of Fig. 3-38. As shown, the parallel output resistance is approximately equal to the collector voltage squared, divided by twice the power output or, for the final amplifier,

$$\frac{28^2}{2 \times 30} = 13.1 \ \Omega$$

This equation, although a guideline, has proved to be within 20 per cent of the actual collector load resistance required for maximum power transfer.

The computed value of parallel output resistance R_p is combined with the parallel equivalent output capacitance C_{OUT} of the final amplifier to form the conjugate of the collector load impedance to be matched to the external am-

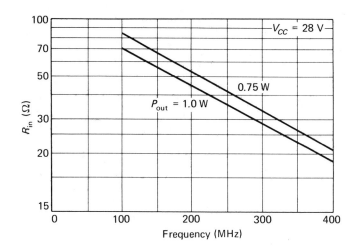

Courtesy Motorola

Fig. 3-48 Parallel equivalent input resistance versus frequency of 2N5160

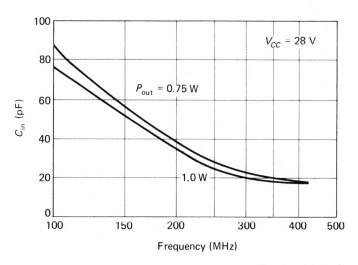

Courtesy Motorola

Fig. 3-49 Parallel equivalent input capacitance versus frequency of 2N5160

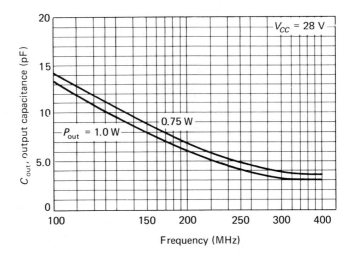

Courtesy Motorola

Fig. 3-50   Parallel output capacitance versus frequency of 2N5160

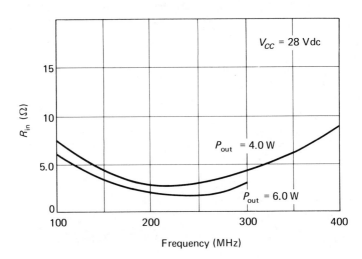

Courtesy Motorola

Fig. 3-51   Parallel equivalent input resistance versus frequency of 2N5161

plifier load. From Fig. 3-56, the output capacitance of the 2N5162 is found to be 36 pF when operating with 30-W output power at 175 MHz. This a reactance of 28 Ω (at 175 MHz).

The impedance to be conjugately matched at the collector is composed of

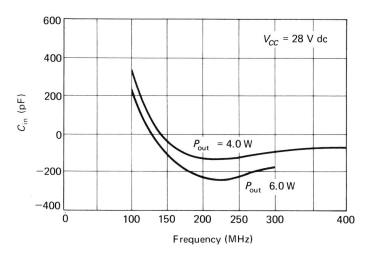

Courtesy Motorola

Fig. 3-52 Parallel equivalent input capacitance versus frequency of 2N5161

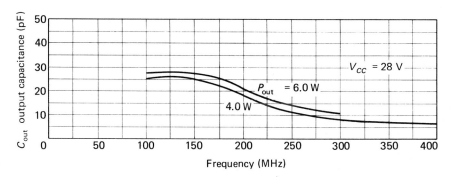

Courtesy Motorola

Fig. 3-53 Parallel output capacitance versus frequency of 2N5161

13.1-Ω resistance in parallel with 28-Ω capacitive reactance. For the output network configuration selected (that of Fig. 3-39 where R_1 is considerably smaller than R_L), it is more convenient to work with the equivalent series components of impedance. The parallel impedances can be converted to series with the equations of Fig. 3-38, as follows:

$$R_S = \frac{13.1}{1 + (13.1/28)^2} = 10.75$$

$$X_S = 10.75 \times \frac{13.1}{-28} = -5.04$$

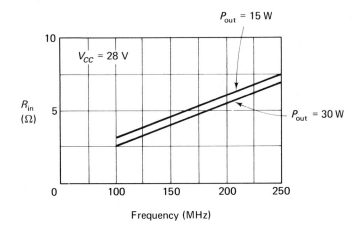

Fig. 3-54   Parallel equivalent input resistance versus frequency of 2N5162

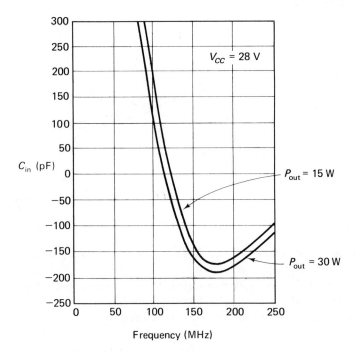

Fig. 3-55   Parallel equivalent input capacitance versus frequency of 2N5162

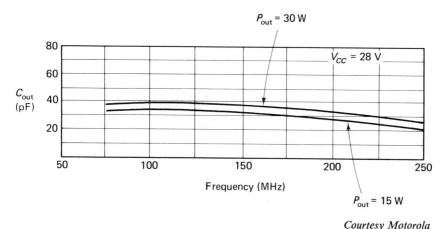

Courtesy Motorola

Fig. 3-56   Parallel output capacitance versus frequency of 2N5162

The series impedance is 10.75-j5.04 (the minus is used since reactance is capacitive).

Since the amplifier is designed to operate into a 50-Ω load, the collector network must transform the amplifier load (50 + j0 Ω) into the required collector load impedance (10.75 + j5.04 Ω). This can be done using the equations of Fig. 3-39. For example, with an assumed Q of 5, the reactance of L_1 is

$$X_{L_1} = (5 \times 10.75) + 5.04 \approx 60 \, \Omega$$

With a frequency of 175 MHz and a reactance of 60, the value of L_1 is approximately 50 nH.

The reactance of C_1 is

$$X_{L_1} = \frac{B}{Q - A} = \frac{280}{5 - 2.14} \approx 100 \, \Omega$$

With a frequency of 175 MHz and a reactance of 100, the value of C_1 is approximately 9 pF.

The reactance of C_2 is

$$X_{C_2} = 2.14 \times 50 \approx 107 \, \Omega$$

With a frequency of 175 MHz and a reactance of 107, the value of C_2 is approximately 8 pF.

The values of A and B on Fig. 3-39 are found as follows:

$$A = \sqrt{\frac{10.75(1 + 5^2)}{50} - 1} \approx 2.14$$

$$B = 10.75(1 + 5^2) \approx 280$$

The values for remaining networks in the amplifier of Fig. 3-44 can be found using similar procedures.

3-11. USING SAMPLED y-PARAMETER TECHNIQUES TO DESIGN RADIO-FREQUENCY AMPLIFIER NETWORKS

This section describes the design of a UHF broadband amplifier using sampled y-parameter techniques, as developed by Motorola. The section also illustrates the use of the Smith chart to calculate y parameters (admittances). The sampled y-parameter technique is an extension of the conventional small-signal y-parameter design described in previous sections of this chapter, and is particularly useful at ultrahigh frequencies.

The amplifier, shown in Fig. 3-57, has a midband gain of 31 dB $\pm$ 1 dB. The lower 3-dB cutoff point occurs at 3.1 MHz and the upper end at 405 MHz. The frequency response is given in Fig. 3-58. The usable sensitivity defined as input signal required to produce a $S + N/N = 10$ dB is 100 μV. The dynamic range is 40 dB, which corresponds to a root-mean-square output voltage of 1 to 100 mV.

3-11.1. Circuit Analysis

As shown in Fig. 3-57, the 400-MHz broadband amplifier consists of three identical stages. Both input and output are terminated into 50-Ω impedances. Since the stages are identical, only one stage is discussed.

A shunt feedback network composed of L_1, L_3, R_1, R_3, and C_2 is used to stabilize gain over the 400-MHz bandwidth. Capacitor C_2 is used for blocking direct current, and could be replaced by an appropriate zener diode. Inductor L_2 is the interstage coupling used to match the output impedance of one stage to the input impedance of the following stage.

The choice of the 2N4957 for this application stems from its high gain and very low noise figure at ultrahigh frequencies. At a bias current of 5 mA, the 2N4957 has a current gain of 50. The input impedance is *approximately* equal to the current gain/transconductance, or 250 Ω. A relatively small feedback loop gain (small feedback conductance) is required to reduce the input impedance to 50 Ω, which is the prescribed source value. Thus, the open-loop gain of the amplifier is not excessively sacrificed, allowing a relatively high closed-loop gain.

If minimum noise figure is a major consideration, the 2N4957 should be biased at 2 mA, which has been found to be an optimum bias point for noise figure. However, if 2 mA is used and the input impedance is held to 50 Ω, a larger feedback loop gain is needed, and a considerable amount of the open-loop gain is sacrificed.

It is not necessary to bias each stage identically. The first stage may be

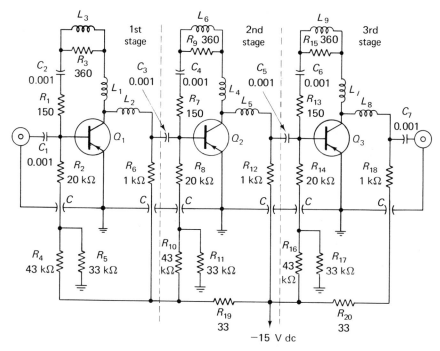

L_1, L_4 and L_7 = 3.5 turns #20 soft-drawn, ID = $\frac{1}{4}$ inch, Length = 1 inch

L_2, L_5 and L_8 = 6 turns #20 soft-drawn, ID = $\frac{1}{4}$ inch, Length = $\frac{3}{4}$ inch

L_3, L_6 and L_9 = 6 turns #24 tinned wire on CF 103, Q_3 toroid with Teflon sleeve

C = 500 pF Erie button capacitor with 0.02 μF ceramic disk. All other
capacitors are ceramic disk.

Q_1, Q_2, Q_3, = 2N4957

Bandwidth: 3.1 - 405 MHz at 3-dB point

Gain: 31 dB ± 1 dB

Sensitivity: 100 μV ($\frac{S + N}{N}$ = 10 dB)

I_C = 5 mA, V_{CE} = 10 V dc for all devices

Courtesy Motorola

Fig. 3-57   A 400-MHz wideband amplifier

biased for minimum noise, and the following stages for high gain to achieve
overall optimum performance. However, the design work is more time consuming for this case.

3-11.2. Design Analysis

The design of a multistage broadband amplifier, without an
overall feedback loop, may be broken down to that of the individual stages.

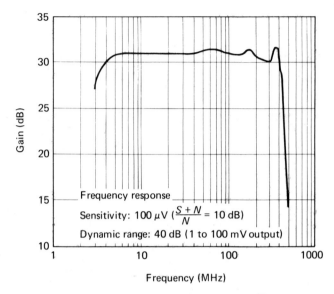

Fig. 3-58   Frequency response of amplifier

To minimize the interaction between stages, an interstage coupling network is needed to match the output impedance of one stage to the input impedance of the next stage. For example, if the individual stages are designed to have input and output terminations of 50 Ω, cascading of individual stages can be achieved without interaction between stages.

Usually, negative feedback is required to stabilize the gain-frequency characteristics. There are four possible feedback schemes, as shown in Fig. 3-59.

Each scheme has its own merit, and can be readily characterized by different two-port parameters: y parameters for shunt–shunt, g parameters for shunt–series, h parameters for series–shunt, and z parameters for series–series.

For the transistor having an input impedance higher than that of the input termination, a shunt–shunt feedback scheme is quite suitable for matching the required source to the input impedance of the device. Moreover, y parameters can be used readily for design calculations. Only the shunt–shunt feedback configuration will be discussed.

The design procedure can be divided into the following steps:

1. Determine the required design specifications of gain, bandwidth, and input and output terminations.
2. Determine the feedback network.
   a. Use a low-frequency transistor model to calculate the value of feedback conductance to meet the gain and input impedance requirements.
   b. Use a constant G_{oo} (which is the power gain when $Y_L = y_{22}$, or the power

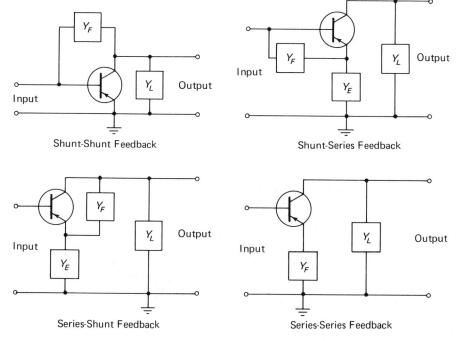

Shunt-Shunt Feedback   Shunt-Series Feedback

Series-Shunt Feedback   Series-Series Feedback

Fig. 3-59   Feedback schemes in RF amplifiers

gain when the load admittance equals the transistor output admittance)
plot to estimate device high-frequency gain capability under the bias
conditions.
  c. Use the constant gain expression to calculate the feedback inductance
     to meet gain and bandwidth requirements.
3. Realize a feedback network.
4. Determine the interstage coupling network.

3-11.3. Design Example

The following is an illustration of how the four basic steps can
be implemented:

1. *Design specifications:*
   Gain: 30 dB $\pm$ 1 dB
   Bandwidth: 400 MHz
   Input termination: 50 Ω
   Output termination: 50 Ω

2a. *Construct a low-frequency model as shown in Fig. 3-60.* Note the gain
    expression and input admittance expression shown on Fig. 3-60.

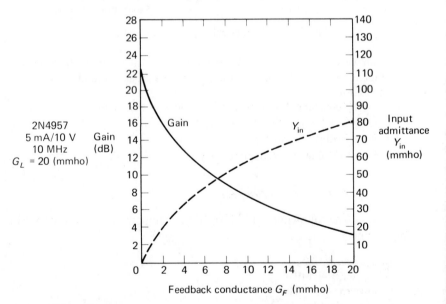

$$G = \frac{(\dfrac{G_m - G_F}{G_F + G_L})^2 \, G_L}{G_1 + G_F \, (1 + \dfrac{G_m - G_F}{G_F + G_L})}$$

$$Y_{in} = G_1 + G_F \, (1 + \frac{G_m - G_F}{G_F + G_L})$$

Courtesy Motorola

Fig. 3-60 Low-frequency model with corresponding gain expression and input impedance expression

Once the load, G_L, is chosen, there is one-to-one correspondence between gain and the value of G_F, and between Y_{IN} and G_F. The choice of G_F for a desired value of gain automatically fixes the input impedance.

The gain and its corresponding input admittance are plotted versus G_F in Fig. 3-61. If the constant gain function is mapped on the Z_F plane of an

2N4957
5 mA/10 V
10 MHz
G_L = 20 (mmho)

Gain (dB)

Gain

Y_{in}

Input admittance Y_{in} (mmho)

Feedback conductance G_F (mmho)

Courtesy Motorola

Fig. 3-61 Gain and corresponding input admittance plotted against G_F for the 2N4957

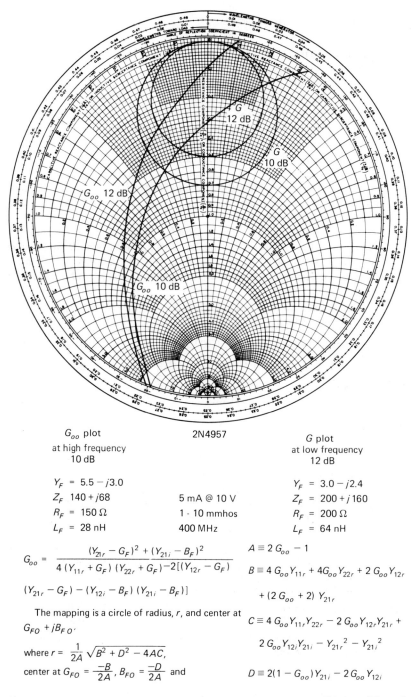

G_{oo} plot	2N4957	G plot
at high frequency		at low frequency
10 dB		12 dB

$Y_F = 5.5 - j3.0$		$Y_F = 3.0 - j2.4$
Z_F $140 + j68$	5 mA @ 10 V	Z_F $200 + j160$
$R_F = 150\,\Omega$	1 - 10 mmhos	$R_F = 200\,\Omega$
$L_F = 28\ \text{nH}$	400 MHz	$L_F = 64\ \text{nH}$

$$G_{oo} = \frac{(Y_{21r} - G_F)^2 + (Y_{21i} - B_F)^2}{4\,(Y_{11r} + G_F)\,(Y_{22r} + G_F) - 2[(Y_{12r} - G_F)}$$

$$(Y_{21r} - G_F) - (Y_{12i} - B_F)\,(Y_{21i} - B_F)]$$

The mapping is a circle of radius, r, and center at $G_{FO} + jB_{FO}$.

where $r = \dfrac{1}{2A}\sqrt{B^2 + D^2 - 4AC}$,

center at $G_{FO} = \dfrac{-B}{2A}$, $B_{FO} = \dfrac{-D}{2A}$ and

$A \equiv 2\,G_{oo} - 1$

$B \equiv 4\,G_{oo}Y_{11r} + 4G_{oo}Y_{22r} + 2\,G_{oo}Y_{12r}$

$\qquad + (2\,G_{oo} + 2)\,Y_{21r}$

$C \equiv 4\,G_{oo}Y_{11r}Y_{22r} - 2\,G_{oo}Y_{12r}Y_{21r} +$

$\qquad 2\,G_{oo}Y_{12i}Y_{21i} - Y_{21r}{}^2 - Y_{21i}{}^2$

$D \equiv 2(1 - G_{oo})Y_{21i} - 2\,G_{oo}Y_{12i}$

Courtesy Motorola

Fig. 3-62 Constant G_{oo} and G for 2N4957 in Smith chart form

impedance chart (known as the Smith chart), it represents a circle of constant R_F, which is equal to the reciprocal of G_F. Figure 3-62 shows two circles of constant gain of 10 and 12 dB plotted on a Smith chart.

2b. *Estimate high-frequency gain capability.* This is done by mapping G_{oo} into the $Y_F = 1/Z_F$ plane of the Smith chart. The equations for calculating G_{oo} are given on Fig. 3-62, along with the values necessary to make the plot. Figure 3-62 shows two 400-MHz constant G_{oo} curves of 10 and 12 dB superimposed on the low-frequency constant gain curves. If the constant G_{oo} curve intersects the corresponding low-frequency gain circle, the device is capable of yielding the gain and bandwidth at the specified input impedance. Maximum gain bandwidth capability occurs when the two circles are tangent to each other.

For this particular example, a 2N4957 biased at 5 mA of collector current and 10-V collector voltage has a capability of 12-dB gain at 400 MHz with an input impedance of 30 Ω, and 10 dB at 400 MHz with a 23-Ω input. Figure 3-63 lists the y parameters of the device at the sampled frequencies.

f MHz	Y_{11r}	Y_{11i}	Y_{12r}	Y_{12i}	Y_{21r}	Y_{21i}	Y_{22r}	Y_{22i}
10	3.77	0.283	0	−0.033	134	−11.7	0.024	0.076
100	4.50	2.80	0	−0.23	53	−35	0.2	0.72
200	5.90	5.9	0	−0.44	50.5	−46.5	0.2	1.5
300	7.80	8.4	0	−0.66	42.5	−54.25	0.3	2.2
400	9.60	10.5	0	−0.90	34.0	−61.0	0.3	3.0

Courtesy Motorola

Fig. 3-63   *Y*-parameters of the FET at sampled frequencies

2c. *Approximate determination of feedback network.* The *intersection* of the constant G_{oo} curve, and its corresponding low-frequency gain, represents a feedback network that at least will yield the gain specified at the two extremes. In this example, for 10-dB gain the possible values of Y_F are 5.5 − j3.0 and 1.0 + j2.5. The latter, being a capacitive network, is discarded. The 5.5 − j3.0 Y_F corresponds to a series feedback network composed of a 150-Ω resistance and a 300-nH inductance. A typical frequency response of the amplifier using this feedback network and $Y_L = y_{22} + Y_F$ at the high-frequency extreme is shown in Fig. 3-64.

2d. *Final determination of feedback network.* In order to flatten the frequency response at the intermediate frequencies, alteration of either Y_F, Y_L, or both should be made. One way of determining the feedback network is by repeating the same steps of 2c at intermediate sampled frequencies. (It is obvious that all these calculations are best accomplished on a computer.)

Figure 3-65 shows the intersections of the 10-dB G_{oo} with the 10-dB low-frequency gain curve at 100, 200, and 300 MHz, with a respective Y_L of 4.2 − j3.5, 4.7 − j3.4, and 5.0 − j3.5 mmhos.

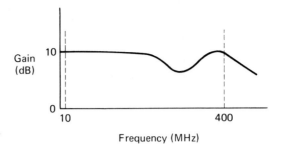

Fig. 3-64 Amplifier frequency response with a series R_L feedback network

Frequency (MHz)

Courtesy Motorola

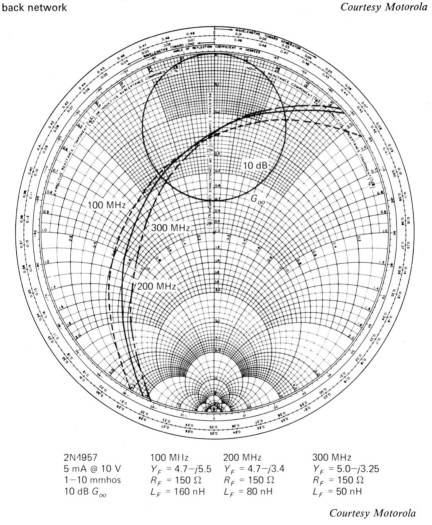

2N4957	100 MHz	200 MHz	300 MHz
5 mA @ 10 V	$Y_F = 4.7 - j5.5$	$Y_F = 4.7 - j3.4$	$Y_F = 5.0 - j3.25$
1–10 mmhos	$R_F = 150\ \Omega$	$R_F = 150\ \Omega$	$R_F = 150\ \Omega$
10 dB G_{oo}	$L_F = 160$ nH	$L_F = 80$ nH	$L_F = 50$ nH

Courtesy Motorola

Fig. 3-65 Intersections of 10-dB G_{oo} and 10-dB low-frequency gain curve at 100, 200, 300 MHz

213

The corresponding series inductance, L_F, of the feedback network is plotted versus frequency as the dashed line of Fig. 3-66. If Y_F is kept constant, in this case $20 + j0$, L_F varies with frequency as shown by the solid line of Fig. 3-66. The solid line is obtained from calculation of the 10-dB constant gain expression.

3. *Realization of the feedback network.* The frequency characteristics of a VHF toroidal inductance are similar to that of the desired L_F, so it is particularly well suited for this application. Since frequency response is highly dependent upon the feedback network, it is best determined with the physical circuit. Figure 3-67 shows the typical responses obtained when the corresponding feedback networks are used. The measured L_F in the final circuit is given in Fig. 3-66.

4. *Determination of interstage coupling network.* As long as Y_L is not frequency selective, Y_L does not affect the frequency response significantly. The interstage coupling network is designed to provide some impedance transformation to improve gain at the high-frequency end. At low frequen-

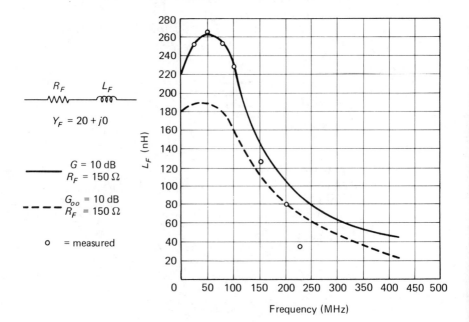

$$G = 10\,dB \quad \frac{(Y_{21} - Y_F)^2 \, \mathrm{Re}\,(Y_L)}{(Y_L + Y_{22} + Y_F)^2 \, \mathrm{Re}\left[Y_{11} + Y_F - \dfrac{(Y_{12} - Y_F)\,(Y_{21} - Y_F)}{Y_{22} + Y_F + Y_L}\right]}$$

Courtesy Motorola

Fig. 3-66   Frequency versus L_F for G and G_{oo}

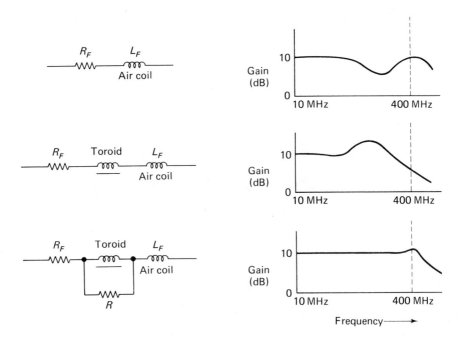

Fig. 3-67   Feedback network response

cies, $Y_L = 20 + j0$ mmhos. A series inductance of 180 nH is used for the interstage coupling in this case.

4. DIRECT-COUPLED
AND COMPOUND
AMPLIFIERS

The amplifiers discussed in Chapters 1 through 3 have one major limitation in certain applications. The amplifiers discussed thus far cannot amplify *direct currents* (or direct voltages). Direct currents will not be passed by coupling capacitors or transformers. Equally as important, amplifiers using transformers and/or coupling capacitors are not well suited for amplification of *low frequencies*. As discussed in Chapter 2, coupling capacitors form a voltage divider with the input impedance of the following stage. Such dividers increase attenuation of the signal as frequency decreases. In the case of transformers, the impedance offered at low frequencies virtually shorts low-frequency signals (since inductive reactance decreases with frequency).

For these reasons, *direct-coupled* amplifiers are used if direct currents and/or low-frequency signals must be amplified. Direct-coupled amplifiers, also known as DC amplifiers, permit a signal to be fed directly to the transistor without the use of any coupling device. Direct-coupled amplifiers are used in many electronic instruments, to operate relays, and for similar applications. Direct-coupled amplifiers may be single stage or multistage. However, direct coupling is generally limited to three stages.

4-1. BASIC DIRECT-COUPLED AMPLIFIER THEORY

The basic circuit of a DC amplifier using a *PNP* transistor in a common-emitter configuration is illustrated in Fig. 4-1. Resistors R_1 and R_2 form a voltage divider to forward bias the emitter–base junction of the transis-

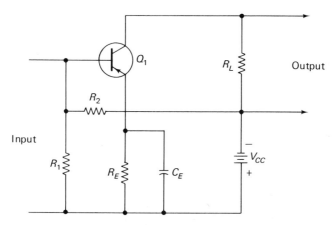

Fig. 4-1   Basic circuit of a direct-coupled amplifier

tor. Resistor R_E is the emitter stabilizing resistor, and C_E is the bypass capacitor. Resistor R_L is the collector load resistor. The voltage drop across R_L, caused by the flow of collector current, is the output voltage.

If the input voltage increases the forward bias of the emitter–base junction, the emitter current is increased. This, in turn, increases the collector current and the output voltage. If the input voltage reduces the forward bias, the collector current is reduced, as is the output voltage.

The range of input voltages which may be applied to a direct-coupled transistor is small. The forward bias must not be increased to the point where the transistor operates in its saturated region, nor can the bias be reduced so that the transistor is cut off.

The circuit of a two-stage direct-coupled amplifier using two *PNP* transistors in common-emitter configuration is shown in Fig. 4-2. Resistors R_1 and

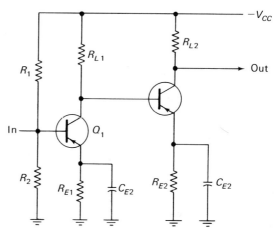

Fig. 4-2 Direct-coupled amplifier with two stages of like transistors (*PNP*)

R_2 form a voltage divider to forward bias the emitter–base junction of Q_1, Resistor R_{E1} is the emitter stabilizing resistor for Q_1, and C_{E1} is the bypass capacitor.

Resistor R_{L1} serves the double purpose of collector load resistor for Q_1 and voltage-dropping resistor to forward bias the emitter–base junction of Q_2. Resistor R_{E2} is the emitter stabilizing resistor for Q_2, and C_{E2} is the bypass capacitor. Resistor R_{L2} is the collector load resistor for Q_2.

The amplifier of Fig. 4-2, where both transistors are *PNP* (or *NPN*), has a tendency to become unstable. For example, if the collector current of Q_1 varies because of power supply or temperature changes, Q_2 will amplify these changes, and will add to them the changes caused by its own variations. For this reason, it is difficult to cascade more than two stages of direct coupling where both transistors are of the same type (*PNP* or *NPN*).

The circuit of a more stable direct-coupled amplifier is illustrated in Fig. 4-3. Here, Q_1 is an *NPN* transistor with the collector coupled directly to the base of Q_2, a *PNP* transistor. Such an arrangement is known as a *complementary* amplifier. The forward bias for the emitter–base junction of Q_1, and the reverse bias for the collector–base junction, are obtained from the $+V_{CC}$ supply in the normal manner. The emitter of Q_2 is connected to $+V_{CC}$ through R_{E2} (which is bypassed by C_{E2}). The collector resistor R_{L1}, the impedance of Q_1, and resistor R_{E1} form a voltage divider across the V_{CC} supply.

With such an arrangement, the base of Q_2 is less positive (or more negative) than the emitter. Thus, the emitter–base junction of Q_2 (*PNP*) is forward biased. The collector of Q_2 is grounded through the output load resistor R_{L2}. Since the negative terminal of the power supply (or $-V_{CC}$) is also grounded, reverse bias for the collector–base junction of Q_2 is established.

The increased stabilization for the complementary direct-coupled amplifier arises from the fact that a change in collector current of Q_1 (due to tempera-

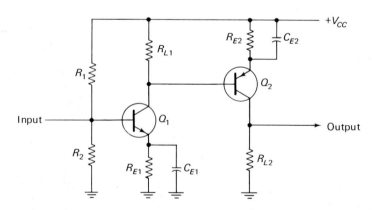

Fig. 4-3   Direct-coupled complementary amplifier (*NPN–PNP*)

ture, power supply variation, etc.) is *opposed by an equal change* in collector current of Q_2 (but in the opposite direction). If more stages are to be added, the complementary system is continued. That is, *NPN* and *PNP* amplifiers are used alternately. The circuit of such a three-stage complementary amplifier is shown in Fig. 4-4.

It is possible to use direct coupling between unlike stages, such as a common-emitter amplifier and an emitter follower. This arrangement is shown in Fig. 4-5. Here, transistor Q_1 is connected in a conventional common-emitter amplifier configuration (with emitter bypass for high gain). The collector of Q_1 is direct coupled to the base of Q_2, which is connected in a common-collector (or emitter-follower) configuration.

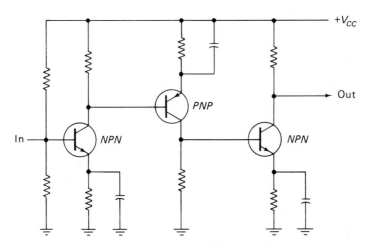

Fig. 4-4   Three-stage direct-coupled complementary amplifier

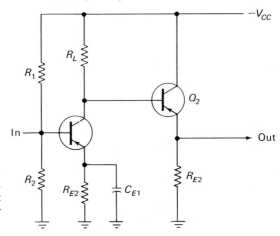

Fig. 4-5   Direct-coupled high-gain common-emitter stage, followed by a low-output impedance common-collector (emitter-follower) stage

Forward bias for the emitter–base junction of Q_2 is obtained from the power supply through resistor R_L, which also serves as the collector load for Q_1. Capacitor C places the collector of Q_2 at ground potential, as far as signal is concerned. Output is taken across R_{E2}, which is the emitter resistor for Q_2. The circuit of Fig. 4-5 is particularly useful when the output must be at a low impedance. The value of R_{E2} sets the approximate output impedance of the circuit.

4-2. PRACTICAL DIRECT-COUPLED TWO-JUNCTION TRANSISTOR AMPLIFIERS

Figure 4-6 is the working schematic of a two-stage direct-coupled complementary amplifier. Figure 4-6 also shows the equations for finding approximate component values, and gives some typical component values and currents.

Note that two- and three-stage amplifiers, similar to that shown in Fig. 4-6, are available commercially as *hybrid circuits*. As discussed in Sec. 2-10, hybrid circuits are complete (or nearly complete) packages similar to integrated circuits. It is generally easier to design with hybrid circuits rather than

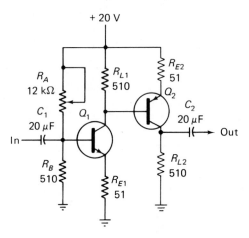

$$Z_{in} \approx R_B \approx 510$$
$$Z_{out} \approx R_{L2} \approx 510$$
$$Q_1 \; V_{gain} \approx \frac{R_{L1}}{R_{E1}} \approx 10$$

$$Q_2 \; V_{gain} \approx \frac{R_{L2}}{R_{E2}} \approx 10$$

$$R_L > 5R_E$$
$$R_L \approx 10R_E$$
$$R_B \approx 10R_E$$

$$R_B < 20R_E$$

$$Q_2 \; V_{collector} \approx 0.5 \times \text{supply (as adjusted by } R_A)$$

Fig. 4-6   Basic direct-coupled amplifier circuit and equations

with individual components, since impedance relationships, Q point, and so forth, have been calculated by the circuit manufacturer. Also, the datasheets supplied with the hybrid circuits provide information regarding source voltage, gain, impedances, and the like.

The datasheet information can be followed to adapt the hybrid circuit for a specific application. However, in some cases, it is necessary to select values of components external to the hybrid circuit package. For this reason, and since it may be necessary to design a multistage direct-coupled amplifier (with individual components) for some special application, the following design considerations and examples are provided.

4-2.1. Design Considerations

Note that the circuit elements for transistor Q_1 in Fig. 4-6 are essentially the same as for the circuit of Fig. 2-20 (the basic single-stage two-junction transistor amplifier) of Chapter 2. Also note that the same circuit arrangement is used for transistor Q_2, except that R_A and R_B are omitted (as is the coupling capacitor between stages).

The design considerations for the circuit of Fig. 4-6 are essentially the same as for the circuit of Fig. 2-20, except for the following:

The input impedance of the complete circuit is approximately equal to R_B (510 Ω). The output impedance is set by R_{L2} (510 Ω).

The value of C_1 is dependent upon the low-frequency limit and the value of R_B. For example, if the low-frequency limit is 30 Hz, an approximate value of 20 μF is required for C_1 to produce a 1-dB (at 30 Hz). The value of C_2 is dependent upon the low-frequency limit and the value of input resistance of the following stage (or the load). Capacitors C_1 and C_2 can generally be eliminated, except in those cases where an ac signal is mixed with direct current.

When two stages are direct coupled in the stabilized circuit of Fig. 4-6, the overall voltage gain is about 70 per cent (possibly higher) of the combined gains of each stage. The gain of each stage is approximately 10, as set by the 10-to-1 ratio of collector emitter resistances. The combined gain of the circuit is theoretically 100, and practically about 70. Thus, an input of 100 mV is increased to about 7 V at the output.

The signal at the base of Q_2 is approximately 10 times the signal at the base of Q_1. Thus, there is an approximate 1-V signal at the base of Q_2. This signal can be a varying direct current or an ac sinewave, or even a pulse. In any event, the base of Q_2 must be biased to accommodate the 1-V signal.

The collector of Q_2 should be approximately 10 V at the Q point. This will allow for the full 7-V output swing. With the collector of Q_2 at 10 V, there is an approximate 20-mA current through R_{L2} and R_{E2}. This produces an approximate 1-V drop across R_{E2}. Assuming that the transistors are silicon, the base of Q_2 should be about 0.5 V from the emitter. Since Q_2 is *PNP*, the base should be more negative (or less positive) than the emitter. The emitter of Q_2 is at about $+19$ V ($+20$-V supply, minus the 1-V drop across R_{E2}). Thus, the base of Q_2 should be $+18.5$ V. This also sets the collector voltage for Q_1 at the Q point. A Q-point collector voltage of 18.5 V will allow for the full 1-V signal swing.

The Q-point voltage of Q_1 is set by the collector current (approximately 3 mA). In turn, the collector current is set by the bias network R_A and R_B in the usual manner. In practice, the resistance values are approximated, and then R_A is adjusted to give the desired Q point. The final adjustment of R_A is made for distortion-free 7-V signal at the collector of Q_2 (with a 100-mV input signal applied to Q_1).

Since gain of the amplifier is set by circuit values, little concern need be given to transistor beta. Of course, the beta of each transistor must be greater than 10 at all frequencies that the circuit must amplify. Note that the circuit of Fig. 4-6 is highly stable, and has a very wide band frequency response (typically from direct current on up to whatever maximum is set by the transistor high-frequency limitations).

If gain must be increased, the emitter resistor can be bypassed. Amplifier gain is then entirely dependent upon transistor beta, rather than circuit values. However, as discussed in Chapter 2, bypassing the emitters creates problems for low-frequency signals. Since one of the prime reasons for using direct coupling is to amplify low-frequency signals, emitter bypassing may defeat the advantage of direct coupling. Likewise, if the circuit is used to amplify dc signals, the emitter bypass will be of little value.

Figure 4-7 shows an alternative method for bypassing the emitters of a direct-coupled amplifier with one capacitor. In effect, collector load resistor R_{L2} is broken up into two resistances (R_{L2A} and R_{L2B}), and connected in series with the emitter of Q_1 as a form of feedback. The bypass capacitor C_3 is connected between the junction of R_{L2A} and R_{L2B}. With such an arrangement,

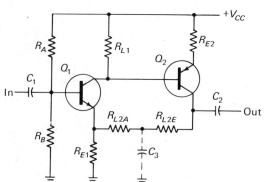

Fig. 4-7 Direct-coupled amplifier with partially bypassed feedback

the signal is bypassed, *as is part of the feedback*. This alters input impedance, as well as voltage gain. In effect, voltage gain is sacrificed for increased input impedance, resulting in increased stability. However, the low-frequency signal-attenuation problem caused by C_3 still exists.

4-3. PRACTICAL DIRECT-COUPLED FET AMPLIFIERS

Both IGFETs and JFETs can be used in direct-coupled amplifiers. However, IGFETs are especially well suited to direct-coupled applications. Since the gate of an IGFET acts essentially as a capacitor, rather than a doide junction, no coupling capacitor is needed between stages. For ac signals, this means that there are no low-frequency cutoff problems, *in theory*. In practical design the input capacitance can form an *RC* filter with the source resistance, and result in some low-frequency attenuation.

Figure 4-8 is the working schematic of an all-IGFET, three-stage amplifier. Note that all three IGFETs are of the same type, and all three drain resistors $(R_1, R_2, \text{ and } R_3)$ are the same value. This arrangement simplifies design. At first glance it may appear that all three stages are operating at zero bias. However, when I_D flows, there is some drop across the corresponding drain resistor, producing a voltage at the drain of the stage, and an identical voltage at the gate of the next stage. The gate of the first stage is at essentially the same voltage as the drain of the last stage, because of feedback resistor R_F.

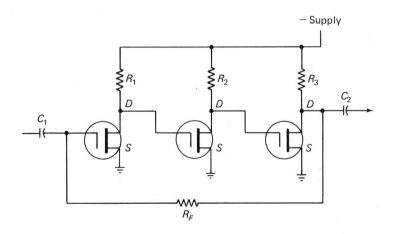

Gain without $R_F \approx (R_1 \times y_{fs})\,(R_2 \times y_{fs})\,(R_3 \times y_{fs})$

$$Z_{in} \approx \frac{R_F}{\text{gain}}$$

Fig. 4-8   Basic IGFET multistage amplifier

There is no current drain through R_F, with the possible exception of reverse gate current (which can be ignored for practical purposes).

Operating point. To find a suitable operating point for the amplifier, it is necessary to trade off between desired output voltage swing, IGFET characteristics, and supply voltage. For example, assume that an output swing of 7 V peak-to-peak is desired, and the supply voltage is 24 V. Furthermore, assume that the I_D is about 0.55 mA when V_{GS} is 7 V. A suitable operating point would be 7 V to accommodate the 7-V output swing without distortion. (The swing would be from 3.5 to 10.5, about the 7-V point.) This requires an 18-V drop from the 24-V supply. With 0.55-mA I_D and an 18-V drop, the values of R_1, R_2, and R_3 should be about 33 kΩ.

Gain. The overall voltage gain is dependent upon the relationship of the gain without feedback and the feedback resistance R_F. Gain without feedback is determined by Y_{fs} and the value of R_1, R_2 and R_3. For example, assume a Y_{fs} of 1000 μmhos (0.001 mho); the gain of each stage is 33 (33,000 × 0.001 = 33). With each stage at a gain of 33, the overall gain (without feedback) is about 36 kΩ.

To find the value of R_F, divide the gain without feedback by the desired gain. Multiply the product by 100. Then multiply the resultant product by the value of R_1. For example, assume a desired gain of 3000 (the gain without feedback is 36,000): 36,000/3000 = 12; 12 × 100 = 1200; 1200 × 33,000 = 39.6 MΩ (use the nearest standard to 40 MΩ).

Input impedance. The input impedance is dependent upon the relationship of gain and feedback resistance. The approximate impedance is R_F/gain.

Since gain is dependent upon Y_{fs}, input impedance is subject to variation from FET to FET, and with temperature.

4-3.1. Direct-Current FET Amplifier

The circuit of Fig. 4-8 requires one coupling capacitor at the input. This is necessary to isolate the input gate from any direct current voltage that may appear at the input generator or other device. This makes the circuit of Fig. 4-8 unsuitable for use as a dc amplifier. The coupling capacitor forms an RC filter with the input resistance. However, since the resistance is so high, a 0.01-μF coupling capacitor will produce less than 1-dB drop, even at frequencies of a few hertz.

The circuit can be converted to a dc amplifier when the coupling capacitor is replaced by a series resistor R_{IN}, as shown in Fig. 4-9.

The considerations concerning operating point are the same for both circuits. However, the series resistance must be terminated at a dc level equivalent to the operation point. For example, if the operation point is −7 V, point A must be at −7 V. If point A is at some other dc level, the operating point is shifted.

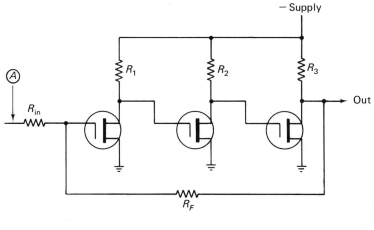

$$\text{Gain} \approx \frac{R_F}{R_{in}}$$

Input impedance $\approx R_{in}$ Output impedance $\approx R_3$

Fig. 4-9   Basic IGFET direct-coupled amplifier

The relationships of input impedance, R_F, and gain still hold. However, input impedance is approximately equal to R_{IN}. Therefore, gain is approximately equal to the ratio of R_F/R_{IN}. This makes it possible to control gain by setting the R_F/R_{IN} ratio. Of course, the gain cannot exceed the gain-without-feedback (open-loop) factor, no matter what the ratio of R_F/R_{IN} (closed-loop). As a general rule, the greater the ratio of open-loop gain to closed-loop gain, the greater the circuit stability.

As an example, assume that the open-loop gain is 36,000, and the desired gain is 6000. This requires a ratio of 6 to 1.

As another example, assume that the desired gain is 5000, R_F is 40 MΩ, and the open-loop gain is 36,000. 5000 is considerably less than 36,000, so the circuit is well capable of producing the desired gain with feedback. To find the value of R_{IN};

$$R_{IN} \approx \frac{40 \times 10^6}{5 \times 10^3} \approx 8 \times 10^3 (8 \text{ k}\Omega)$$

4-3.2. Amplification from Grounded Sources

The circuit of Fig. 4-9 requires that the signal source be at a dc level equal to the operating point. In many cases, it is necessary to amplify dc signals at the zero or ground level. This can be accomplished by using a depletion-mode JFET at the input, as shown in Fig. 4-10.

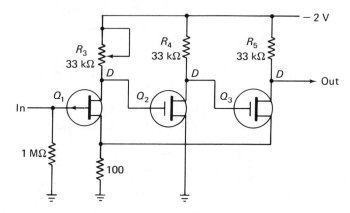

Fig. 4-10   Direct-coupled FET amplifier with JFET input, followed by two IGFETs

Feedback is introduced by connecting the sources of both Q_1 and Q_3 to a common source resistor R_2. The source of Q_2 is not provided with source resistor, but there is some bias on Q_2 produced by the I_D drop across R_3. The input impedance of the Fig. 4-10 circuit is set by the value of R_1 and the gate–drain capacitance of Q_1.

Gain can be sacrificed for stability by increasing the value of R_2. With the values shown and typical FETs, the gain should be on the order of 3000 to 5000. The bias and operating point for Q_2 and Q_3 is set by R_3, shown as 33-kΩ. In practice, the value of R_3 is approximated by calculation, and then adjusted for a desired operating point at the output (drain) of Q_3.

4-4. HYBRID DIRECT-COUPLED AMPLIFIERS

In certain applications, FET stages and two-junction transistor stages can be combined to form hybrid amplifiers. The classic example is where a single FET stage is used at the input, followed by two two-junction amplifier stages. Such an arrangement takes advantage of both the FET and two-junction transistor characteristics.

A FET is essentially a voltage-operated device permitting large voltage swings with low currents. This makes it possible to use high resistance values (resulting in high impedance) at the input and between stages. In turn, these high resistance values permit the use of low-value coupling capacitors and eliminate the need for bulky, expensive electrolytic capacitors. If operated at the 0TC point, the FET is highly temperature stable, tending to make the overall amplifier equally stable. However, FETs have the characteristic of operating at low currents, and are therefore considered as low-power devices.

Two-junction transistors are essentially current-operated devices permitting large currents at about the same voltage levels as the FET. Thus, with equal supply voltage and signal voltage swings, the two-junction transistor can supply much more current gain (and power gain) than the FET. Since currents are high, the impedances (input, interstage, and output) must be low in two-junction transistor amplifiers. This requires large-value coupling capacitors if low frequencies are involved. The low impedances also place a considerable load on devices feeding the amplifier, particularly if the devices are high impedance. On the other hand, a low output impedance is often a desirable characteristic for an amplifier.

When a FET is used as the input stage, the amplifier input impedance is high. This places a small load on the signal source and allows the use of a low-value input coupling capacitor (if required). If the FET is operated at the $0TC$ point, the amplifier input is temperature stable. (Generally, the input stage is the most critical in regards to temperature stability.) When two-junction transistors are used as the output stages, the output impedance is low, and current gain (as well as power gain) is high.

Hybrid amplifiers can be direct coupled or capacitor coupled, depending on requirements. The direct-coupled configuration offers the best low-frequency response, permits dc amplification, and is generally simpler (uses less components). The capacitor-coupled hybrid amplifier permits a more stable design and eliminates the voltage-regulation problem common to all direct-coupled amplifiers. (That is, a direct-coupled amplifier cannot distinguish between changes in signal level and changes in power-supply level. This problem is discussed further in Sec. 4-5.)

The FET can be combined with any of the classic two-stage two-junction transistor amplifier combinations. The two most common combinations are the Darlington pair (for no voltage gain, but high current gain and low output impedance), and the *NPN–PNP* complementary amplifier pair (for both voltage gain and current gain). (The Darlington pair is discussed in Sec. 4-6.)

4-4.1. FET Input, Two-Junction Transistor Output Amplifier

Figure 4-11 is the working schematic of a direct-coupled amplifier using a FET input stage, and a two-junction transistor pair as the output. Note that *local feedback* is used in the FET stage (provided by source resistor R_S), as well as *overall feedback* (provided by resistance R_4).

The design considerations for the FET portion of the circuit are essentially the same as described in Chapter 2, with certain exceptions. Input impedance is set by the value of R_2, as usual. Output impedance is set by the combination of R_4 and R_S. However, since R_S is quite small in comparison to R_4, the output impedance is essentially equal to R_4.

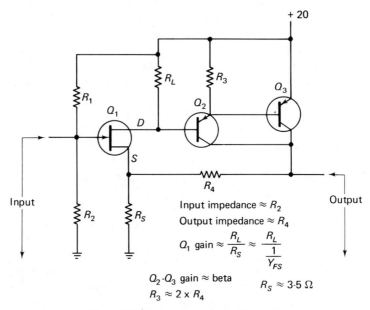

Fig. 4-11   Direct-coupled hybrid amplifier

The gain of the FET stage is set by the ratio of R_L to R_S, plus the $1/Y_{fs}$ factor. However, since R_s is quite small, the FET gain is set primarily by the ratio of R_L to $1/Y_{fs}$. The gain of the two-junction transistor pair is set by the beta of the two transistors and the feedback. Thus, the gain can only be estimated.

Note that the drop across R_3 is the normal base–emitter drop of a transistor (about 0.5 to 0.7 V for silicon, and 0.2 to 0.3 V for germanium). The drop across R_L is twice this value (about 1 to 1.5 V for silicon, and 0.4 to 0.6 V for germanium). Thus, for a typical silicon transistor, the base of Q_2 and the drain of Q_1 operate at about 1 V removed from the supply. In a practical experimental circuit, R_L must be adjusted to give the correct bias for Q_2 (and operation point for Q_1). The same is true for R_3. However, as a first trial value, R_3 should be about twice the value of R_4.

Design starts with a selection of I_D for the FET. If maximum temperature stability is desired, use the $0TC$ level of I_D. This usually requires a fixed bias, as described in Chapters 1 and 2. If temperature stability is not critical, the FET can be operated at zero bias by omitting R_1. There is some voltage developed across R_S. However, since R_S is small, the V_{GS} is essentially zero, and the I_D is set by the $0V_{GS}$ characteristics of the FET.

With the value of I_D set, select a value of R_L that produces approximately a 1- to 1.5-V drop to bias Q_2.

The input impedance is set by R_2, with the output impedance set by R_4. The value of R_3 is approximately twice that of R_4. The value of R_S is less than 10 Ω, typically on the order of 3 to 5 Ω.

As a brief design example, assume that the circuit of Fig. 4-11 is to provide an input impedance of 1 MΩ, an output impedance of 500 Ω, and maximum gain. Temperature stability is not critical.

Under these conditions, the values of R_2 and R_4 are set at 1 MΩ and 500 Ω but the voltage drop across R_S can be ignored. FET Q_1 operates at $V_{GS} = 0$, for practical purposes. Assume that I_D is 0.2 mA under these conditions. With a required drop of 1.5 V and 0.2 mA I_D, the value of R_L is approximately 7.5 kΩ. Since R_4 is 500 Ω, R_3 should be 1 kΩ.

The key component in setting up this circuit is R_L. With the circuit operating in experimental form, adjust R_L for the desired Q-point voltage at the output (collectors of Q_2 and Q_3).

4-4.2. Nonblocking Direct-Coupled Amplifier

Generally, a direct-coupled amplifier does not require any coupling capacitors. One exception is a coupling capacitor at the input to isolate the amplifier from direct current (when the signal is composed of direct current and alternating current). When a coupling capacitor is used at the gate of a JFET (or at the base of some two-junction transistors) a condition known as *blocking* can possibly occur.

Blocking is produced by the fact that the gate junction of a JFET is similar to that of a diode. That is, the diode acts to rectify the incoming signal. If a capacitor is connected in series with the diode (gate junction), large signals can charge the capacitor. On one half-cycle, the diode is forward biased and charges rapidly. On the opposite half-cycle, the diode is reverse biased and discharges slowly. If the signal and charge are large enough, the amplifier can be biased at or beyond cutoff, until the capacitor discharges. Thus, the amplifier can be blocked to incoming signals for a period of time.

One method of eliminating the blocking problem is to use an IGFET at the input. Such a circuit is shown in Fig. 4-12, where an IGFET drives a two-junction transistor pair. Since the gate of an IGFET acts essentially as a capacitor (rather than a diode junction), there is no rectification of the signal and no blocking.

The input impedance is set by R_{IN}, with the output impedance set by R_7. With the resistance ratios as shown by the equation of Fig. 4-12, the voltage gain is approximately 10 when capacitor C_2 is *out of the circuit*, and about 1000 when C_2 is in the circuit. Keep in mind that feedback is reduced or removed (and gain increased) when C_2 is in the circuit, since capacitor C_2 functions to bypass feedback signals to ground. With C_2 removed, the full feedback is applied, and gain is minimum (stability is maximum).

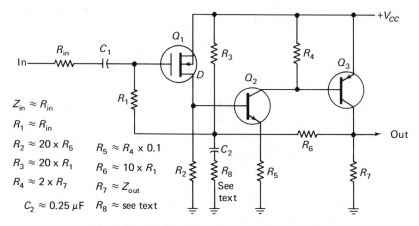

Fig. 4-12   Nonblocking, direct-coupled amplifier

If it is desired to operate the amplifier at some gain level between 10 and 1000, use a resistance in series with C_2 (shown in phantom as R_8).

4-5. STABILIZATION PROBLEMS IN DIRECT-COUPLED AMPLIFIERS

One of the main problems with any direct-coupled amplifier is that the circuit responds the same way to a change in power-supply voltage (due, for example, to temperature drift) as to a change in the dc signal level. Although high feedback can stabilize the gain of a capacitor-coupled amplifier to a point where the gain is determined almost entirely by the resistors in the feedback network, the level of a direct-coupled amplifier is not so easily stabilized.

Temperature-drift problems can be stabilized (or the effects of temperature drift minimized) in several ways. If FETs are involved, one or more of the FET stages can be operated at the $0TC$ point. If two-junction transistors are used, thermistors and/or temperature-compensating diodes can be used, as discussed in Chapter 1. Likewise, complementary circuits tend to be more stable in the presence of temperature and power-supply variations. However, none of these methods can compensate for a constantly changing or drifting power supply (from whatever cause) when the signal is also direct current. For that reason, several techniques have been developed to stabilize direct-coupled amplifiers.

4-5.1. Chopper Stabilization

A widely used method for circumventing the voltage-change problems of direct-coupled amplifiers is to convert the dc signal to an equivalent ac signal (through modulation). The ac signal is amplified in a gain-stabilized ac amplifier and then reconverted to direct current (through demodulation). During amplification, the signal is represented by the difference between the maximum and minimum excursions of the ac waveform, and is not affected by drift in the absolute voltage levels within the amplifier.

One method used to convert the dc signal to an ac signal is to switch the amplifier input alternately to both sides of a transformer, as shown in Fig. 4-13. This periodically inverts the polarity of the signal applied to the amplifier. The switches illustrated may be mechanical, transistor, or photoconductive. Another pair of contacts at the output establishes the ground level for a storage capacitor in series with the output. The output storage capacitor becomes charged to a level corresponding to the amplitude of the output squarewave. Synchronous detection preserves the polarity of the input voltage, and recovers both positive and negative voltages with the correct polarity. The synchronous modulation and demodulation is known as *chopping*, or *chopper stabilization*. A direct-coupled amplifier with a chopper circuit offers drift-free amplification of low-level signals in the microvolt region.

Another modulation technique uses two photoconductors—one in series with and one parallel to the amplifier input, as shown in Fig. 4-14. The photoconductor resistance is proportional to the illumination. By illuminating the photoconductors alternately (with flashing neon lights in this case), the amplifier input is connected to the signal and to ground. By synchronizing the output photoconductors with the input photoconductors as shown, the output storage capacitor charges to a level that corresponds to the output squarewave amplitude. This level is equivalent to the amplifier dc signal.

Typically, the illumination for all the photoconductors is from a bi-stable neon-tube relaxation oscillator. Note that when V_1 is illuminated, V_2 is dark,

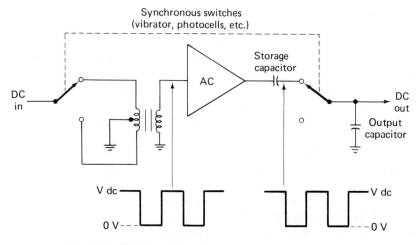

Fig. 4-13   Modulated amplifier technique for dc amplification

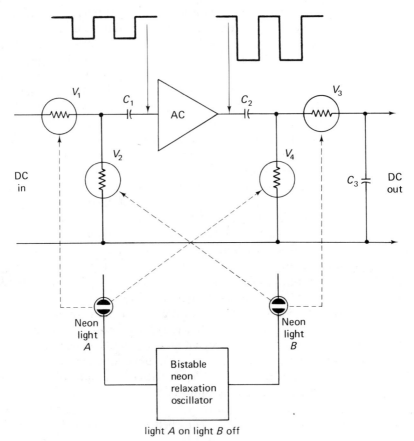

Fig. 4-14   Direct-current amplification circuit with photoconductive modulation

and vice versa. When V_1 is illuminated and V_2 is dark, the signal passes and charges C_1. On the alternate cycle, V_1 is dark and offers a high resistance to the signal. Likewise, V_2 is illuminated, allowing C_1 to discharge to ground. V_3, V_4, and C_2 operate in a similar manner at the amplifier output. Capacitor C_3 functions as a low-pass filter for the amplified signal.

One problem with any type òf modulated amplifier (chopper stabilized) is frequency response. If the input is pure direct current, there is no problem using modulation. However, if the input is very low frequency alternating current, the modulating frequency must be higher than the signal frequency. If not, the input signal waveform can be distorted or completely lost. For example, if both the input signal and modulating frequency are 100 Hz, and if the amplifier input is shorted at the same instant as the positive swing of the input signal, the amplifier will see only the negative portion of the input. As a rule of thumb, the modulating frequency should be *four times that of the highest* ac input signal to be amplified.

Another method of chopper stabilization is shown in Fig. 4-15. Here, a chopper-modulated amplifier is used to correct the direct-coupled amplifier for voltage change or drift. Note that the input signal direct current is amplified through a conventional dc amplifier. A portion of the amplified output signal is tapped off from a divider network and compared with the original input signal. The divider network reduces the output by the same

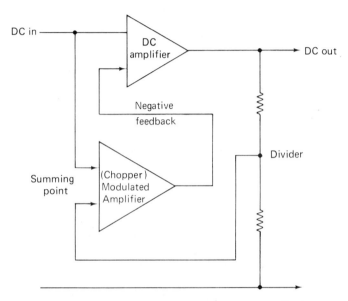

Fig. 4-15   Chopper-stabilized technique for dc amplifications

amount that is amplified in the dc amplifier. Therefore, the divider output should be equal to the input level at the summing point. Any difference at the summing point (caused by voltage change, drift, etc.) is amplified through the modulated amplifier, then applied to the main channel dc amplifier as *negative feedback* to cancel the drift.

4-5.2. Dual Amplification

Techniques other than modulation can be used when it is necessary to amplify both direct and alternating current (up to about 100-kHz). One such method is shown in Fig. 4-16, where two parallel amplifiers are used. One amplifier is direct-coupled for direct current and low-frequency alternating current. The other amplifier is a conventional ac amplifier. Appropriate networks separate the two frequency bands.

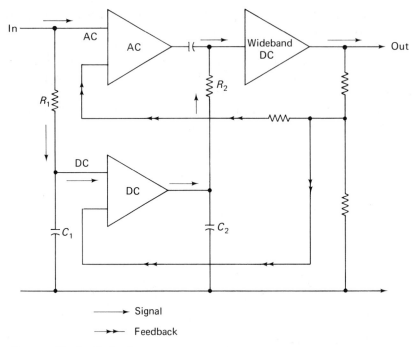

Courtesy Hewlett-Packard

Fig. 4-16   Dual amplification circuit

For example, direct current is rejected from the ac amplifier branch by coupling capacitors, but appears at the dc amplifier input and output as a charge across capacitors C_1 and C_2. High-frequency alternating current is attenuated at the dc amplifier input by resistors R_1 and R_2. Note that feedback is used in both branches to assure uniform gain. Such an amplifier will provide stable gain (1 per cent or better) from direct current up to about 100 kHz.

4-6. DARLINGTON COMPOUNDS

Figure 4-17 shows the basic Darlington circuit (known as the Darlington compound), together with two practical versions of the circuit. As shown, the Darlington compound is an emitter follower (or common collector) driving a second emitter follower. Going back to the basic amplifier theory of Chapters 1 and 2, an emitter follower provides no voltage gain, but can provide considerable power gain.

The main reason for using a Darlington compound (especially in audio work) is to produce high current (and power) gain. For example, Darlington compounds are often used as audio *drivers* to raise the power of a signal from a voltage amplifier to a level suitable to drive a final power amplifier. Darling-

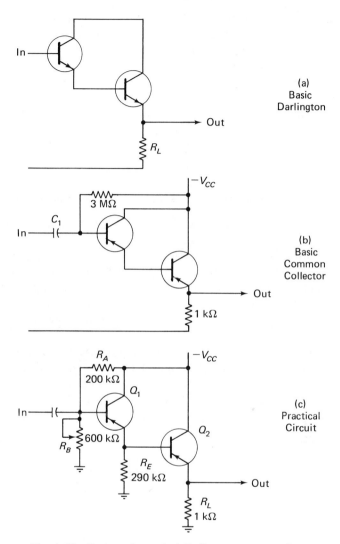

Fig. 4-17   Basic and practical Darlington compounds

tons are also used as a substitute for a driver section (or to eliminate the need for a separate driver).

4-6.1. Darlingtons as Basic Common Collectors (Emitter Followers)

When the Darlington is used as a common collector, as shown in Fig. 4-17, the output impedance is approximately equal to the load resistance R_L. The input impedance is approximately equal to beta^2 $\times$ R_L. The current gain is approximately equal to the average beta of the two transistors, squared. However, in most common collectors, power gain is of primary concern. That is, the designer is interested in how much the signal power can be increased across a given output load.

As an example, assume that the value of R_L (in Fig. 4-17c) is 1 kΩ and that the average beta is on the order of 15. This results in an input impedance of about 225 kΩ ($15^2 \times 1000$), and an output impedance of 1 kΩ. Now assume that a 2.5 V signal is applied at the input, and an output of 2 V appears across R_L. This input power is $2.5^2/225$ kΩ ≈ 0.028 mW. The output power is $2^2/1$ kΩ ≈ 4 mW. The power gain is $4/0.028 \approx 140$.

4-6.2. Darlingtons as Basic Common-Emitter Amplifiers

Darlington compounds can be used as common-emitter amplifiers to provide voltage gain. This can be accomplished by simply adding a collector resistor to any of the circuits in Fig. 4-17, and taking the output from the collector rather than the emitter. In effect, Q_1 then becomes a common-collector driving Q_2, which appears as a common-emitter amplifier. The entire circuit then appears as a common-emitter amplifier and can be used to replace a single transistor. Such an arrangement is often used where high voltage gain is desired.

A more practical method of using a Darlington as a common-emitter amplifier is to eliminate R_B and R_E (of Fig. 4-17c), ground the emitter of Q_2, and transfer R_L to the collector of Q_2. Such an arrangement is shown in Fig. 4-18. The circuit of Fig. 4-18 is stabilized by the collector feedback through R_A, which holds both collectors at a potential somewhat less than 0.5 V from the

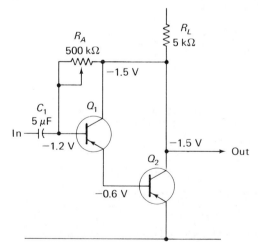

Fig. 4-18 Darlington compound with collector feedback and common-emitter output

base of Q_1. Note that both collectors are at the same voltage, and that this voltage is approximately equal to two base–emitter voltage drops (or about 1.5 V for two silicon transistors).

With the circuit of Fig. 4-18, the current gain is approximately equal to the ratio of R_A/R_L. Both the input and output voltage swings are somewhat limited in the Fig. 4-18 circuit. The input is biased at approximately 1 to 1.5 V. However, voltage gains of 100 (or more) are possible, since input impedance (or resistance) is approximately equal to the ratio of R_A/current gain, or equal to R_L. (With input and output impedances approximately equal to R_L, the voltage gain follows the current gain.)

4-6.3. Multistage Darlingtons

Darlington compounds need not be limited to two transistors. Three (and even four) transistors can be used in the Darlington circuit. A classic example of this is the General Electric circuit of Fig. 4-19 (which is available in hybrid form, in a TO-5 style package). This circuit is essentially a common-collector and common-emitter Darlington, followed by a common-emitter amplifier.

With R_I out of the circuit, both the input and output impedances are set by R_L. (In practice, the input impedance is slightly higher than R_L, typically on the order of 700 to 800 Ω). With R_I removed, the voltage gain is about 1000. When R_I is used, the input impedance is approximately equal to R_I, and the voltage gain is reduced accordingly. For example, if R_I is 10 kΩ, the 1000 voltage gain drops to about 50.

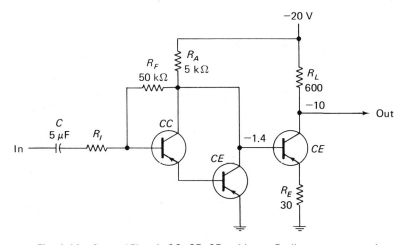

Fig. 4-19   General Electric CC–CE–CE multistage Darlington compound

4-7. SPECIAL DIRECT-COUPLED AMPLIFIER CIRCUITS

In addition to the basic direct-coupled amplifier circuits described in previous sections, there are three special direct-coupled amplifiers in common use: the emitter-coupled circuit, the cascode amplifier, and the power complementary circuit.

4-7.1. Emitter-Coupled Amplifier

Figure 4-20 is the basic schematic of an emitter-coupled amplifier. This circuit, or one of its many variations, is similar to that of the *phase inverter* or *phase splitter* described in Sec. 2-10 in that two 180° out-of-phase signals or outputs can be taken from one input. Unlike the single-stage inverter of Sec. 2-10, the emitter-coupled amplifier of Fig. 4-20 provides considerable gain. Therefore, an emitter-coupled amplifier can be used in a design when a low-voltage input must be amplified to drive a push–pull output stage.

The emitter resistor R_E is common to both Q_1 and Q_2, which are biased at the Q point (typically at about one half the supply voltage) in the normal manner. The signal is applied to Q_1 and appears in amplified, but phase inverted, form at the collector of Q_1. An in-phase signal is also developed

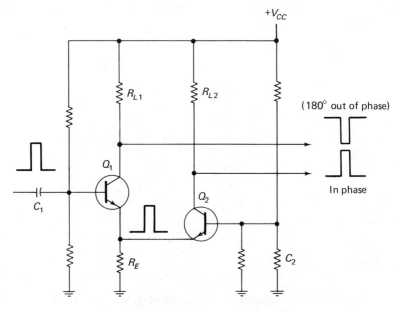

Fig. 4-20   Basic emitter-coupled amplifier

across R_E and appears at the emitters of both Q_1 and Q_2 simultaneously. Since the base of Q_2 remains fixed by the bias, signal variations on the emitter of Q_2 produce an in-phase signal at the collector of Q_2. If the values of R_{L1} and R_{L2} are equal, the two output collector signals will be identical in amplitude but opposite in phase. Capacitor C_2 provides a signal path for the base of Q_2.

The values for components of the Fig. 4-20 circuit are found using the guidelines established in previous chapters. However, keep in mind that R_E will pass *twice the current* normally found in a single-stage amplifier, all other factors being equal. This is because the emitter–base and emitter–collector currents of both transistors must pass through R_E. Thus, R_E will normally be one half the value of a single-stage design for a given base–emitter voltage relationship.

4-7.2. Cascode Amplifier

The cascode amplifier is a form of direct coupling used primarily in RF circuits. The cascode circuit is quite popular in vacuum-tube amplifiers, but not widely used in solid-state RF equipment. The basic circuit is shown in Fig. 4-21.

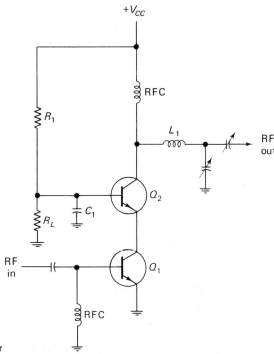

Fig. 4-21 Cascode RF amplifier

The input signal is fed to the base of Q_1, which is a conventional common-emitter amplifier. Variations in the signal at the base of Q_1 appear as amplified variations in voltage at the collector of Q_1.

The collector of Q_1 is coupled directly to the emitter of Q_2, which operates as a common-base amplifier. The base is grounded as far as the signal is concerned through C_2. A grounded or common-base amplifier does not require neutralization. (Refer to Chapter 3.)

One problem in solid-state cascode amplifiers is the bias network. The collector of Q_1 and the emitter of Q_2 are at the same potential. Generally, this potential is less than the supply voltage, but can still be well above ground. To function properly, the base of Q_2 should be within about 1 V of the emitter. This bias is provided by the network of R_1 and R_2.

4-7.3. Power Complementary Amplifier

The basic complementary amplifier described in Sec. 4-1 can be used in audio power applications when a third power transistor is added. Such an arrangement is shown in Fig. 4-22 where a *PNP–NPN* pair is followed by a power *PNP* transistor. Using the values shown and assuming that Q_3 is capable of handling about 5 W, the circuit of Fig. 4-22 is capable of delivering about 5 W of stable power at audio frequencies. Although the voltage gain is low (about 10), the power gain is high (about 100,000), since there is a large difference between input and output impedances.

The input impedance is approximately equal to the value of R_B. The output impedance is equal to the impedance of the transformer. If the loudspeaker has an 8-Ω impedance, and T_1 has a 1-to-1 ratio, the output impedance is 8 Ω.

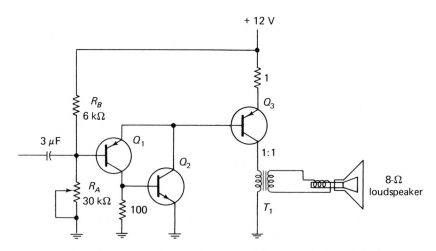

Fig. 4-22   Power complementary amplifier for audio frequencies

The circuit is adjusted for class A operation by the value of R_A. In practice, the value shown (30 kΩ) is used as a starting point and is adjusted for the desired class A operation and/or a specific Q point (collector voltage of Q_3) under no-signal conditions. Note that Q_3 must be operated with a heat sink since the power dissipation is in excess of 5 W.

4-8. TRANSFORMERLESS DIRECT-COUPLED AUDIO AMPLIFIER

As discussed, audio amplifiers can be *RC* coupled and/or transformer coupled. Both methods have limitations. One way to get around these limitations is to use a *transformerless audio amplifier circuit*. There are three such basic circuits: the series-output amplifier, the quasi-complementary amplifier, and the full-complementary amplifier

4-8.1. Transformerless Series-Output Amplifier

Figure 4-23 shows two typical transformerless series-output amplifiers used in audio systems. One configuration requires two power supplies, but omits the coupling capacitor to the load. This configuration provides better low-frequency response (since there is no capacitor), but can be inconvenient because of the two power supplies. The configuration with a single power supply has reduced low-frequency response since the coupling capacitor forms a high-pass filter with the load resistance.

Either configuration of series-output has two drawbacks. A phase inverter (Sec. 2-10) is required to drive the series-output stage, even if gain is not required. Also, an additional driver stage (or possibly an emitter-coupled amplifier, Sec. 4-7.1) may be necessary to bring the power up from the output of a voltage amplifier to a level required by the series-output stage.

These problems are overcome by means of a *complementary* circuit, either quasi-complementary or full complementary.

4-8.2. Quasi-Complementary Amplifier

Figure 4-24 is the schematic of a quasi-complementary output. This circuit consists of a Darlington compound using *NPN* transistors, and a direct-coupled complementary pair using an input *PNP* and an output *NPN*. Both base signals can be in phase (although they are often at different voltage levels) so that a phase inverter or emitter-coupled amplifier is not needed.

For example, if the input is positive going, Q_1 is forward biased, as is Q_2. A positive-going input at Q_3 reverse biases Q_3, since Q_3 is *PNP*. This produces a negative-going output from Q_3 to Q_4 (an *NPN*) and reverse biases Q_4.

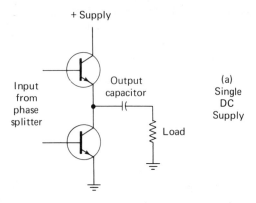

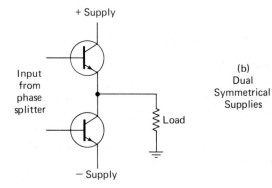

Fig. 4-23 Transformerless series output amplifier

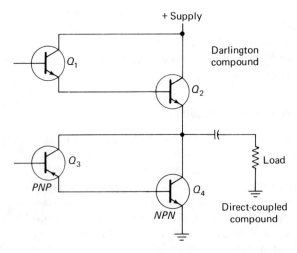

Fig. 4-24 Basic quasi-comple-mentary amplifier

When the signal at the bases of Q_1 and Q_3 is negative going, the condition is reversed (Q_3 is reverse biased, Q_4 is forward biased).

4-8.3. Full-Complementary Amplifier

Figures 4-25 and 4-26 show two versions of the full-complementary output. Either version has an advantage over the quasi-complementary in that both halves of the circuit are identical. This makes it easier to match both halves (for positive and negative signals) to minimize distortion that could be caused by uneven amplification of the signal.

The circuit of Fig. 4-25 uses two Darlington compounds, and is often called a dual-Darlington output. The amplifier of Fig. 4-25 is used when *power gain* is needed.

The circuit of Fig. 4-26 has two direct-coupled complementary compounds in the output, and is used when *voltage gain* is most needed.

Phase inversion is not required for either circuit. A positive-going input will forward bias Q_1 and Q_2 and reverse bias Q_3 and Q_4. A negative-going input will produce opposite results.

As in the case of series-output circuits (Sec. 4-8.1), the load-coupling capacitor can be omitted if two power supplies are used (one positive and one negative with respect to ground or common). The tradeoff between the in-

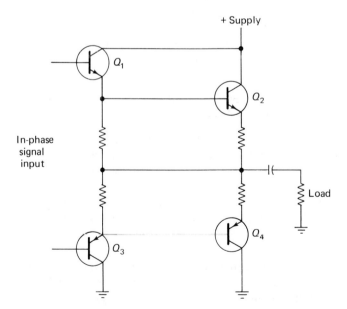

Fig. 4-25   Full-complementary amplifier with dual-Darlington output

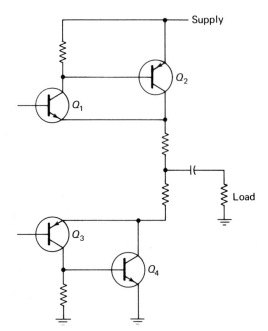

Fig. 4-26   Full-complementary amplifier with direct-coupled output

convenience of two power supplies versus improved low-frequency response must be decided by circuit requirements.

As a guideline, a 2000-μF capacitor working into a 4-Ω load (such as a 4-Ω loudspeaker) produces an approximate 3-dB drop at 20 Hz (a 10-V output drops to 7 V).

4-8.4. Examples of Transformerless Direct-Coupled Audio Amplifiers

The following examples are used to illustrate the problems involved with transformerless amplifiers. Two circuits are discussed. Both circuits were developed by Motorola using their complementary plastic transistors.

A four-transistor circuit, capable of delivering 3 to 5 W of audio power with an approximate 0.25-V input signal, is shown in Fig. 4-27. A six-transistor circuit, delivering between 7 and 35 W with 0.1- to 0.45-V input signals, is shown in Fig. 4-28.

As shown, the most significant difference between the circuits of Figs. 4-27 and 4-28 is the output configuration. The circuit of Fig. 4-27 is essentially the series-output amplifier (with single power supply) described in Sec. 4-8.1. The output configuration of the six-transistor circuit (Fig. 4-28) is a composite complementary pair described in Sec. 4-8.3. The complementary pairs serve

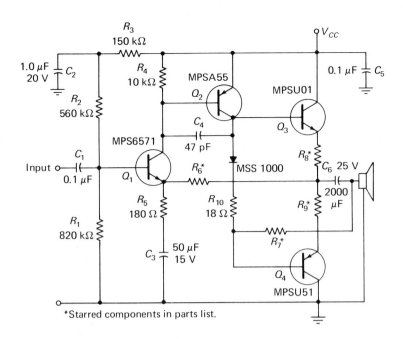

*Starred components in parts list.

Parts List

	3 W	5 W
V_{CC}	18 V	22 V
R_6	3.9 kΩ	4.7 kΩ
R_7	470 Ω	560 Ω
R_8	0.75 Ω	0.47 Ω
R_9	0.75 Ω	0.47 Ω

*See text.

Amplifier Output		
	3 W	5 W
Idle current, nominal	25 mA	25 mA
Current drain at rated power output	290 mA	370 mA
Nominal sensitivity for rated power output	0.25 V	0.25 V
THD at rated maximum power output, 50 Hz to 15 kHz, nominal	2%	2%
THD at rated power output, nominal	0.25 %	0.25%
Input impedance, typical	300 kΩ	300 kΩ
Maximum power output at 5% THD, without current-limiting diodes	4.5 W	6.7 W
Maximum power output at 5% THD with current-limiting diodes*	4.0 W	6.0 W

Courtesy Motorola

Fig. 4-27   Four-transistor transformerless audio amplifier

245

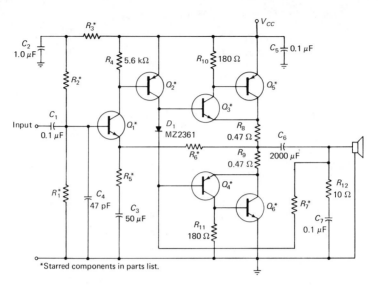

*Starred components in parts list.

Parts List

	7 W	10 W	15 W	20 W	25 W	30 W	35 W
V_{CC}	26 V	30 V	35 V	43 V	46 V	50 V	54 V
R_1	2.7 MΩ	2.7 MΩ	560 kΩ	560 kΩ	220 kΩ	220 kΩ	220 kΩ
R_2	1.2 MΩ	1.0 MΩ	330 kΩ	330 kΩ	150 kΩ	150 kΩ	150 kΩ
R_3	390 kΩ	820 kΩ	120 kΩ	120 kΩ	47 kΩ	47 kΩ	47 kΩ
R_5	100 Ω	82 Ω	100 Ω	75 Ω	220 Ω	270 Ω	270 Ω
R_6	8.2 kΩ	8.2 kΩ	10 kΩ	10 kΩ	10 kΩ	10 kΩ	10 kΩ
R_7	3.9 kΩ	3.9 kΩ	4.7 kΩ	5.6 kΩ	5.6 kΩ	6.8 kΩ	8.2 kΩ
Q_1	MPS6571	MPS6571	MPS6571	2N5088	2N5088	2N5088	2N5088
Q_2	2N5087	2N5087	2N5087	2N5087	MPSA56	MPSA56	MPSA56
Q_3	MPSA05	MPSA05	MPSU05	MPSU05	MPSU05	MPSU05	MPSU05
Q_4	MPSA55	MPSA55	MPSU55	MPSU55	MPSU55	MPSU55	MPSU55
Q_5	MJE371	MJE371	MJE105	MJE105	MJE2901	MJE2901	MJE2901
Q_6	MJE521	MJE521	MJE205	MJE205	MJE2801	MJE2801	MJE2801

	Amplifier Output (W)						
	7	10	15	20	25	30	35
Idle current, nominal (mA)	25	25	25	25	25	25	25
Current drain at rated output power (mA)	420	500	580	700	750	850	950
Sensitivity at rated output power (V)	0.1	0.1	0.1	0.1	0.45	0.4	0.45
THD at rated output power, typical (%)	2	2	1	1	0.5	0.5	0.5
THD at 1/2 rated output power, typical (%)	0.5	0.5	0.25	0.25	0.1	0.1	0.1
Input impedance, typical (kΩ)	800	700	200	200	90	90	90

Courtesy Motorola

Fig. 4-28 Six-transistor transformerless audio amplifier

as extra driver transistors to supply the higher drive current necessary for higher power.

Four-transistor circuit. As shown in Fig. 4-27, the output circuit is formed by Q_3 and Q_4. To ensure maximum output signal swing, the dc voltage at the emitters of Q_3 and Q_4 must be one half of V_{CC}. To prevent clipping on either half-cycle of output signal, the dc voltage at the Q_3–Q_4 emitters must "track" with any variations in V_{CC}. The dc feedback arrangement meets these requirements, in that the amplifier is essentially a unity-gain- voltage follower for direct current.

The voltage at the base of Q_1 is approximately one half of V_{CC} and is set by the divider of resistors R_1, R_2, and R_3. The Q_1 emitter voltage "follows" the base voltage, so the dc output voltage is approximately equal to the emitter voltage. The Q_1 base is actually slightly greater than one half of V_{CC} (since R_1 is larger than R_2 and R_3). This compensates for the small drop across resistor R_6 and the $V_{BE(on)}$ of Q_1.

The ac gain of the Fig. 4-27 circuit is set by the ratio of R_6 to R_5:

$$A_V = \frac{R_5 + R_6}{R_5} = \frac{R_6}{R_5}, \qquad \text{when } R_6 \gg R_5$$

Capacitor C_3 allows the bottom of R_5 to be at signal ground, and provides dc isolation for the emitter of Q_1.

Elimination of crossover distortion is desirable, especially at low listening levels. The collector current of Q_2 through D_1 and R_{10} produces a dc voltage that slightly forward biases the base–emitter junctions of output transistors Q_3 and Q_4. With the output transistors forward biased, the amplifier operates in class A for small output signals and class B for large output swings. Keep in mind that Q_3 and Q_4 are complementary (*PNP* and *NPN*). Thus, one transistor is conducting with the other transistor cut off (or near cutoff) on each half-cycle.

Objectionable power-supply hum is filtered from the input circuit by resistor R_3 and capacitor C_2. Capacitor C_5 protects the V_{CC} line against oscillations that could occur when high current transients are present.

High-frequency oscillations may occur with varying source and load impedances. Capacitor C_4 provides a high-frequency rolloff at approximately 30 kHz to prevent such oscillations.

Six-transistor circuit. As shown in Fig. 4-28, transistors Q_3 and Q_5 form the composite *NPN* transistor, and Q_4 and Q_6 form the composite *PNP* transistor. Driver transistors Q_3 and Q_4 operate as emitter follower to establish the output voltage; at the same time they are common-emitter amplifiers supplying drive current to output transistors Q_5 and Q_6.

Both ac and dc feedback is applied to the composite pairs by resistors R_8 and R_9. To prevent thermal runaway, which can occur with an increase in collector–base leakage current due to high temperature, resistors R_{10} and R_{11}

are placed between the base and emitter of the output transistors. The resistance of R_{10} and R_{11} is sufficiently low to prevent forward biasing of the output transistors by temperature-induced leakage currents.

High-frequency oscillations are suppressed by capacitor C_4. The network composed of resistor R_{12} and capacitor C_7 prevents oscillations that arise due to the increased impedance of the loudspeaker at high frequencies. The functions of the remaining circuits in Fig. 4-28 are identical to those of the Fig. 4-27 circuit.

Keeping direct current out of the loudspeaker. The circuits of Figs. 4-27 and 4-28 use the output capacitor as a bootstrap to provide driving current to the *PNP* side on the negative half-cycle. This method requires the collector current of the predriver transistor Q_2 to flow through the loudspeaker. Some designers consider direct current in the loudspeaker as objectionable. One objection is that there can be a loud "pop" in the loudspeaker when power is first applied. Another objection is that a fixed direct current in the loudspeaker can change the acoustic characteristics, resulting in poor reproduction of sound. Some designers consider that a constant direct current tends to burn out the loudspeaker windings.

If the bootstrapping method of Figs. 4-27 and 4-28 is considered as objectionable, for whatever reason, the circuit can be modified as shown in Fig.

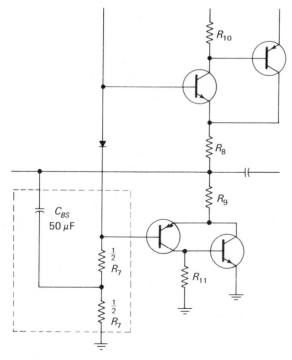

Courtesy Motorola

Fig. 4-29 Bootstrapping R_7 with a separate capacitor to keep direct current from the loudspeaker

4-29. The capacitor C_{BS} in conjunction with the top resistor labeled $\frac{1}{2} R_7$ supplies the required drive current needed to ensure a full negative signal swing.

Overload protection. The amplifiers of Figs. 4-27 and 4-28 are not generally intended for use in sealed systems, so provision for overload protection is desirable. One method is to add two diodes D_2 and D_3 to the output circuit, as shown in Fig. 4-30. This method uses the forward voltage drop of diodes

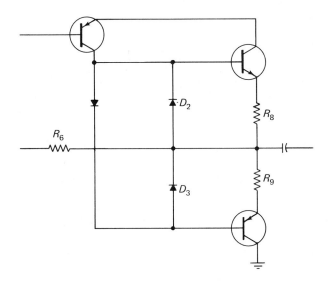

Amplifier Version (W)	Short-Circuit Current (A)	Diode D1	Diodes D2 and D3	R8 (Ω)	R9 (Ω)
35	2.1	MZ2361	MZ2361	0.47	0.39
30	2.0	MZ2361	MZ2361	0.47	0.39
25	1.6	MZ2361	MSS1000	0.47	0.39
20	1.55	MZ2361	MSS1000	0.47	0.39
15	1.35	MZ2361	MSS1000	0.47	0.47
10	1.3	MZ2361	MSS1000	0.47	0.47
7.0	1.25	MZ2361	MSS1000	0.47	0.47
5.0	1.0	MSS1000	MSS1000	0.47	0.47
3.0	0.8	MSS1000	MSS1000	0.47	0.47

Fig. 4-30   Connection of diodes D_2 and D_3 for overload protection. The diodes and circuit values are shown in the table

D_1, D_2, and D_3 to provide a clamp at the base of the output transistors (Fig. 4-27) or the drivers (Fig. 4-28), which limits the maximum current in the output transistors.

Figure 4-31a shows the equivalent circuit of the clamp on the positive half-cycle, and Fig. 4-31b shows the action on the negative half-cycle. The chart in Fig. 4-30 shows the dc level at which the various amplifiers limit under short-circuit conditions, along with parts values and diode types used in the overload protection circuitry. The 20- and 35-W versions use a 0.39-Ω resistor for R_9, rather than the 0.47 Ω, to prevent clipping, which can occur on the negative half-cycle with the larger resistor.

The overload network protects the amplifier against accidental short circuits in the output terminals, and against an overload due to connecting extra loudspeakers. The problem of overloads in audio amplifiers is discussed further in Sec. 4-9.

Specifying transistors. One of the most important aspects of amplifier design is determining which transistor parameters are of greatest importance in each function, and then specifying these parameters to best take advantage of the inherent characteristics of the device, while still maintaining the desired circuit performance.

One of the most important "parameters" to the audio designer is cost, and since testing is one of the costliest steps in manufacturing any semiconductor,

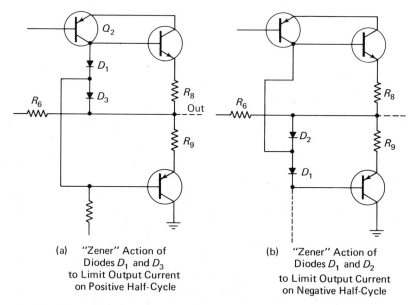

(a) "Zener" Action of Diodes D_1 and D_3 to Limit Output Current on Positive Half-Cycle

(b) "Zener" Action of Diodes D_1 and D_2 to Limit Output Current on Negative Half-Cycle

Courtesy Motorola

Fig. 4-31   Use of diodes for overload protection

both vendor and user should eliminate as many unnecessary tests as possible. Although this may not apply to all laboratory applications, it certainly does apply to manufacturing. For example, in production-line manufacturing it is ridiculous to test noise figure in a power output transistor or safe area in a preamplifier device.

Let us examine several sections in an amplifier circuit to illustrate one method of approaching the problem. The basic circuit of Fig. 4-28 with a 25-W 8-Ω power rating is referred to in the following discussion.

In a power amplifier the logical place to begin design is the output terminal. For 25 W rms to 8 Ω, the peak-to-peak output voltage, or signal voltage swing, is given by

$$V_{PP} = 2\sqrt{(R)(2)(P_{\text{rms}})}$$
$$= 2\sqrt{(8)(2)(25)}$$
$$= 40 \text{ V}$$

where V_{pp} is the peak-to-peak output voltage, and P_{rm} is the rms output power.

The power supply voltage must be

$$V_{CC} = V_{PP} + 2I_{Pk}R_E + V_{CE}(Q_5) + V_{CE}(Q_6)$$

I_{Pk} is found from the equation

$$I_{Pk} = \frac{(1/2)V_{PP}}{R_L} = \frac{(1/2)(40)}{8} = 2.5 \text{ A}$$

Figure 4-28 shows R_E (R_8 or R_9) to be 0.47 Ω, and the recommended power transistors are the MJE2801 and MJE2901. The datasheets for these devices lists a minimum h_{FE} of 25 at 3 A, and V_{CE} of 2 V for the output units. These numbers do not change significantly at 2.5 A, so the V_{CE} of Q_5 and Q_6 can be taken as 2 V. Note that the $V_{CE(sat)}$ measurement found on many datasheets is meaningless in this application because the output transistors are operated as large-signal class B emitter followers, and do not saturate.

Using these figures to find V_{CC},

$$V_{CC} = 40 + 2(2.5)(0.47) + 2 + 2 \approx 46 \text{ V}$$

Adding 10 per cent for a high line voltage, the maximum V_{CC} is 50 V. Examination of the important parameters indicates that the following are *minimum specifications for the output transistors:*

$$BV_{CEO} = 50 \text{ V min at 200 mA}$$
$$h_{FE} = 30 \text{ min at 2.5 A and } V_{CE} \text{ of 2 V}$$

$$V_{BE(on)} = 1.2 \text{ V max at 3 A and } V_{CE} \text{ of 2 V}$$

$$I_{CBO} = 0.1 \text{ mA max at 25°C and } V_{CB} \text{ of 50 V}$$

$$I_{CBO} = 2 \text{ mA max at 150°C and } V_{CB} \text{ of 50 V}$$

The MJE2801 and MJE2901 meet these requirements satisfactorily.

The collector current of the driver transistors is the base current of the output units. The peak output current is 2.5 A, and h_{FE} of the output units is 30, minimum. The maximum collector current of each driver (Q_3 or Q_4) is then

$$I_{max} = \frac{I_{\text{load (Pk)}}}{h_{FE} \min(Q_5 \text{ or } Q_6)} = \frac{2.5}{30} = 83 \text{ mA}$$

The recommended drivers for the 25-W amplifiers are the MPS-UO5 (NPN) and MPS-U55 (PNP). The datasheet has a minimum h_{FE} specification of 50 at collector currents of 10 mA and 100 mA, and BV_{CEO} is 50 V.

The designer may wish to increase or decrease certain specifications. For example, it may be desirable to use a transistor with a 55-V BV_{CEO}, or perhaps a minimum h_{FE} of 30 or 40 for the driver and 20 for the outputs. In doing this, the overall effects must be considered. For instance, if the h_{FE} specifications on the output and driver units are relaxed, the designer must ensure sufficient collector current in the predriver (Q_3) to provide full drive to the output stages. The predriver will then be operated at a higher collector current, which would increase the base current, making additional compensation in the preamplifier (Q_1) necessary. Such compensation can take the form of a higher h_{FE} preamplifier, or a decrease in the impedance of the base biasing resistors of the preamplifier. Another obvious solution is to specify a higher h_{FL} for the predriver, thus making a change in the preamplifier unnecessary.

Intelligent use of device specifications enables the designer to safely make numerous changes in the circuits of Fig. 4-27 and 4-28. For example, input impedances can be lowered or raised, power levels can be changed, or the circuits can be used to drive different load impedances. It is only necessary to decide exactly what the circuit requires; then the exact device can be specified. For example, if the circuit is to operate into a 16-Ω loudspeaker, the value of I_{PK} is changed from 2.5 to 1.25 A, reducing the V_{CC} to about 43 V. In turn, this reduces the required BV_{CEO} for the output transistors and reduces the maximum collector current I_{max} of the driver transistors.

Power supplies. When the amplifier circuits are used in consumer applications, the associated power supply does not need to be complex. The circuit of Fig. 4-32 is a typical example. Capacitor C_1 is usually 1000 to 4000 μF. Resistor R_1 and capacitor C_2 are selected to provide voltage V_2, which may be necessary for operation of other circuitry, such as preamplifiers, AM or FM tuners, multiplex decoders, and so on.

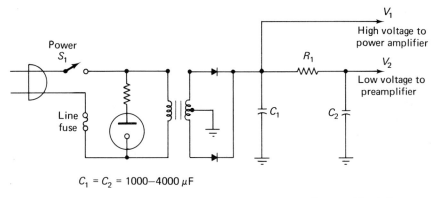

$C_1 = C_2 = 1000-4000\ \mu F$

Courtesy Motorola

Fig. 4-32   Typical power supply for transformerless amplifiers

Voltage V_1 is the V_{CC} for the power amplifier. The regulation of the power supply determines its ability to maintain V_1 at its nominal design level. Regulation is defined as the ratio of the drop in voltage with load to full-load voltage:

$$\% \text{ regulation} = \frac{V_M - V_0}{V_0} \times 100$$

where V_M is no-load voltage and V_0 is full-load voltage.

For a more comprehensive discussion of solid-state power supplies, see the author's *Handbook of Simplified Solid-State Circuit Design* (Prentice-Hall, Inc., Englewood Cliffs, N.J., 1971).

Heat sinks. The output transistors of all circuits shown in Figs. 4-27 and 4-28 require heat sinks. In addition, the driver transistors of the 15- to 35-W versions should be provided with heat sinks. These heat sinks should allow the dissipation capability of the drivers to be increased to about 10 per cent of the rated output power of the amplifier, which protects the devices under short-circuit conditions.

4-9. SHORT-CIRCUIT PROTECTION AND SECONDARY BREAKDOWN

One problem with any of the output circuits described in Sec. 4-8 is that one (or more) of the collectors will operate without a load resistance. In effect, the output load (usually the loudspeaker) forms the load for the collector. Although the working load does present a load similar to that of a collector resistor, there is one major difference. The working load in any of the output circuits is not in series with the collector, as is the load resistor in a basic class A amplifier.

In the basic *RC*-coupled class A amplifiers, the collector load resistor reduces the collector voltage when heavy current is flowing, and raises the voltage only when a light current is present. If there should be a short in the output, heavy current may flow, but the collector voltage will drop.

In the various direct-coupled series- and complementary-output circuits (as well as transformer-coupled circuits), a short circuit of the load will produce heavy current without reducing the collector voltage. This can result in the destruction of the transistors. Thus, some short-circuit protection should be considered for any output configuration involving circuits without collector resistors.

The destructive condition is usually known as *secondary breakdown* or *second breakdown*, and results from a sudden channeling of collector current into a localized area of the transistor. Secondary breakdown is usually prevented by limiting the collector current–voltage product.

Another overload problem is presented when the load is a loudspeaker. High-fidelity loudspeaker systems can appear *capacitive* or *inductive*, as well as resistive. The current and voltage appearing in the amplifier will thus be out of phase when the load appears *reactive*. A 60° phase shift is not uncommon. At a 60° phase shift, *half* the source voltage and *all* the load current can appear *simulaneously* at the output transistor, or the full source voltage and one half the load current can appear, depending upon whether the load is capacitive or inductive.

This condition of a large, simultaneous voltage–current combination can result in secondary breakdown, just as if it were caused by a short circuit in the load.

As a matter of reference, the *minimum peak-power level* to which the short-circuit dissipation can be limited is the product of peak current and voltage appearing at the output transistor under the worst-case allowable phase shift. Assuming a 60° phase shift, the short-circuit power dissipation is set by

$$P_{PD \text{ (short circuit)}} \frac{V_{CC} \times I_{\text{peak}}}{2}$$

where P_{PD} is the peak dissipation for each *output transistor* (also the total average power dissipation for the amplifier).

The average power dissipation of each transistor is

$$P_{AD \text{(short circuit)}} = \frac{(1/2)V_{CC} \times I_{\text{peak}}}{2}$$

The worst-case average power dissipation in driver transistors (such as Q_3 and Q_4 of Fig. 4-28) is the power dissipation expressed as P_{AD} divided by the current gain of the output transistor.

With a typical short-circuit protection network, the safe operating requirements (based on an overload of 1-s minimum) for the output transistors occurs at V_{CC} and is the same as the peak dissipation expressed as P_{PD}.

The maximum thermal resistance (or the minimum power dissipation rating) required for each output transistor is found by

$$\theta_{JC(max)} = \frac{T_{J(max)} - T_A - \theta_{CA} \times P_{AD}}{P_{AD}}$$

where $T_{J(max)}$ is the maximum junction temperature rating of the transistor, T_A is the maximum ambient temperature, and θ_{CA} is the thermal resistance of the heat sink including any insulating washer.

The minimum power dissipation rating of the transistor is found by

$$P_{DM} = \frac{T_{J(max)}}{\theta_{JC(max)}}$$

Refer to Sec. 2-3.1.3 of for a further discussion of thermal resistance, heat sinks, and temperature design problems.

4-9.1. Overload-Protection Circuit

A good heat sink will protect a transistor from most temporary overloads, and some prolonged overloads, due to short circuits, simultaneous high voltage and current, and so on. However, if there is a possible danger of prolonged overload (typically longer than a few seconds), a well-designed amplifier should include some form of overload-protection circuit.

Figure 4-33 illustrates such a circuit. Before going into the short-circuit protection portion of the circuit, let us discuss overall operation of the amplifier.

Transistors Q_1, Q_2, Q_4, and Q_6 through Q_{10}, along with their associated components, comprise the standard full-complementary circuit. Transistors Q_1 and Q_2 are used in a differential amplifier configuration, which, when used with the dual power supply, provide a convenient means for setting the dc voltage level to be direct coupled to the loadspeaker. (Differential amplifiers are discussed in Chapter 5.)

Resistor R_F provides 100 per cent dc feedback from the output line to the input (at the base of Q_2), resulting in excellent dc stability. (This is a form of *overall feedback* or *loop feedback* since it involves more than one stage.) The resistance ratio of R_F to R_1 determines the *closed-loop* ac voltage gain of the amplifier. (The term closed-loop gain refers to the gain with feedback. Open-loop gain is gain without feedback.)

Transistor Q_4 functions as a high-gain common-emitter driver. Since the output transistors (Q_7, Q_8, Q_9, Q_{10}) function as emitter followers (no voltage gain), Q_4 must be capable of handling the full-load voltage swing.

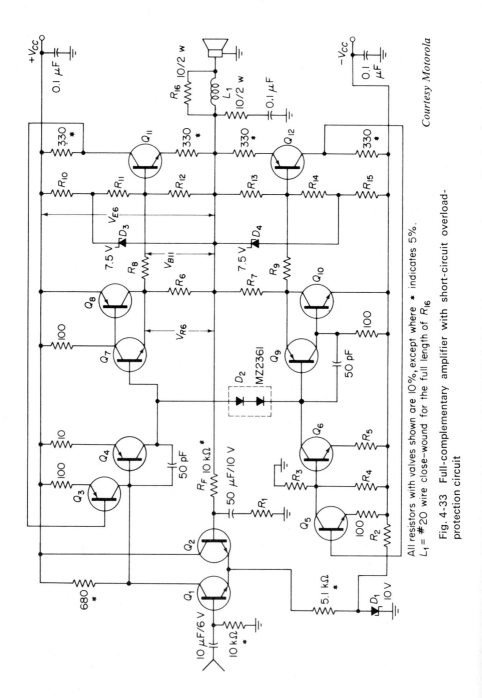

All resistors with valves shown are 10%, except where * indicates 5%.

L_1 = #20 wire close-wound for the full length of R_{16}

Courtesy Motorola

Fig. 4-33 Full-complementary amplifier with short-circuit overload-protection circuit

Transistor Q_6 serves as a constant-current source for the dc bias current, which flows through Q_4 and the bias diode D_2. (Q_6 eliminates the need for the large bootstrap capacitor found in some complementary audio-amplifier designs.)

Transistors Q_7 and Q_8 form a compound pair, which functions as an emitter follower with high current gain and unity voltage gain, for the positive portion of the output signal. Transistors Q_9 and Q_{10} function similarly for the negative porrion of the output signal.

Zener diode D_1 is used to set the dc voltage through the differential amplifier and provide ac hum rejection from the negative power supply.

Transistors Q_3, Q_5, Q_{11}, Q_{12}, and semiconductors D_3 and D_4, along with their associated resistors, *comprise the short-circuit protection network.*

Resistors R_8, R_{10}, R_{11}, and R_{12} form a voltage-summing network. The voltage appearing at the base of transistor Q_{11} is determined by the collector current of Q_8 flowing through R_6, and the voltage appears from the source $(+V_{CC})$ to the output. This summing network, since it detects both the voltage and current of Q_8, effectively senses the peak-power dissipation occurring in Q_8.

At a predetermined power level in Q_8, the summing network can be chosen so that transistor Q_{11} conducts sufficiently to turn on transistor Q_3. Transistor Q_3 then steals the drive current from the base of Q_4, and thus limits power dissipated in Q_8. Diode D_3 is used to prevent Q_{11} from turning on, under normal load conditions, when the output signal swings negative. Resistors R_9, R_{13}, R_{14}, and R_{15}, along with transistors Q_{12}, Q_5 and diode D_4, similarly limit the power dissipation occuring in the output transistor Q_{10}.

4-10. DIRECT-COUPLED INTEGRATED-CIRCUIT AMPLIFIERS

Most ICs (integrated circuits) use some form of direct coupling. It is difficult to fabricate a capacitor on an IC chip, and it is next to impossible to form a coil or transformer within any IC. These problems are eliminated in ICs by direct coupling between stages.

Direct-coupled IC amplifiers operate at all frequency ranges, from direct current to radio frequency. Of course, coils, capacitors, and transformers are external components connected to the IC. The transistors and resistors associated with the amplifier circuit are part of the IC.

The following paragraphs include descriptions of commercial IC amplifiers using some form of direct coupling. Additional IC amplifiers (those using the differential-amplifier principle) are discussed in Chapter 5. It should be noted that the majority of IC amplifiers use the differential-amplifier system.

4-10.1. Motorola MC1550 Amplifier

Figure 4-34 shows the full schematic and simplified schematic of the amplifier. As shown, the input is applied to the base of Q_1, and the output is taken from the collector of Q_3. The combination of Q_1 and Q_3 acts as a common-emitter common-base pair (Q_1 is common emitter, Q_3 is common base). This combination reduces internal feedback. Automatic gain control is introduced at Q_2, with a reference voltage V_R at Q_3. The amount of AGC is determined by the difference between V_{AGC} and V_R. When V_{AGC} is at least 114 mV greater than V_R, Q_3 is turned off and ac gain is at a minimum. If V_{AGC} is less than V_R by 114 mV or more, Q_3 is turned on and ac gain is maximum. Changing the AGC voltage has little effect on Q_1 operating point, and the input impedance of Q_1 remains constant.

Voltage V_S and resistance R_S establish the current in diode D_1, which is on the same IC chip as Q_1. The emitter current of Q_1 remains within 5 per cent of the diode current. This biasing technique takes advantage of the matching characteristics that are available with ICs. Resistors R_1 and R_2 bias the diode and D_1 also establish a base voltage for transistor Q_3. Resistors R_3 and R_4 serve to widen the AGC voltage from 114 mV to about 0.86 V. This is necessary in some applications so that the AGC line will be less susceptible to external noise.

Figure 4-35 shows the MC1550 used as a tuned, narrowband amplifier. Note that with an AGC of 0 V and a 30-dB power gain, the bandwidth is about 0.5 MHz, centered on 60 MHz.

Figure 4-36 shows the MC1550 used as a tuned wideband amplifier. Both the external circuit values and the performance characteristics are given. Note that with an AGC of 0 V and a 30-dB power gain, the bandwidth is about 15 MHz, centered on 45 MHz.

Figure 4-37 shows the MC1550 used as a video amplifier. Note that the voltage gain remains essentially flat to about 3 MHz, and remains above the 3-dB down point to about 20 MHz.

As shown in Figs. 4-35 through 4-37, the characteristics of the basic IC can be altered considerably by external components and connections. Of course, the IC is not limited to the applications shown. These configurations are presented to show what is available in typical IC amplifiers.

4-10.2. RCA CA3023 Amplifier

Figure 4-38 shows the schematic of the RCA CA3023 amplifier. Gain is obtained by use of transistors Q_1, Q_3, Q_4, and Q_6, which are amplifiers having a voltage gain of about 60 dB. The common-collector configuration provides the necessary impedance transformation (high-impedance

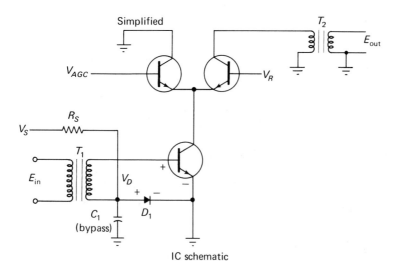

IC schematic

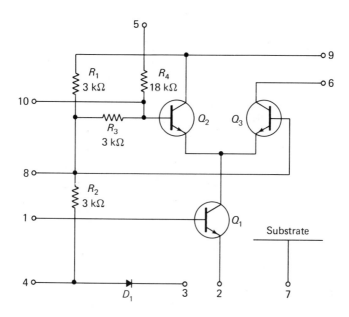

Fig. 4-34   Motorola MC1550 RF–IF amplifier

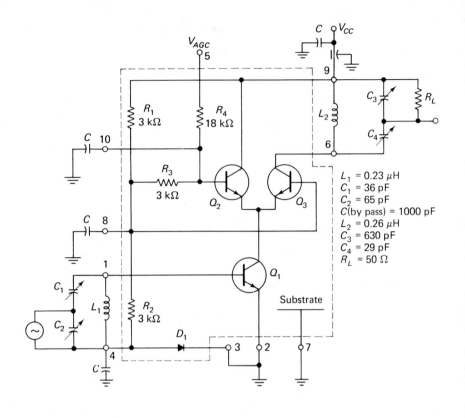

$L_1 = 0.23\ \mu H$
$C_1 = 36\ pF$
$C_2 = 65\ pF$
C(by pass) = 1000 pF
$L_2 = 0.26\ \mu H$
$C_3 = 630\ pF$
$C_4 = 29\ pF$
$R_L = 50\ \Omega$

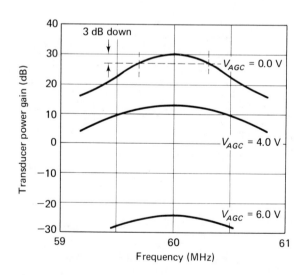

Courtesy Motorola

Fig. 4-35 Sixty-MHz tuned amplifier

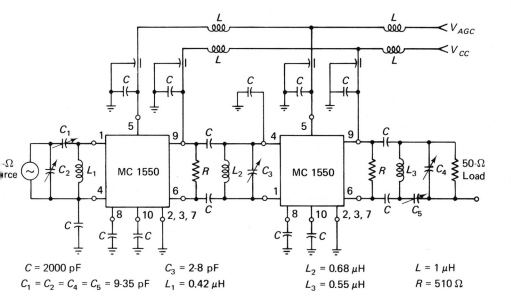

| $C = 2000$ pF | $C_3 = 2\text{-}8$ pF | $L_2 = 0.68\,\mu H$ | $L = 1\,\mu H$ |
| $C_1 = C_2 = C_4 = C_5 = 9\text{-}35$ pF | $L_1 = 0.42\,\mu H$ | $L_3 = 0.55\,\mu H$ | $R = 510\,\Omega$ |

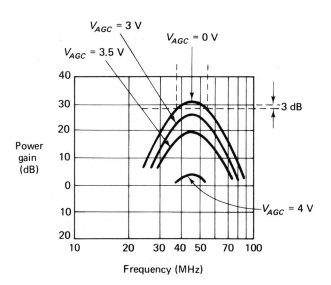

Courtesy Motorola

Fig. 4-36 Wideband tuned amplifier

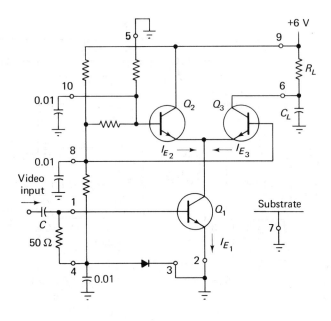

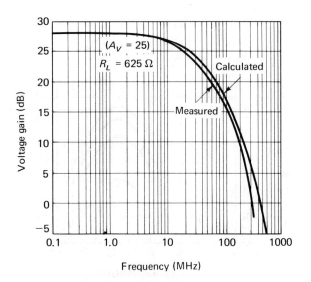

Courtesy Motorola

Fig. 4-37 Video amplifier

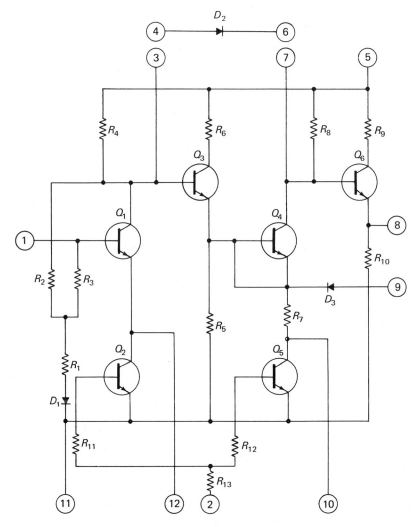

Fig. 4-38   RCA CA3023 amplifier

input and low-impedance output) for wide bandwidth. The output transistor Q_6 provides the low output impedance. The circuit must be capacitively coupled, and should have a low-impedance source for best operation.

Figure 4-39 shows typical connections for wideband and bandpass applications, with and without AGC, and for limiter applications.

An external feedback resistor R_F, or a tuned circuit, can be added between terminals 3 and 7 for desired bandwidth and gain performance. Linear operat-

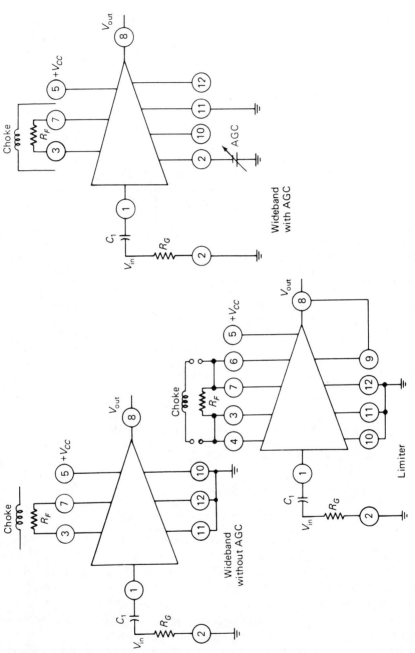

Fig. 4-39 Wideband and limiter amplifiers

Courtesy RCA

ing conditions are maintained by the bias applied between the collector and base of Q_1, supplied from the resistor–diode network R_1, R_2, R_3, and D_1. Because the collector of Q_1 is held at a fixed potential that is relatively independent of supply, IC characteristics, and temperature, dc coupling to the remainder of the circuit can be used.

For applications in which gain control is desired, terminals 10 and 12 are left floating and AGC is applied to terminal 2. For maximum gain, terminal 2 is operated at a positive voltage not larger than the supply voltage applied to terminal 5. In the positive voltage condition, transistors Q_2 and Q_5 are saturated and the impedance in the emitters of Q_1 and Q_4 is low. When the gain-control voltage becomes negative, Q_2 and Q_5 come out of saturation and provide high emitter feedback to reduce the gain.

In limiting applications, diodes D_2 and D_3 are connected in the feedback loops. The diodes provide clamping for sufficient input-signal swing. Limiting can be achieved with input signal swings up to 2.5 V.

Figure 4-40 shows two CA3023 amplifiers used as a 28-MHz limiter amplifier. Terminals 3 and 7 are connected to terminals 4 and 6, respectively. Terminal 8 is connected to terminal 9 to provide limiting action. A self-resonant coil in parallel with a 2-kΩ resistor is inserted in the feedback loop of each amplifier to provide gain and stability. The bandwidth of the system, before limiting, is 3.8 MHz, and the effective Q is 7.35. The total gain is 61 dB (30.5 dB per stage), and the power dissipation is 66 mW. Full limiting occurs at an input of about 300 μV.

4-10.3. RCA CA3018 Transistor Array

An IC transistor array provides several transistors on a single semiconductor chip. Transistor arrays are particularly suitable for amplifier applications in which closely matched device characteristics are required, or in which a number of active devices must be interconnected with external parts not practical to fabricate in IC form (tuned circuits, large-value resistors, variation resistors, large-value capacitors, etc.).

Figure 4-41 shows the schematic of the CA3018. The IC provides four silicon transistors on a single chip. The four active devices include two isolated transistors plus two transistors with an emitter–base common connection. Because it is necessary to provide a terminal for connection to the substrate, the two transistor terminals must be connected to a common lead. The particular configuration is useful in emitter-follower and Darlington circuit connections. Also, the four transistors can be used almost independently if terminal 2 is grounded, or ac grounded, so that Q_3 can be used as a common-emitter amplifier and Q_4 as a common-base amplifier.

In pulse video amplifiers and line drivers, Q_4 can be used as a forward-biased diode in series with the emitter of Q_3. Likewise, transistor Q_3 can be

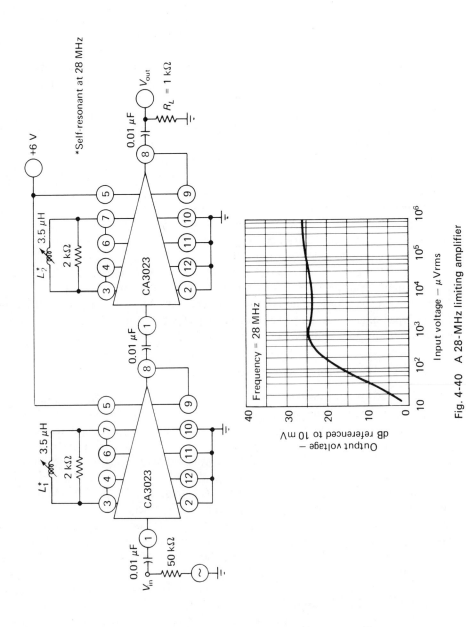

Fig. 4-40 A 28-MHz limiting amplifier

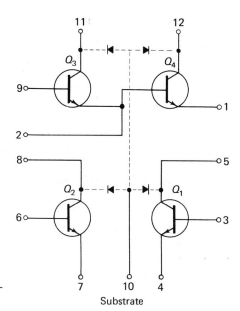

Fig. 4-41 RCA CA3018 transistor array

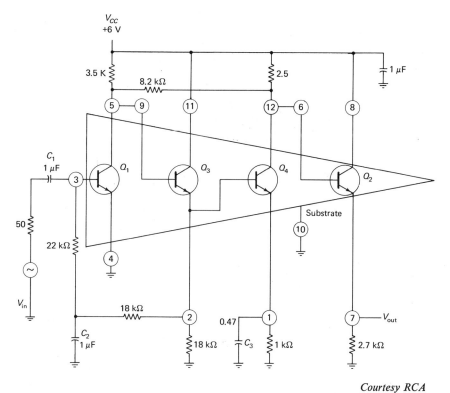

Courtesy RCA

Fig. 4-42   Broadband video amplifier using IC transistor array

267

used as a diode collected to the base of Q_4, or, in a reverse-biased connection, Q_3 can serve as a protective diode in RF circuits connected to operational antennas. The presence of Q_3 does not inhibit the use of Q_4 in a large number of circuits.

In transistors Q_1, Q_2, and Q_4, the emitter load is interposed between the base and collector leads to minimize package and lead capacitances. In Q_3 the substrate lead serves as the shield between base and collector. This lead arrangement reduces feedback capacitance in common-emitter amplifiers, and thus extends video bandwidth and increases tuned-circuit amplifier gain stability.

Figure 4-42 shows a broadband video amplifier design using the four-transistor array. This amplifier can be considered as two direct-coupled stages, each consisting of a common-emitter, common-collector configuration. The common-collector transistor provides a low-impedance source to the input of the common-emitter transistor, and a high-impedance low-capacitance load at the common-emitter output.

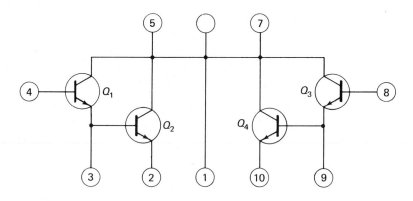

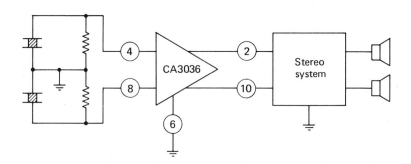

Fig. 4-43   RCA CA3036 Darlington array used as stereo preamp

Two feedback loops provide dc stability of the broadband video amplifier and exchange gain for bandwidth. The feedback loop from the emitter of Q_3 to the base of Q_1 provides dc and low-frequency feedback. The loop from the collector of Q_4 to the collector of Q_1 provides both dc feedback and ac feedback at all frequencies.

4-10.4. RCA CA3036 Dual Darlington Array

As in the case of transistor arrays, IC Darlington arrays provide several Darlington pairs on a single chip. Figure 4-43 shows both the schematic and typical application for the CA3036. The IC can be used to provide two independent low-noise wideband amplifier channels, and is particularly useful for preamplifier and low-level amplifications in single-channel or stereo systems.

As shown in Fig. 4-43, the array consists of four transistors connected to form two independent Darlington pairs. The block diagram illustrates the use of the array in a typical stereo phonograph. The IC can be mounted directly on a stereo cartridge. Because of the low noise, high input impedance, and low output impedance of the array, only minimum shielding is required from the pickup to the amplifier. The buffering action of the IC also substantially reduces losses and decreases hum pickup.

5. DIFFERENTIAL AMPLIFIERS

The differential amplifier is similar to the emitter-coupled amplifier (Fig. 4-20), except that the two output signals are the result of a *signal difference* between the two inputs. In a theoretical differential amplifier, no output is produced when the signals at the inputs are identical. That is, an output is produced *only when there is a difference* in signals at the input.

One of the main uses for differential amplifiers is as the input stage for an *operational amplifier*, as is described in Chapter 6. Another use for a differential amplifier in laboratory work is that of an amplifier for meters, oscilloscopes, recorders, and the like. Such instruments are operated in areas where many signals may be radiated (power-line radiation, stray signals from generators, etc.). Test leads connected to the input terminals will pick up these radiated signals, even when the leads are shielded. If a single-ended input is used, the undesired signals will be picked up and amplified along with the desired signal input. If the amplifier has a differential input, both leads will pick up the same radiated signal at the same time. Since there is no difference between the radiated signals at the two inputs, there is no amplification of the undesired inputs.

Signals common to both inputs (such as radiated signals) are known as *common-mode signals*. The ability of a differential amplifier to prevent conversion of a common-mode signal into a difference signal (which produces an output) is expressed by its *common-mode rejection ratio* (CMR or CMRR).

5-1. BASIC DIFFERENTIAL-AMPLIFIER THEORY

Figure 5-1 is the schematic of a basic differential amplifier. The circuit responds differently to common-mode signals than it does to single-ended signals.

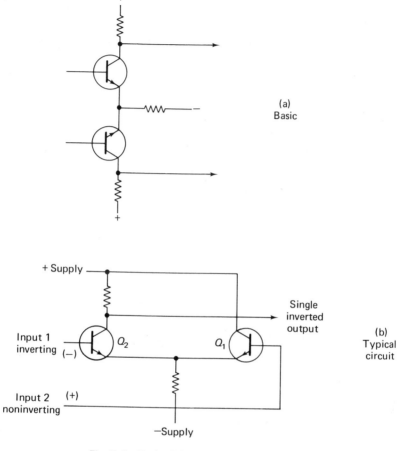

Fig. 5-1   Basic differential-amplifier circuits

A common-mode signal (like power-line pickup) drives both bases in phase with equal-amplitude ac voltages, and the circuit behaves as though the transistors are in parallel to cancel the output. In effect, one transistor cancels the other.

Normal signals are applied to either of the bases (Q_1 or Q_2). The *inverting input* is applied to the base of transistor Q_2, and the *noninverting* input is applied to the base of Q_1. With a signal applied only to the inverting input and the noninverting input grounded, the output is an amplified and inverted version of the input. For example, if the input is a positive pulse, the output is a negative pulse. If the noninverting input is used with the inverting input grounded, the output is an amplified version of the input (without inversion).

The emitter resistor introduces emitter feedback to both transistors simultaneously. This reduces the common-mode signal gain without reducing the differential signal gain in the same proportion.

Figure 5-2 is the schematic of a more practical differential amplifier. The amplifier of Fig. 5-2 is packaged as an integrated circuit (the RCA CA3000). The circuit is basically a single-stage differential amplifier (Q_2 and Q_4) with input emitter followers (Q_1 and Q_5) and *constant-current source* Q_3 in the emitter-coupled leg. Note that the single emitter resistor of the Fig. 5-1 circuit is replaced by Q_3 and its associated components in the circuit of Fig. 5-2.

The use of a transistor such as Q_3 is typical for many differential amplifiers, particularly those found in operational amplifiers (as the first stage). The circuit of transistor Q_3 is known as a *temperature-compensated constant-current source*. All current for the differential amplifier is fed through Q_3

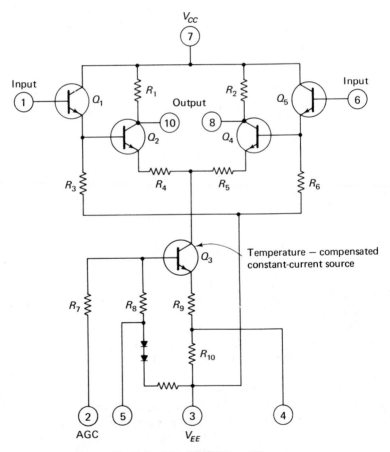

Fig. 5-2   RCA CA3000 amplifier

(an *NPN*) connected between the emitters of the differential amplifier and V_{EE} (the negative power supply). If there is an increase in current, a larger voltage is developed across the current-source Q_3 emitter resistor. This larger voltage acts to reverse bias the base emitter, thus reducing current through Q_3. Since all current for the differential amplifier is passed through Q_3, current to the amplifier is also reduced. If there is a decrease in current, the opposite occurs, and the amplifier current increases. Thus, the differential amplifier is maintained at a constant current level.

Transistor Q_3 is also *temperature compensated* by the diodes connected in the base–emitter bias network. These diodes have the same (approximate) temperature characteristics as the base–emitter junction, and offset any change in Q_3 base–emitter current flow that results from temperature change.

The circuit of Fig. 5-2 has an input impedance of approximately 100 kΩ (for each input), and a gain of approximately 30 dB at frequencies up to about 1 MHz. The useful frequency response can be increased by means of external resistors and coils. The use of degenerative feedback resistors R_4 and R_5 in the emitter-coupled pair of transistors increases the linearity of the circuit. The low-frequency output impedance between each output (terminals 8 and 10) and ground is essentially the value of the collector resistors R_1 and R_2 in the differential stage.

5-2. FET VERSUS TWO-JUNCTION TRANSISTORS

Figure 5-3 shows the basic circuits involved for both two-junction and FET differential amplifiers. With either circuit, *floating signals* not referenced to ground can be amplified and, since large values of common-mode rejection can be achieved, small differential signals can be discriminated from large common-mode signals upon which the differential signals may be riding.

Also, because of the matching and temperature tracking that can be achieved between the two transistors in the differential amplifier (particularly in ICs and in transistors on the same chip), the circuit can have very good dc stability over a very wide temperature range.

In certain applications, the FET has advantages over the two-junction transistor. The main advantage of the FET is the high input impedance. As a comparison, whereas differential-amplifier input impedances of 100 kΩ to 1 MΩ can be obtained with two-junction transistors, values ranging upwards of 10^9 are easily within reach of FET amplifiers.

5-2.1. Matching Requirements

In designing a differential amplifier, either two junction or FET, some degree of matching is required between certain characteristics of the

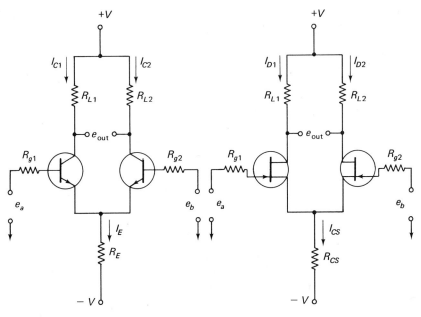

(a) Two-Junction Transistor   (b) Field-Effect Transistor

Courtesy Motorola

Fig. 5-3   Comparison of FET and two-junction differential-amplifier circuits

two transistors. In two-junction amplifiers, the dc current gain (h_{FE}) and the base–emitter voltage (V_{BE}) of the two transistors should be matched. The collector–base leakage currents (I_{CBO}) should also be nearly equal, although this parameter is generally not too critical, especially in silicon transistors where I_{CBO} is very small.

The FET requires a match of the transconductance (y_{fs}) and the gate–source voltage (V_{GS}) of the two devices. In addition, if the impedance of the driving source is high (1 MΩ or higher), I_{GSS}, the gate–source leakage current, should be matched.

5-2.2. Temperature Coefficients

The base–emitter voltage of a two-junction transistor has a *negative temperature coefficient*, the magnitude of which is some function of the emitter current (typically around 2 to 2.5 mV/°C in most applications). If two matched chips are selected from the same silicon wafer, the two base–emitter junctions will track over a fairly wide current range.

The situation is quite different for FET devices. Figure 5-4 shows a typical plot of I_D (drain current) versus V_{GS} for a FET at three different temperatures. For low values of I_D, V_{GS} has a positive temperature coefficient (TC), and the higher values of I_D yield a negative TC. At point 0, the temperature coefficient is zero. Thus, by selecting the proper drain current, a single FET can be made to have zero temperature coefficient ($0TC$). Refer to Sec. 1-11.4.

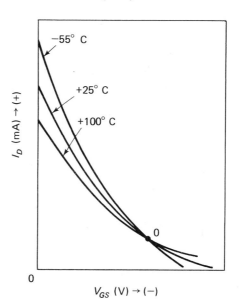

Fig. 5-4 Drain current versus gate–source voltage

Unfortunately, all the temperature curves of a single FET do not always intersect at one point. The problem becomes more pronounced when two FETs are involved. In practical terms, *perfect* compensation is not possible, although changes in V_{GS} of a few millivolts can be achieved. If two well-matched FETs are used in a differential-amplifier configuration, as shown in Fig. 5-3b, these minor changes can be balanced out quite effectively.

5-2.3. Finding the Zero Drift Point for Differential-Amplifier FETs

Theoretically, if the V_{GS} versus I_D characteristics of the two devices in a FET differential amplifier are well matched, the differential amplifier can be operated at any current level (within the limits of the V_{GS}–I_D curve) with little or no drift. Unfortunately, the degree of matching required is generally not practical from an economical point of view.

In order to minimize drift in a FET differential amplifier, it becomes very desirable to operate the FETs at or near their zero drift point. Matching can be fairly easily obtained at this one point, and because the FETs are operating with essentially $0TC$, the effects of any mismatch are greatly minimized.

At the zero drift point, V_{GS} and I_D are given by

$$V_{GS(Z)} = V_P - 0.63 \tag{5-1}$$

$$I_{D(Z)} = I_{DSS} \left(\frac{0.63}{V_p}\right)^2 \tag{5-2}$$

where V_p is the pinch-off voltage, and I_{DSS} is the drain current at zero gate–source voltage ($V_{GS} = 0$). The pinch-off voltage is defined as that point where a further increase in V_{DB} causes little change in I_D. (See Fig. 5-5, and refer to Chapter 1.)

5-2.4. Varying I_{CS} to Achieve Minimum Drift

Equations (5-1) and (5-2) are based on some approximations, and as such are subject to some degree of error. The underlying assumption is that the voltage drops across the input resistances R_g due to I_{GSS} either completely cancel or are so small as to be negligible. Usually, the latter is the case. However, even when the possible errors are considered, the equations are useful in arriving at values of V_{GS} and I_D that are reasonably close to the zero drift point.

In practical design, additional testing (over temperature) then becomes necessary to establish the minimum drift point. By varying the current I_{CS}

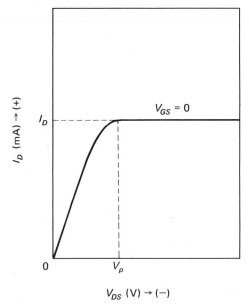

Fig. 5-5   Drain current versus drain–source voltage

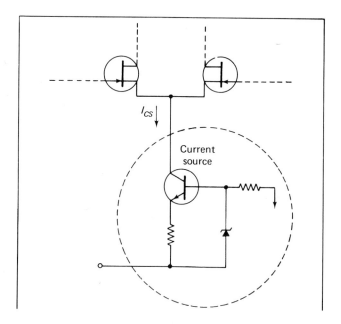

Fig. 5-6   Two-junction transistor current source for FET differential amplifier

(the sum of the two drain currents), value of drift on the order of a few hundred microvolts per degree Celsius (referred to input) can be obtained.

In general, the technique of varying I_{cs} for drift compensation is not satisfactory. The compensation is usually not as good as is desired. More important, the resistor R_{cs} of Fig. 5-3b is replaced by a current source for most applications. (Refer to Sec. 5-1.) A typical circuit is shown in Fig. 5-6. The zener diode voltage is chosen such that it compensates for variations in the transistor base–emitter junction over temperature. As discussed in Sec. 5-1, the current source provides higher common-mode rejection and greatly reduces the effects of power-supply variations on drift. It is very difficult to adjust the value of I_{cs} if a current source is used. A better way of compensation is available.

5-2.5. Drift and Offset Compensation Circuit

Figure 5-7 shows an expanded view of the region around the $0TC$ point of the typical FET differential amplifier. For clarity, only two temperatures are shown, and the differences between the two devices have been greatly exaggerated. The $0TC$ points are at A and B. Assuming negligible drop across the input resistances R_{g1} and R_{g2}, the gate-to-source voltage of the two FETs is equal.

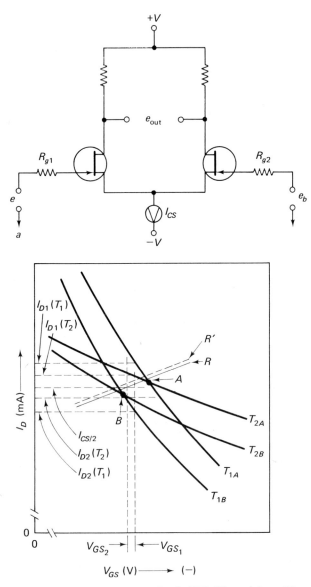

Fig. 5-7 Temperature compensation in FET differential amplifier

At temperature T_1, the value of $V_{GS}(V_{GS1})$ is established by the circuit such that the sum of the two drain currents I_{D1} and I_{D2} is equal to I_{CS}. At temperature T_2, V_{GS} shifts to a new value (V_{GS2}) so that, again, I_{D1} plus I_{D2} equals I_{CS}. This change in I_D results in a drift at the differential-amplifier output, since I_{D1} has decreased and I_{D2} has increased.

The *condition for zero drift* at the differential amplifier output is

$$\Delta I_{D1} R_L = \Delta I_{D2} R_L \tag{5-3}$$

since

$$I_{D1} + I_{D2} = I_{CS} \tag{5-4}$$

and

$$\Delta I_{CS} = 0 \quad \text{(constant current)} \tag{5-5}$$

Then, for zero drift

$$\Delta I_{D1} = \Delta I_{D2} = 0 \tag{5-6}$$

In order to compensate the circuit for drift, operating points must be found such that the drift in V_{GS} for one FET compensates the drift in V_{GS} of the other FET at constant values of I_D. The solid line R of Fig. 5-7 defines such a condition.

As the temperature changes from T_1 to T_2, the operating point moves from the solid line R to the dashed line R'. The change in V_{GS} is equal for both FETs, and there is no change in I_D. This results in a condition of no drift.

The stable operating points are accomplished by adding a resistor in the source of the proper FET. The added resistor goes in the source of the FET in which the current *must be decreased*. In the case of the Fig. 5-7 circuit, the resistor goes in the left side, corresponding to I_{D1}. The value of R can be calculated from

$$R = \frac{V_{GS1} - V_{GS2}}{I_D} \tag{5-7}$$

where V_{GS1} and V_{GS2} are the value of gate-to-source voltage at the desired operating points, and I_D is the current through the FET that has the added R.

5-2.6. A Practical Drift–Offset Compensation Circuit

In most practical applicctions there is already some resistance in the source, as shown in Fig. 5-8. The source resistances R_S are included to stabilize the amplifier gain. To produce drift–offset compensation, one of the source resistors is shunted. The shunted resistor is the one in the *side opposite* that in which a compensating resistor is added. For example, in the Fig. 5-7 circuit the source resistor is to be added at the left side. In the Fig. 5-8 circuit the shunt resistor is added across R_{S2} (the right side). The *difference* between the shunted resistor (R_{S2} in this case) and the unshunted resistor (R_{S1}) is the value of the compensating resistor that would be added [using equation (5-7)].

Although the value of compensating resistance required can be approximated using equation (5-7) and the curves of Fig. 5-7, the best results are

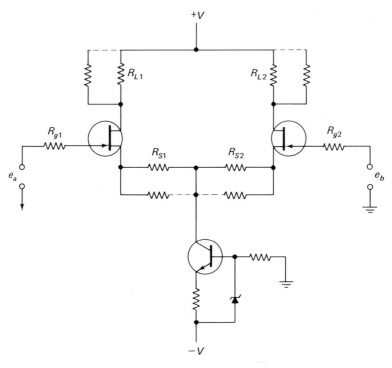

Courtesy Motorola

Fig. 5-8   Drift and offset compensation

obtained by a trial-and-error selection. Drifts of a few microvolts per degree Celsius can be obtained over a wide (-55 to $+100°$C) temperature range. If the temperature range is narrowed (0 to $+50°$C), almost perfect compensation is possible with trial and error selection.

5-2.7. *Voltage Gain of Compensated Differential Amplifier*

Voltage gain of the FET differential amplifier is the same as that for a single common-source stage, if both sides are well matched. The gain is expressed by

$$A_{dd} = \frac{y_{fs}R_L}{1 + y_{os}R_L} \tag{5-8}$$

where A_{dd} is the differential-input to differential-output gain, y_{fs} is the device transadmittance, y_{os} is the device output admittance, and R_L is the load resistor (equal for each side) of the circuit.

In most applications, 1 is much greater than $y_{os}R_L$, so the gain becomes

$$A_{dd} \approx y_{fs}R_L \qquad (5\text{-}9)$$

The *common-mode gain* is given by

$$A_{cc} = \frac{y_{fs}R_L}{1 + y_{os}R_L + 2R_{CS}(y_{os} + y_{fs})} \qquad (5\text{-}10)$$

where A_{cc} is the common-mode-input to differential-mode-output gain, and R_{CS} is the value of the resistance in the common-source circuit of the two FETS. This indicates the desirability of using a constant current source in order to minimize the common-mode gain.

Since, in most cases, y_{fs} is much greater than $y_{os}R_L$, equation (5-10) can be reduced to

$$A_{cc} \approx \frac{y_{fs}R_L}{1 + 2R_{CS}y_{fs}} \qquad (5\text{-}11)$$

Furthermore, since $2R_{CS}y_{fs}$ will generally be much larger than 1, the common-mode gain becomes

$$A_{cc} \approx \frac{R_L}{2R_{CS}} \qquad (5\text{-}12)$$

Again, it must be emphasized that the foregoing equations for calculating gain are based on the assumption of well-matched devices. The accuracy of the equations is dictated to a large extent by the degree of matching.

5-2.8. Eliminating Offset Voltage in the Differential Amplifier

After the drift compensation has been accomplished, there will arise the need for eliminating the *offset voltage* at the differential amplifier *output*. That is, if one source has a different resistance value than the other source, there is a fixed difference in dc voltage at the outputs (drains), as well as a difference in gains. This difference can be offset by shunting one of the load (drain) resistors (R_{L1} or R_{L2}), as shown in Fig. 5-8.

This shunt resistance causes a slight, but usually negligible, variation in gain between the two sides, and offsets the difference produced by the different source resistances. All other factors being equal, the gain of each side is set by the R_L/R_S ratio. If R_{S2} is shunted to provide drift compensation, the value of R_{S2} is reduced and the gain is increased, so that the right side of the differential amplifier produces more gain than the left side. This can be offset by shunting R_{L2} to restore the R_L/R_S ratio, and make the gain for both sides equal.

5-3. COMMON-MODE DEFINITIONS

The terms *common mode* and *common-mode rejection* are used frequently in differential-amplifier applications. All manufacturers do not agree on the exact definition of common-mode rejection.

One manufacturer defines common-mode rejection (CMR, or sometimes listed as CM_{rej}), or the common-mode rejection ratio (CMRR), as the ratio of differential gain (usually large) to common-mode gain (usually a fraction). That is, the amplifier may have a large gain of differential signals (different signals at each input terminal, or one input terminal grounded and the opposite input terminal with a signal), but little gain (or possibly a loss) of common-mode signals (same signal at both terminals).

Another manufacturer defines CMR as the relationship of *change* in output voltage to the *change* in the input common-mode voltage producing it, divided by the open-loop gain (amplifier gain without feedback).

For example, using the latter definition, assume that the common-mode input (applied to both terminals simultaneously) is 1 V, the resultant measured output is 1 mV, and the open-loop gain is 100. The CMR is then

$$\frac{(\text{output/input})}{\text{open-loop gain}} = \text{CMR}$$

$$\frac{(0.001/1)}{100} = 100,000 = 100 \text{ dB}$$

Another method by which to calculate CMR is to divide the output signal by the open-loop gain to find an *equivalent differential input signal.* Then the common-mode input signal is divided by this equivalent differential input signal. Using the same figures as in the previous CMR calculation,

$$\frac{0.001}{100} = 0.00001, \quad \text{equivalent differential input signal}$$

$$\frac{1}{0.000001} = 100,000 \quad \text{or} \quad 100 \text{ dB}$$

No matter what basis is used for calculation, CMR is an indication of the *degree of circuit balance* of the differential stages, since a common-mode input signal should be amplified identically in both halves of the circuit. A large output for a given common-mode input is an indication of large unbalance or poor CMR. If there is an unbalance, a common-mode signal becomes a differential signal after it passes the first stage (or at the output in a single-stage differential amplifier).

As with amplifier gain, CMR usually decreases as frequency increases. However, as a rule of thumb, the CMR of any differential amplifier should be at least 20 dB *greater* then the open-loop gain at any frequency (within the limits of the amplifier).

5-4. FLOATING INPUTS AND GROUND CURRENTS

Since a differential amplifier is sensitive only to the difference between two input signals (in theory), the signal source need not be grounded and can be *floating*. Therefore, differential amplifiers are often used in test-equipment applications where the signal source is from a bridge (such as a bridge-type transducer) and the power supply is grounded.

The differential amplifier also allows injection of a fixed dc voltage into either channel to permit establishment of a new voltage reference level at the output (some point other than 0 V). This is commonly referred to as *zero suppression*, and is discussed further in Chapter 6.

A floating-input circuit can create problems. When the input is floating, cable shielding between the amplifier and signal source may be connected to chassis ground rather than to signal ground. However, both ac and dc voltages can exist between two widely separated earth grounds, causing current to flow. (Such currents are known as *ground currents*, and the circuits producing the current flow are known as *ground loops*) The condition is shown in Fig. 5-9, where a bridge-type transducer is used with a differential amplifier.

Note that the signal source is connected to the transducer earth ground (local ground or physical ground, as it may sometimes be called). This ground point is connected to the amplifier ground through the cable shielding. The amplifier ground is connected to one of the differential inputs through the

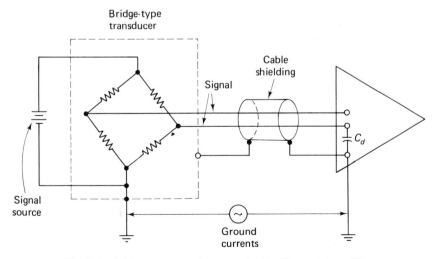

Fig. 5-9   Bridge-type transducer used with differential amplifier

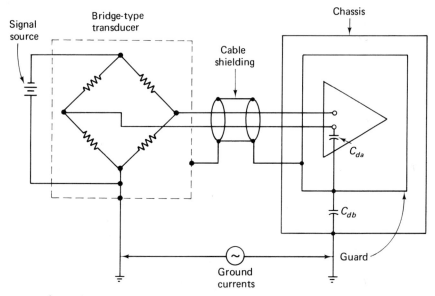

Fig. 5-10 Guard-shield technique used to reduce capacitance between signal leads and ground

internal capacitance (represented as C_d) of the amplifier, even though there may be no dc connection between ground and the input terminal of a floating-input amplifier. The same differential input terminal is connected to the signal source through the signal leads and the transducer elements (bridge resistors in this case). Thus, the ac ground currents are mixed with the signal currents. This can result in an unbalance of the differential amplifier. Also, radiated signals picked up by the shield appear as undesired differential signals, rather than common-mode signals, and produce an undesired output.

One method used to minimize this condition is shown in Fig. 5-10. Here, a guard shield is placed around the input circuits of the differential amplifier. This not only shields the differential amplifier from radiated signals, but also provides an electrostatic shield to break the internal capacitance C_d into two series capacitances C_{da} and C_{db}. A much higher impedance is then presented to the flow of ac ground signals. This type of amplifier is termed a *floated-input* and *guarded* amplifier.

5-5. DIFFERENTIAL AMPLIFIERS IN LABORATORY TEST EQUIPMENT

Because of their ability to reject common-mode signals, differential amplifiers are often used in laboratory test equipment. Figure 5-11 is the diagram of a differential amplifier used in a typical laboratory instru-

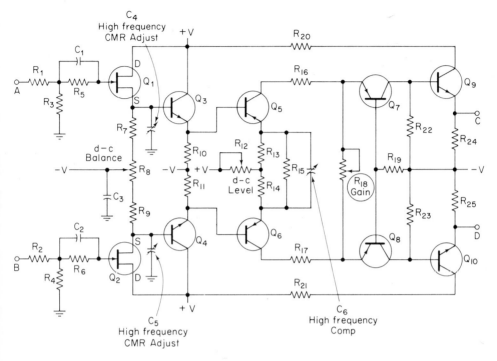

Fig. 5-11   Differential amplifier used in laboratory instrument

ment. Note that FETs are used at the input (for high input impedance), whereas the output uses the emitter–follower configuration (for low output impedance). Direct coupling is used throughout to ensure good low-frequency response.

In either design or troubleshooting of any differential amplifier, a consideration of major importance is that both halves of the amplifier be *electrically symmetrical*. That is, the input impedance at point A should be exactly equal to the input impedance at point B. R_1 and R_3 form an input attenuator with an impedance of $R_1 + R_3$. The voltage attenuation is equal to the ratio $R_3/R_1 + R_3$. (Similarly, R_2 and R_4 form a division ratio of $R_4/R_2 + R_4$.) The sum of $R_1 + R_3$ is typically on the order of 1 MΩ. Resistors R_5 and R_6 (typically in the range of 100 kΩ) limit the input current to protect Q_1 and Q_2. Capacitors C_1 and C_2 couple ac signals directly to the gates of Q_1 and Q_2, and thus give high-frequency peaking. (At high frequencies, the impedance offered by R_5 and R_6 could reduce the input signals.)

5-5.1. Adjustment Controls in Differential Amplifiers

Note that there are many adjustment controls in the circuit of Fig. 5-11. To properly analyze the circuit for troubleshooting or design, the effects of these controls must be known. For example, in troubleshooting it may be possible to eliminate an apparent major fault simply by adjustment of the controls. The following descriptions are applicable to the specific controls of Fig. 5-11. However, similar controls (possibly identified by different names) appear on most differential amplifiers in laboratory equipment.

Capacitors C_4 and C_5 (high-frequency CMR adjust) shunt the high-frequency signals to ground. Capacitor C_6 (high-frequency compensation) shunts high-frequency signals between the emitters of Q_5 and Q_6. These three controls form the high-frequency compensation adjustments for the amplifier. When properly adjusted, the high-frequency signals are attenuated so that the response is flat over the desired range.

The dc balance potentiometer R_8 compensates for any inherent difference between two halves of the input circuit. It is impossible to match the two halves perfectly. Also, component values will change with age and thus produce an unbalance. Generally, R_8 is a very temperature stable potentiometer, and is mechanically secure so as not to lose its setting with any vibration. (Sometimes the potentiometer shift is provided with a locknut to prevent a change in setting due to vibration.)

The function of R_{12} (dc level) and R_{18} (gain) are often confused. Potentiometer R_{12} sets the *level* of the dc voltage at the output, whereas R_{18} sets the *overall gain* of the amplifier. It is quite possible for the amplifier to operate at the correct output level, but not to provide the necessary gain, and vice versa.

In order to prevent temperature from causing the amplifier dc levels to drift, transistors Q_1 through Q_4 are mounted on the same temperature heat sink. Sometimes Q_1 through Q_4 are enclosed by a metal can, so they will all remain at the same temperature. In some cases, Q_5 and Q_6 may also share the same heat sink, but usually they are separate from Q_1 through Q_4. However, transistors Q_1 through Q_4 invariably are *matched pairs* and should be replaced (for troubleshooting) or selected (for design) as such.

Transistors Q_5 and Q_7 (as well as corresponding transistors Q_6 and Q_8) form a cascode amplifier (Sec. 4-7.2), with the result that the input to Q_5 is voltage, whereas the output of Q_5 into Q_7 is current. The output of Q_7 is voltage again, since the input impedance of Q_7 is quite low, whereas the Q_7 output impedance is quite high. Transistors Q_9 and Q_{10} are emitter followers used to reduce the high impedance from Q_7 to a low output impedance.

5-5.2. Failure Patterns for Differential Amplifiers

The four most common failures of a differential amplifier are loss of gain (voltage or current), poor common-mode rejection, dc unbalance, and output signal dc drift.

Loss of gain (voltage or current). If there is a lack of gain in a particular design, the cause is usually one of incorrect component values (low R_L/R_E

ratio, for example), or possibly a problem of low beta (or poor y_{fs} in FETs). Such conditions can only be corrected by trial-and-error substitution in experimental circuits.

In an existing differential amplifier, if a loss of gain has been gradual, component aging is usually the fault. Always try to correct a loss-of-gain problem, frequency-response problem, or an improper-output-level problem by adjustment. Gradual losses are not common in well-designed equipment. Also, usually *only-half* the circuit will lose gain, thus unbalancing the amplifier. Any differential amplifier that requires continued adjustment of the dc balance control is suspect. If there is a total breakdown or extreme unbalance, check the active elements in the usual manner. Refer to Chapter 7.

Poor common-mode rejection. Poor CMR is due most commonly to nonsymmetrical gain on both halves of the amplifier. This is one of the reasons for including the CMR adjustment capacitors (C_4 and C_5 of Fig. 5-11).

Always try to correct a CMR problem by adjustment first. Do not forget to include the *balance* adjustments, since any unbalance is reflected as a poor CMR ratio. Another point to remember is that *attenuator probes, attenuator networks*, and the like, must be perfectly matched with regard to both alternating and direct current to get a good CMR ratio. (Some laboratory instruments are provided with attenuator networks and probes, usually at the input circuit. Although usually not part of the amplifier, their operation has a direct effect on amplifier performance. Try to avoid using external attenuators in *calibrating* a differential amplifier for CMR. Unmatched external calibrators just add one more unknown.)

Direct-current unbalance. Problems of dc unbalance are almost always due to component aging or mismatched temperature coefficients of symmetrical components. That is, the components can age faster in one half of one stage, thus unbalancing the entire amplifier circuit. The same applies to an unbalance introduced by a mismatched temperature coefficient.

If many stages are used in the amplifier, start by balancing the *last stage first*. Make the inputs to the last stage equal (tie both halves together, ground both inputs, or apply the same signal to both inputs, depending on the specific circuit), and adjust the stage's dc balance control. Sometimes the last stage will not have a balance control, but can be checked for balance. For example, in Fig. 5-11 the voltage drop across R_{20} should be the same as across R_{21}, the drop across R_{24} should be the same as R_{25}, and so on.

Continue this checking process, moving toward the first stage and always monitoring the *main output*, until the trouble has been located. Usually, this sequence can isolate a badly unbalanced section.

If the circuit is part of an oscilloscope (differential amplifiers are often found in laboratory oscilloscopes), an unbalance is easily detected by grounding the two inputs and rotating the oscilloscope gain control (often called the *vernier* control). An unbalance will show up in the form of a trace shift. That is, the oscilloscope trace will shift vertically as the gain control is changed, even though both inputs are at zero. This is a result of dc signal being developed across the gain control due to unbalance. The amplifier can then be balanced by adjusting the amplifier dc balance potentiometer until no trace shift is observed when the gain control is rotated.

Output signal drift. A slow drift of any kind at the output is usually linked to a dc unbalance. Temperature compensation (or a lack thereof) plays a major role in such cases. A well-designed circuit is very seldom plagued by temperature problems. However, occasionally circuits may be overheated, and solid-state components can be partially damaged (changing their temperature coefficients, but not destroying their function).

When replacing parts (for troubleshooting) or selecting parts (for design) always make sure that both halves of the amplifier are well matched in *all characteristics* (gain, temperature coefficient, etc.), especially in the *very first stage*. Minor unbalances in the first stage of a differential amplifier are amplified by each stage thereafter, possibly resulting in a major unbalance at the output.

5-6. DIFFERENTIAL AMPLIFIERS AS TUNED AMPLIFIERS AND CASCADED AMPLIFIERS

The differential amplifier of Fig. 5-2 (Sec. 5-1) is found in IC form. There are a variety of uses for such IC differential amplifiers. The following paragraphs describe two typical applications.

5-6.1. Cascaded RC Coupled Feedback Amplifier

Figure 5-12 shows the IC differential amplifier used in a two-stage, feedback, cascade amplifier. (Note that two ICs are required.) The cascade amplifier produces a typical open-loop (without resistor R_F) midband gain of 63 dB. The circuit of Fig. 5-12 uses a 100-pF capacitor C_1 to shunt the differential outputs of the first stage. Capacitor C_1 staggers the high-frequency roll-offs of the amplifier, and thus improves stability (prevents oscillations).

The low-end rolloff of the amplifier is determined by the interstage coupling. Because AGC may be applied to the first stage (terminal 2 at the base of the current source transistor, Fig. 5-2), the amplifier of Fig. 5-12 can be used in high-gain video AGC applications under open-loop conditions. If resistor R_F is included and feedback is used to control the gain, AGC may still be applied successfully.

If three or more ICs are cascaded, the low-frequency rolloffs must be staggered (by using different values of interstage coupling capacitors). Likewise, the high-frequency rolloffs should be staggered (by using different values of

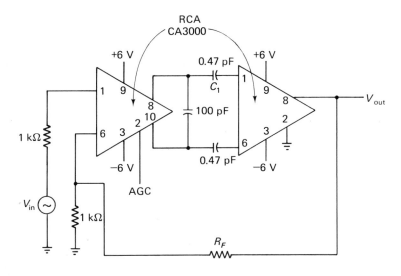

AGC = 0

R_F	Gain (dB)
∞	63 (Open loop, without RF)
100 kΩ	40
9 kΩ	20

Courtesy RCA

Fig. 5-12   Cascaded, *RC*-coupled, feedback amplifier

C_1). This staggering of rolloff frequencies will minimize the possibility of oscillation. Three cascaded stages produce about 94-dB gain.

5-6.2. Narrowband Tuned Amplifier

Because of its high input and output impedances, the IC is suitable for use in parallel tuned-input and tuned-output applications. There is a comparative freedom in selection of circuit Q because the differential amplifier exhibits inherently low feedback qualities, provided the following conditions are met: (1) the collector of the driven transistor is returned to ac ground, and the output is taken from the nondriven side, and (2) the input is adequately shielded from the output by a ground plane.

The tuned amplifier circuit is shown in Fig. 5-13. By comparison of this circuit against the IC schematic of Fig. 5-2, it will be seen that the driven transistor Q_2 has the collector (terminal 10) returned to ac ground through

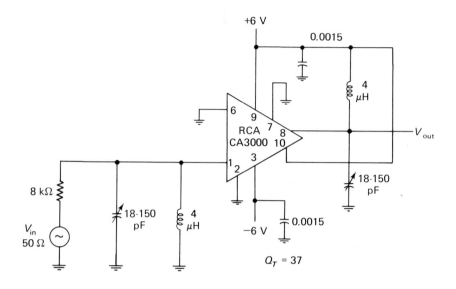

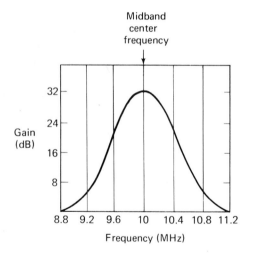

Courtesy RCA

Fig. 5-13   Narrowband tuned amplifier

the 0.0015-μF capacitor, and that the output is taken from the nondriven side (terminal 8, with input terminal 6 grounded).

The circuit of Fig. 5-13 is a narrowband tuned-input tuned-output configuration for operation at 10 MHz, with an input Q of 26 and an output Q of 25. A typical response curve is also shown. The 10-MHz voltage gain is about 30 dB, and the total effective Q is 37. The CA3000 can be used in tuned-amplifier applications at frequencies up to the 30-MHz range.

5-7. DIFFERENTIAL INPUTS AND OUTPUTS

Differential amplifiers can have differential inputs with differential outputs, differential inputs with single-ended output, or single-ended input with differential output, depending upon the particular application. The differential input is usually required where there are high common-mode signals, or where the source is floating (no reference to ground). The differential output is required when both amplitude and polarity must be considered.

A laboratory *differential voltmeter* is a classic example of this latter requirement. Such meters must amplify very small differential signals in the presence of large common-mode signals. The output of the meter amplifier is fed to the meter movement, where both the amplitude and polarity of the differential signal are indicated (as a plus or minus voltage readout), but common-mode signals are rejected.

A differential output amplifier can be made by using two amplifiers. However, to be really effective, both amplifiers must be *identical* in characteristics (gain, temperature drift, etc.). This is best accomplished using a dual-channel amplifier, in differential form, with both channels fabricated on the same semiconductor chip. Such an arrangement is shown in Fig. 5-14, which is the schematic of a Motorola MC1520 IC differential amplifier.

5-7.1. Circuit Description

Transistors Q_2 and Q_4 differentially amplify the input signal applied between pin 9 and pin 10. The emitter follower input buffering, provided by Q_1 at the noninverting input and Q_3 at the inverting input, is used to obtain the high input impedance, to minimize the input capacitance, and to reduce the dependence of offset and impedance variations upon device current gains.

Transistor Q_5 serves as the temperature-compensated current source, used primarily to provide a high common-mode rejection ratio. A common-mode feedback loop is incorporated around the input stage to provide Q-point operating stability for the input devices, and thus ensure a more predictable input common-mode voltage swing. The base of Q_7 and collector of Q_8 are wire bonded to pins 1 and 7 (as are the base of Q_9 and collector of Q_{10} to pins 2 and 6) to provide for open-loop frequency compensation capacitors.

A resistive level shifter follows the input stage to maintain the desired voltage levels at the overall device output. (Level shifters are discussed further in Sec. 5-8.) The second differential gain stage also contains emitter-follower

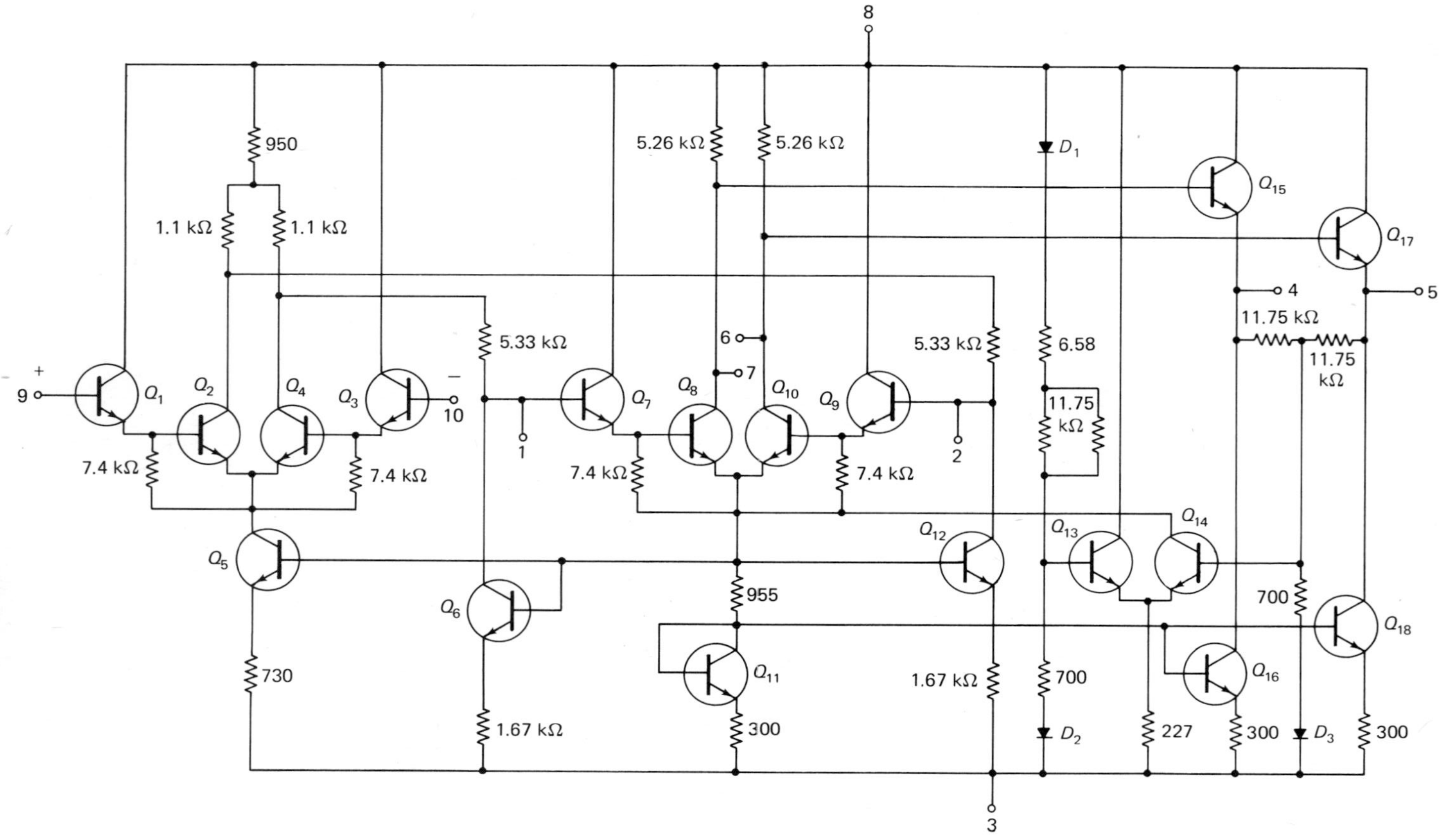

Fig. 5-14 Motorola MC1520 differential output amplifier

buffering to minimize losses in the resistive level shifter and loading of the first stage collectors.

Low output impedance and a large output swing are provided at the output by differential emitter followers Q_{15} and Q_{17}. To improve peak negative swing of the amplifier, a current source Q_{18} (rather than the usual emitter resistor) is used to bias the emitter follower.

Stabilization of output Q point is obtained by using a *local* common-mode feedback loop incorporating a separate differential amplifier stage (Q_{13} and Q_{14}). This circuit holds average voltage output at ground potential for split-power-supply operation, and at $V_{CC}/2$ for single-power-supply operation. Changes in output Q point (also known as *no-signal drift* or *zero drift*) are minimized by using temperature compensation diodes D_2 and D_3 in the base circuits of the feedback differential amplifier.

5-7.2. Typical Connections and Applications

The IC differential amplifier shown in Fig. 5-14 can be used wherever a high-gain wide-frequency-range differential amplifier is required.

Balanced input - differential output

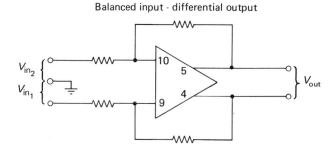

Single-ended input - differential output

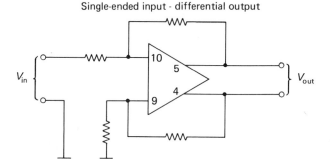

Fig. 5-15 Typical connections for differential output amplifier

Figure 5-15 shows the connections for both balanced-input and unbalanced-input operation. The recommended values for the external resistors and capacitors, together with corresponding gain and frequency ranges, are given in the IC datasheet. The IC can also be used as a single-ended operational amplifier (Chapter 6) if desired.

5-8. DIFFERENTIAL AMPLIFIERS AS LINEAR INTEGRATED-CIRCUIT OPERATIONAL AMPLIFIERS

The most common form of linear IC is the operational amplifier, or *op-amp*. As discussed in Chapter 6, such amplifiers are high-gain direct-coupled circuits where the *gain* and *frequency response* are controlled by external feedback networks. With these feedback networks, the op-amp can be used to produce a broad range of intricate transfer functions, and thus may be adapted for use in many widely different applications.

Although the op-amp was originally designed to perform various mathematical functions (differentiation, integration, analog comparisons, and summation) the op-amp may also be used for many other applications. For example, the same op-amp, by modification of the feedback network, may be used to provide the broad, flat frequency-gain response of video amplifiers, or the peaked response of various types of shaping amplifier. This capability makes the op-amp the most versatile configuration used for linear ICs.

5-8.1. Balanced Differential Operational Amplifier

The most common type of linear IC op-amp uses a *balanced differential amplifier* circuit. Figure 5-16 shows a typical circuit. The complete circuit shown is contained in a ceramic flat pack (less than $\frac{1}{2}$ in. square) with 14 terminal leads. Note that no internal capacitors are used, only resistors and transistors.

The basic purpose of such a circuit is to produce an output signal that is linearly proportional to the *difference between two signals applied to the input*. The circuit shown will provide an over all open-loop gain of approximately 2500.

As discussed, in a theoretical differential amplifier, no output is produced when the signals at the inputs are identical. Thus, differential amplifiers are particularly useful for op-amps, because common-mode signals are eliminated or greatly reduced.

5-8.2. Operational-Amplifier Circuit Description

The following is a brief description of the circuit shown in Fig. 5-16. As illustrated, the operational amplifier consists basically of the two

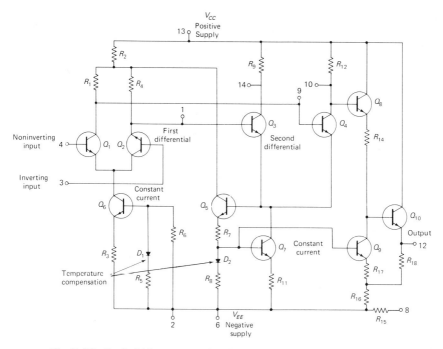

Fig. 5-16   Typical IC op-amp using balanced differential-amplifier circuit

differential amplifiers and a single-ended output circuit in cascade. The pair of cascaded differential amplifiers provides most of the gain.

The inputs to the op-amp are applied to the bases of the pair of emitter-coupled transistors Q_1 and Q_2 in the first differential amplifier. The inverting input is applied to the base of transistor Q_2, and the noninverting input is applied to the base of Q_1. With the noninverting input grounded, and signals applied only to the inverting input, the output is an *amplified* and *inverted* version of the input. For example, if the input is a positive pulse, the output is a negative pulse. If the noninverting input is used, with the inverting input grounded, the output is an amplified version of the input (without inversion). Transistors Q_1 and Q_2 develop driving signals for the second differential amplifier.

Transistor Q_6, a *constant-current* device, is included in the first differential stage to provide bias stabilization for Q_1 and Q_2. If there is an increase in supply voltage, which will normally increase the current through Q_1 and Q_2, the reverse bias on Q_6 increases. This reduces the Q_1–Q_2 current, and offsets the effect of the initial supply voltage increase.

Diode D_1 provides *thermal compensation* for the first differential stage. If there is an increase in operating temperature of the IC, which would normally increase the current through Q_1 and Q_2, there is a corresponding increase in current flow through D_1, since the diode is fabricated on the same silicon chip as the transistors. The increase in D_1 current also increases the reverse bias of Q_6, thus offsetting the initial change in temperature.

The emitter-coupled transistors, Q_3 and Q_4, in the second differential amplifier are driven push–pull by the outputs from the first differential amplifier. Bias stabilization for the second differential amplifier is provided by current transistor Q_7. Compensating diode D_2 provides the thermal stabilization for the second differential amplifier, and also for the current transistor Q_9 in the output stage.

Transistor Q_5 develops the *negative feedback* to reduce common-mode error signals that are developed when the same input is applied to both input terminals of the operational amplifier. Transistor Q_5 samples the signal that is developed at the emitters of transistors Q_3 and Q_4. Because the second differential-amplifier stage is driven push–pull, the signal at this point (the emitters) is zero when the first differential-amplifier stage and the base–emitter circuits of the second stage are matched, and there is no common-mode input.

A portion of any common-mode, or error, signal that appears at the emitters of transistors Q_3 and Q_4 is developed by transistor Q_5 across resistor R_2 (the common collector resistor for transistors Q_1, Q_2, and Q_5) in the proper phase to reduce the error. The emitter circuit of transistor Q_5 also reflects a portion of the same error signal into current transistor Q_7 in the second differential stage so that the initial error signal is further reduced.

Transistor Q_5 also develops feedback signals to compensate for common-mode effects produced by variations in the supply voltage. For example, a decrease in the positive supply voltage results in a decrease of the voltage at the emitters of Q_3 and Q_4. This negative-going change in voltage is reflected by the emitter circuit of Q_5 to the bases of current transistors Q_7 and Q_9. Less current then flows through these transistors.

The decrease in collector current of Q_7 results in a reduction of the current through Q_3 and Q_4, and the collector voltages of these transistors tend to increase. This tendency partially cancels the decrease that occurs with the reduction of the positive supply voltage. The partially canceled decrease in the collector voltage of Q_4 is coupled directly to the base of Q_8, and is transmitted by the emitter of Q_8 to the base of Q_{10}. At this point, the decrease in voltage is further canceled by the increase in the collector voltage of transistor Q_9.

In a similar manner, transistor Q_5 develops the compensating feedback to cancel the effects of an increase in the positive supply (or of variations in the negative supply voltage). Because of the feedback stabilization provided by transistor Q_5, the IC op-amp circuit of Fig. 5-16 provides high common-mode rejection, has excellent open-loop stability, and has a low sensitivity to power-supply variations. All these characteristics are critical to IC op-amps, as is discussed in Chapter 6.

In addition to their function in the cancellation of supply voltage variations, transistors Q_8, Q_9, and Q_{10} are used in an emitter-follower type of
single-ended output circuit. The output of the second differential amplifier
is directly coupled to the base of Q_8, and the emitter circuit of transistor Q_8
supplies the base-drive input for output transistor Q_{10}.

A small amount of signal gain in the output circuit is made possible by the
connection from the emitter of the output transistor Q_{10} to the emitter circuit
of transistor Q_9 (at the junction of R_{16} and R_{17}). Without this connection, the
transistor could be considered as merely a constant-current sink for drive
transistor Q_8. Because of the connection, however, the output circuit can
provide a signal gain of 1.5 from the collector of differential-amplifier transistor Q_4 to the output. Although this small amount of gain may seem insig-
nificant, it does increase the output swing capabilities of the op-amps.

The output from the op-amp circuit is taken from the emitter of the output
transistor Q_{10} so that the dc level of the output signal is substantially lower
than that of the differential amplifier output at the collector of transistor Q_4.
In this way, the output circuit shifts the dc level at the output so that it is
effectively the same as the at the input when no signal is applied. This problem of dc level shifting is discussed further in Sec. 5-8.3.

Resistor R_{15} in series with terminal 8 increases the ac short-circuit load
capability of the operational amplifier, when terminal 8 is shorted to terminal
12, so that R_{15} is connected between the output and the negative supply.

5-8.3. Direct-Current Level-Shifting Problems

In any cascade direct-coupled amplifier, either discrete component or IC, the dc level rises through successive stages toward the supply
voltage. In linear ICs, the dc voltage builds up through the *NPN* stages in
the positive direction, and must be shifted negatively if large output signal
swings are to be obtained. For example, if the supply voltage is 10 V and the
output is at 9 V under no-signal conditions, the maximum output voltage
swing is limited to less than 1 V.

In multistage high-gain ICs, such as op-amps and special-purpose multi-
function circuits, which use external feedback, it is especially important to
provide for compensation of the dc level shift. Such amplifiers must have
equal (and preferably zero) input and output dc levels so that the dc coupling
of the feedback connection does not shift any bias point. For example, if the
input is at 0 V, but the output is at 3 V, this 3 V is reflected back through an
external feedback resistor, and changes the input to a 3-V level.

The use of an output stage, such as shown in Fig. 5-16, is a commonly
used technique to prevent a shift in dc level between the output and input of
an IC. Transistor Q_8 operates as an input buffer, and transistor Q_9 is essentially a current sink for Q_8. The shift in dc level is accomplished by the volt-

age drop across resistor R_{14} produced by the collector current of transistor Q_9. The emitter of the output transistor Q_{10} is connected to the emitter of Q_9. Feedback through R_{18} results in a decrease in the voltage drop across R_{14} for negative-going output swings, and an increase in this voltage drop for positive-going output swings.

If properly designed, the circuit shown in Fig. 5-16 can provide substantial voltage gain, high input impedance, low output impedance, and an output swing *nearly equal* to the supply voltages, in addition to the desired shift in dc level. Moreover, feedback may be coupled from the output to the input to compensate for dc common-mode effects that result from variations in the supply voltages.

5-8.4. Summary of Advantages for Differential Integrated-Circuit Op-Amps

The balanced differential amplifier is considered the optimum configuration for general-purpose linear ICs by most manufacturers. This circuit configuration is generally preferred over other possible types (a feedback pair for example) for the following reasons:

Advantage can be taken from the exceptional balance between the differential inputs that results from the inherent match in base-to-emitter voltage and short-circuit current gain of the two (differential-pair) transistors, which are processed in exactly the same way and are located very close to each other on the same very small silicon chip.

The differential-amplifier circuit uses a minimum number of capacitors. Generally, no internal capacitors are used in IC differential amplifiers.

The use of large capacitors can usually be avoided, and the gain of the differential amplifier circuit is a function of *resistance ratios*, rather than of actual resistance values.

The differential amplifier is much more versatile than other possible circuit configurations, and can be readily adapted for use in a variety of component applications. For example, in the circuit of Fig. 5-16 there are many connections to internal circuit components, in addition to the inputs, output, and power supplies. This is typical of differential amplifier ICs. The additional connections (such as the collectors of Q_1, Q_2, Q_3, Q_4, and Q_9) provide a variety of input–output points for the *phase-lead* and *phase-lag compensation* techniques described in Chapter 6.

6. OPERATIONAL AMPLIFIERS

The subject of operational amplifiers (or *op-amps*) is so broad that it is difficult to cover in one book, much less one chapter. Op-amps are discussed fully in the author's *Manual for IC Users* (Reston Publishing Company, Inc., Reston, Va., 1973). The following sections summarize theory and use of op-amps.

The designation "op-amp" was originally adopted for a series of high-performance dc amplifiers used in analog computers. These amplifiers were used to perform mathematical operations applicable to analog computation (summation, scaling, subtraction, integration, etc.). Today, the availability of inexpensive IC amplifiers has made the packaged op-amp useful as a replacement for any low-frequency amplifier.

In the first sections of this chapter we discuss the basic IC op-amp, concentrating on how to interpret IC op-amp datasheets, design considerations for frequency response and gain, as well as a specific design example for a typical IC op-amp. In later sections of the chapter we discuss some typical applications for the op-amp.

Most of the basic design information for a particular IC op-amp can be obtained from the datasheet. Likewise, a typical op-amp datasheet may describe a few specific applications for the op-amp. However, the datasheets generally have two weak points. First, they do not show how the listed parameters relate to design problems. Second, they do not describe the great variety of applications for which a basic op-amp can be used.

In any event, it is always necessary to interpret datasheets. Each manufacturer has its own system of datasheets. It is impractical to discuss all datasheets here. Instead, we shall discuss typical information found on IC op-amp datasheets, and see how this information affects design and use.

299

6-1. TYPICAL INTEGRATED-CIRCUIT OP-AMP

Integrated-circuit op-amps generally use several *differential* stages in cascade to provide common-mode rejection and high gain. The IC op-amp described in Sec. 5-8 is typical. Integrated circuit op-amps generally require both positive and negative power supplies. Since a differential amplifier has two inputs, it provides phase inversion for degenerative (or negative) feedback, and can be connected to provide in-phase or out-of-phase amplification.

In typical use an op-amp requires that the output be fed back to the input through a resistance or impedance. The output is fed back to the negative or inverting input so as to produce degenerative feedback (to set the desired gain and frequency response). As in any amplifier, the signal shifts in phase as it passes from input to output. This *phase shift is dependent upon frequency*. When the phase shift approaches 180° (generally as frequency increases), the through-amplifier phase shift adds to (or cancels out) the 180° feedback phase shift. Thus, the feedback is in phase with the input (or nearly so) and causes the amplifier to oscillate. This condition of phase shift with increased frequency limits the bandwidth of the op-amp. That is, there is some frequency at which any op-amp will produce enough through-amplifier phase shift to oscillate. The condition can be compensated by the addition of a phase-shift network (usually an *RC* circuit, but sometimes a single capacitor) to the internal amplifier circuit or the input.

Linear IC op-amps are available in the three basic IC package types (flat pack, T0-5 style, and dual in-line). In most cases, the packages have terminals that provide for connecting external phase compensation networks to internal amplifier circuits. (See Fig. 5-16, and refer to Sec. 5-8.4.)

6-2. DESIGN CONSIDERATION FOR FREQUENCY RESPONSE AND GAIN

Most of the design problems for IC op-amps are the result of tradeoffs between gain and frequency response (or bandwidth). The open-loop (without feedback) gain and frequency response are characteristics of the basic IC package, but can be modified with external *phase compensation* networks. The closed-loop (with feedback) gain and frequency response are primarily dependent upon *external feedback* components.

The two basic op-amp configurations, *inverting feedback* and *noninverting feedback*, are shown in Figs. 6-1 and 6-2, respectively. The equations on Figs. 6-1 and 6-2 are classic guidelines. The equations do not take into account the fact that open-loop gain is not infinitely high, and the output or load imped-

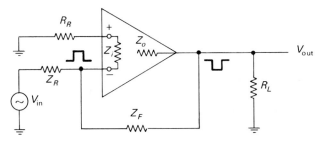

$$V_{out} \approx V_{in} \times \frac{Z_F}{Z_R}$$

$$\text{closed-loop gain} \approx \frac{Z_F}{Z_R}$$

$$\text{loop gain} \approx \frac{\text{open-loop gain}}{\text{closed-loop gain}}$$

$$Z_{in} \approx Z_R , \qquad Z_{out} \approx \frac{Z_o}{\text{loop gain}}$$

Fig. 6-1 Theoretical inverting feedback op-amp

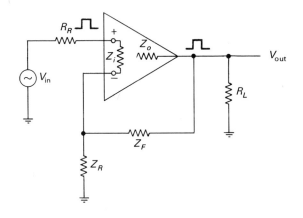

$$V_{out} \approx V_{in} \times \frac{Z_F}{Z_R} + 1$$

$$\text{closed-loop gain} \approx \frac{Z_F}{Z_R} + 1$$

$$\text{loop gain} \approx \frac{\text{open-loop gain}}{\text{closed-loop gain}}$$

$$Z_{in} \approx Z_i + \frac{\text{open-loop gain} \times Z_i}{\text{closed-loop gain}}$$

Fig. 6-2 Theoretical noninverting feedback op-amp

$$Z_{out} \approx \frac{Z_o}{\text{loop gain}}$$

ance is not infinitely low. Thus, the equations contain built-in inaccuracies, and must be used as guides only.

When *loop gain* is large, the inaccuracies in the equations decrease. Loop gain is defined as the ratio of open-loop gain to closed-loop gain, as shown in Fig. 6-3.

The relationships in Fig. 6-3 are based on a theoretical op-amp. That is, the open-loop gain rolls off at 6 dB per octave, or 20 dB per decade. (The term 6 dB per octave means that the gain drops by 6 dB each time frequency is doubled. This is the same as a 20-dB drop each time the frequency is increased by a factor of 10.)

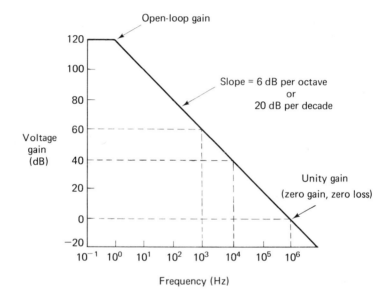

$$\text{Loop gain} \approx \frac{\text{open-loop gain}}{\text{closed-loop gain}}$$

Fig. 6-3   Frequency-response curve of theoretical op-amp

If the open-loop gain of an amplifier is as shown in Fig. 6-3, any stable, closed-loop gain can be produced by the proper selection of feedback components, provided the closed-loop gain is less than the open-loop gain. The only concern is a tradeoff between gain and frequency response.

For example, if a voltage gain of 40 dB (10^2) is desired, a feedback resistance 10^2 times larger than the input resistance is used. The gain is then flat to 10^4 Hz, and rolloff is 6 dB per octave to unity gain at 10^6 Hz. If 60-dB (10^3) gain is required, the feedback resistance is raised to 10^3 times the input re-

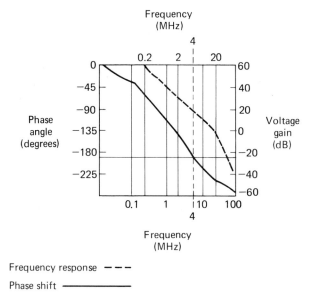

Frequency response – – –

Phase shift ————

Fig. 6-4   Frequency-response and phase-shift characteristics of practical IC op-amp

sistance. This reduces the frequency response. Gain is flat to 10^3 Hz (instead of 10^4 Hz), followed by rolloff of 6 dB per octave down to unity gain.

The open-loop frequency response curve of a *practical* IC op-amp more closely resembles that shown in Fig. 6-4. Here, gain is flat at 60 dB to about 200 kHz, then rolls off at 6 dB per octave to 2 MHz. As frequency increases, rolloff continues at 12 dB per octave (40 dB per decade) to 20 MHz (where gain is about unity or 0), then rolls off at 18 dB per octave (60 dB per decade).

Some IC op-amp datasheets provide a curve similar to that shown in Fig. 6-4. If the curves are not available, it is possible to test the op-amp under laboratory conditions, and draw an actual response curve (frequency response and phase shift). The necessary procedures are described in Chapter 7.

The sharp rolloff at high frequencies, in itself, is not a problem in op-amp use (unless the op-amp must be operated at a frequency very near the high end). However, note that the *phase reponse* (phase shift between input and output) changes with frequency. The phase response of Fig. 6-4 shows that a negative feedback (at low frequency) can become positive (and cause the amplifier to be unstable) at high frequencies (possibly resulting in oscillation). In Fig. 6-4, a 180° phase shift occurs at approximately 4 MHz. This is the frequency at which open-loop gain is about $+20$ dB.

As a guide, when a selected closed-loop gain is equal to, or less than, the open-loop gain at the 180° phase-shift point, the amplifier is *unstable*. For example, if a closed-loop gain of 20 dB or less is selected, a circuit with the curves of Fig. 6-4 will be unstable. (Note the point where the $-180°$ phase angle intersects the phase-shift line. Then draw a vertical line up to cross the open-loop-gain line.)

The closed-loop gain must be *more* than the open-loop gain at the frequency where the 180° phase shift occurs, but *less* than the maximum open-loop gain. Using Fig. 6-4 as an example, the closed-loop gain must be greater than 20 dB, but less than 60 dB.

6-2.1. Phase-Compensation Methods

Op-amp design problems created by excessive phase shift can be solved by the use of compensating techniques that *alter response* so that excessive phase shifts no longer occur. There are three basic methods of phase compensation.

The closed-loop gain can be altered by means of capacitors and/or inductances in the feedback circuit. These elements change the pure resistance to an impedance that changes with frequency, thus providing a different amount of feedback at different frequencies, and a shift in phase of the feedback signal. This offsets the undesired through-amplifier phase shift.

Compensation of phase shift by closed-loop feedback methods is generally not recommended, since they create impedance problems at both the high- and low-frequency limits. Feedback compensation is not considered here, except where the op-amp is to be used in a *bandpass* function, as described in later sections of this chapter.

The open-loop input impedance can be altered by means of a resistor and capacitor, as shown in Fig. 6-5. The input impedance of the series C and R decreases as frequency increases, thus altering through-amplifier (open-loop) gain. As shown in Fig. 6-5, this arrangement causes the rolloff to start at a lower frequency, but produces a stable rolloff similar to that of the "ideal" curve of Fig. 6-3. With the circuit properly compensated, the desired closed-loop gain is produced by selection of external resistors.

The open-loop gain can be altered by one of several phase-compensation schemes, as shown in Figs. 6-6, 6-7, and 6-8.

In Fig. 6-6 (*known as phase-lead compensation*) the open-loop gain is changed by an external capacitor (usually connected between collectors in one of the high-gain stages).

In Fig. 6-7 (*generally referred to as RC rolloff, or straight rolloff, or phase-lag compensation*) the open-loop gain is altered by means of an appropriate external RC network connected across a circuit component. With this method, the rolloff starts at the "corner" frequency produced by the RC network.

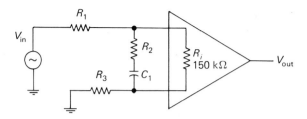

R_i = input impedance of IC

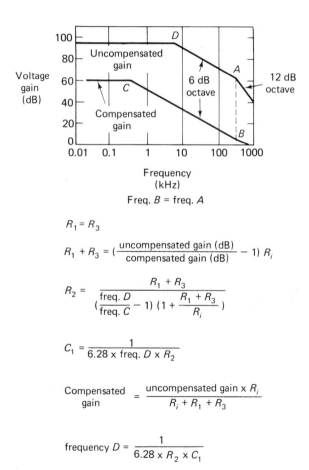

Freq. B = freq. A

$R_1 = R_3$

$$R_1 + R_3 = \left(\frac{\text{uncompensated gain (dB)}}{\text{compensated gain (dB)}} - 1\right) R_i$$

$$R_2 = \frac{R_1 + R_3}{\left(\dfrac{\text{freq. } D}{\text{freq. } C} - 1\right)\left(1 + \dfrac{R_1 + R_3}{R_i}\right)}$$

$$C_1 = \frac{1}{6.28 \times \text{freq. } D \times R_2}$$

$$\frac{\text{Compensated}}{\text{gain}} = \frac{\text{uncompensated gain} \times R_i}{R_i + R_1 + R_3}$$

$$\text{frequency } D = \frac{1}{6.28 \times R_2 \times C_1}$$

Fig. 6-5 Frequency-response compensation by modification of open-loop input impedance of op-amp

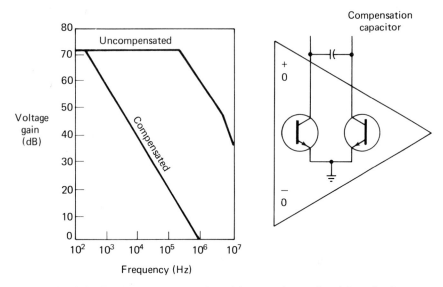

Fig. 6-6   Frequency compensation with external capacitor (phase-lead compensation)

In Fig. 6-8 (*known as Miller-effect rolloff, or Miller-effect phase-lag compensation*) the open-loop gain is altered by an *RC* network connected between the input and output of an inverting gain stage in the op-amp.

6-2.2. Selecting a Phase-Compensation Scheme

A comprehensive IC op-amp datasheet will recommed one or more methods for phase compensation, and will show the relative merits of each method. Usually, this is done by means of response curves for various values of the compensating network.

The recommended phase-compensation methods and values should be used in all cases. Proper phase compensation of an op-amp is a difficult, trial-and-error job at best. By using the datasheet values, it is possible to take advantage of the manufacturer's test results on production quantities of a given IC op-amp.

If the datasheet is not available, or if the datasheet does not show the desired information, it is still possible to design a phase-compensation network using the rule-of-thumb equations.

The first step in phase compensation (when not following the datasheet) is to connect the op-amp to an appropriate power source as discussed in Sec. 6-4. Then test the op-amp for open-loop frequency response and phase shift as described in Chapter 7.

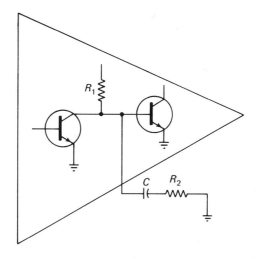

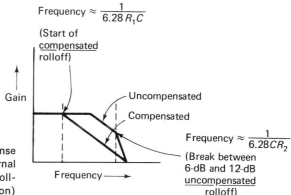

Fig. 6-7   Frequency-response compensation with external capacitor and resistor (*RC* roll-off, or phase-lag compensation)

Draw a response curve similar to Fig. 6-4. On the basis of actual open-loop response and the equations of Figs. 6-5 through 6-8, elect trial values for the phase-compensating network. Then repeat the frequency-response and phase-shift tests. If the response is not as desired, change the values as necessary.

A careful inspection of the equations on Figs. 6-5 through 6-8 show that it is necessary to know certain internal characteristics of the op-amp before an accurate prediction of the compensated frequency response can be found. For example, in Figs. 6-7 and 6-8 the values of *R* and *C* of the compensation network are based on the uncompensated open-loop *frequency at which gain changes* from a 6 dB per octave drop to a 12 dB per octave drop. This can be found by test of the uncompensated op-amp. However, to predict the fre-

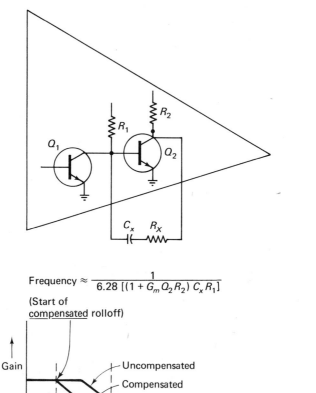

$$\text{Frequency} \approx \frac{1}{6.28\,[(1 + G_m Q_2 R_2)\,C_x R_1]}$$

(Start of
compensated rolloff)

$$\text{Frequency} \approx \frac{1}{6.28 C_x R_x}$$

(Break between 6-dB
and 12-dB uncompensated
rolloff)

Fig. 6-8 Frequency-response compensation with external capacitor
and resistor (Miller-effect rolloff)

quency at which the compensated response will start to roll off (the "corner"
frequency), or the gain after compensation, requires a knowledge of internal
stage transconductance (or gain) and stage load. This information is usually
not available and cannot be found by simple test.

An exception to this is modification of the open-loop input impedance, as
shown in Fig. 6-5. The only op-amp characteristic required here is the input
impedance (or resistance). This is almost always available on the datasheet.
If not, it can be found by a simple test, as described in Chapter 7.

6-2.3. Design Example for Modification of Open-Loop Input Impedance of an Op-Amp

The connections, frequency plots, and equations for modification of open-loop input impedance are shown in Fig. 6-5.

The first step is to note the frequency at which the uncompensated rolloff changes from 6 to 12 dB (point A). The compensated rolloff should be zero (unity gain, point B) at the same frequency.

Draw a line up to the left from point B that increases at 6 dB per octave. For example, if point B is at 350 kHz, the line should intersect 35 kHz as it crosses the 20-dB gain point. (In a practical circuit, the rolloff starts at a slightly lower frequency, about 28 to 30 kHz at 20-dB gain, since the rolloff point is rounded, rather than a sharp "corner.")

Any combination of compensated gain and rolloff starting frequency (point C) can be selected along the line. For example, if the rolloff starts at 10 kHz, the gain is about 30 dB, and vice versa.

Assume that the circuit of Fig. 6-5 is used to produce a compensated gain of 60 dB, with rolloff starting at 280 Hz and dropping to zero (unity gain at 350 kHz). The op-amp to be used has an uncompensated gain of about 94 dB, with a rolloff pattern similar to that of Fig. 6-5. The typical input impedance is 150 kΩ.

Using the compensated gain equation of Fig. 6-5, the relationship is

$$60 \text{ dB} = \frac{(94 \text{ dB})(150{,}000)}{150{,}000 + R_1 + R_3}$$

Therefore,

$$R_1 + R_3 = \left(\frac{94}{60} - 1\right) \times 150{,}000$$

$$= 0.57 \times 150{,}000$$

$$= 85{,}500$$

If $R_1 = R_3$, then $R_1 = R_3 = 42{,}750$. The nearest standard value is 43 kΩ. Using the equation of Fig. 6-5, the value of R_2 is

$$R_2 = \frac{85{,}500}{\left(\dfrac{5000}{280} - 1\right)\left(1 + \dfrac{85{,}500}{150{,}000}\right)}$$

$$= \frac{85{,}500}{16.85 \times 1.57}$$

$$= 3000 \text{ Ω nearest standard value}$$

The value of C_1 is

$$C_1 = \frac{1}{(3000)(6.28)(5000)} = 0.01 \ \mu\text{F, nearest standard value}$$

If the circuit of Fig. 6-5 shows any instability in the open-loop or closed-loop condition, try increasing the values of R_1 and R_3 (to reduce gain); then select new values for R_2 and C_1.

6-3. INTERPRETING INTEGRATED-CIRCUIT OP-AMP DATASHEETS

Most of the basic design information for a particular IC op-amp can be obtained from the datasheet. There are some exceptions to this rule. For certain applications it may be necessary to test the op-amp under simulated operating conditions. However, it is always necessary to interpret datasheets. Each manufacturer has its own system of datasheets. It is impractical to discuss all datasheet formats here. Instead we shall discuss typical information found on datasheets.

6-3.1. Open-Loop Voltage Gain

The open-loop voltage gain (A_{VOL} or A_{OL}) is defined as the ratio of a change in output voltage to a change in input voltage at the input terminals. Open-loop gain is always measured without feedback, and usually without compensation.

Open-loop gain is frequency dependent (gain decreases with increased frequency). Open-loop gain is also temperature dependent (gain decreases with increased temperature, as shown in Fig. 6-9).

As discussed, open-loop gain can be modified by several compensation methods. A typical op-amp datasheet will show the results of such compensation, usually by means of graphs, such as shown in Fig. 6-10.

After compensation is applied, the IC can be connected in the closed-loop configuration. The voltage gain under closed-loop conditions is dependent upon external components (the ratio of feedback resistance to input resistance). Thus, closed-loop gain is usually not listed, as such, on IC op-amp sheets. However, the datasheet may show some typical gain curves with various ratios of feedback (Fig. 6-11). If available, such curves can be used directly to select values of feedback components.

When a capacitor is used to compensate or modify voltage gain, the *slew rate* (or *slewing rate*) of an op-amp can be affected. The slow rate is the maximum rate of change of the output voltage (V_{OUT} or E_{OUT}), with *respect to time,*

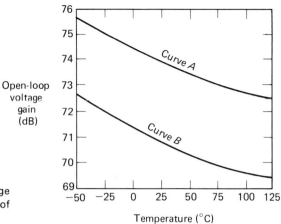

Open-loop voltage gain (dB)

Temperature (°C)

Fig. 6-9   Open-loop voltage gain of op-amp as a function of temperature

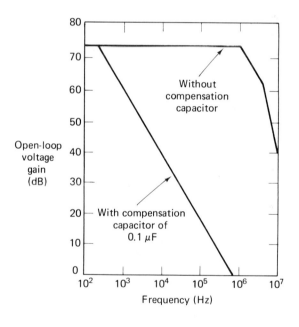

Open-loop voltage gain (dB)

Frequency (Hz)

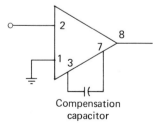

Fig. 6-10   Open-loop voltage gain of op-amp with and without compensation

Compensation capacitor

311

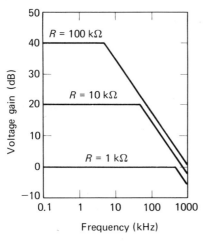
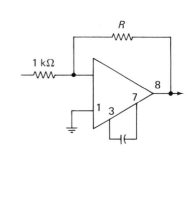

Fig. 6-11   Closed-loop voltage gain of op-amp versus frequency

that the op-amp is capable of producing, while maintaining its linear characteristics.

Slew rate is expressed in terms of

$$\frac{\text{difference in output voltage}}{\text{difference in time}} = \frac{\Delta V_{\text{OUT}}}{\Delta t}$$

Usually, slew rate is listed in terms of volts per microseconds. For example,
if the output voltage from an op-amp is capable of changing 100 V in 1 μs,
then the slew rate is 100. If, after compensation, the output changes to a
maximum of 50 V in 1 μs, the new slew rate is 50.

Slew rate of an op-amp is a direct function of the compensation capacity.
At higher frequencies, the current required to charge and discharge the capacitor can limit available current to succeeding stages or loads, and result in
lower slew rates. (This is one reason why IC op-amp datasheets usually recommend the compensation of early stages in the amplifier where signal
levels are still small and little current is required.)

The major effect of slew rate on design is in output power of the amplifier.
All other factors being equal, a lower slew rate results in lower power output.

Slew rate decreases as compensation capacitance increases. This is shown
in the datasheet curve of Fig. 6-12. Therefore, where high frequencies are involved, the *lowest value* of compensation capacitor should be used.

Figure 6-13 shows the minimum compensating capacitor value that can be
used with different closed-loop gain levels. The curves of Fig. 6-13 are typical
of those found on op-amp datasheets where slew rate is of particular importance.

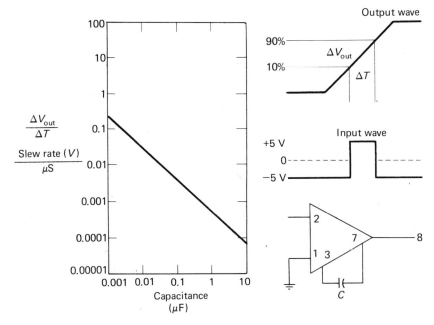

Fig. 6-12   Slew rate of op-amp versus rolloff compensation capacitance

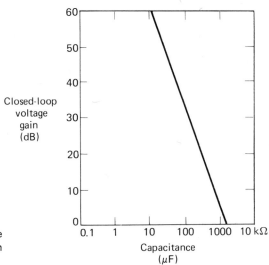

Fig. 6-13   Closed-loop voltage
gain of op-amp versus minimum
rolloff capacitance

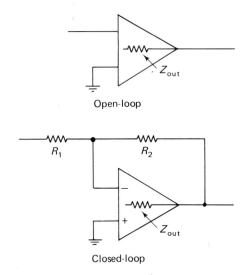

$$\text{Closed-loop } Z_{out} \approx \frac{\text{open-loop } Z_{out}}{1 + \text{open-loop gain } (\dfrac{R_1}{R_1 + R_2})}$$

Fig. 6-14 Open-loop versus closed-loop output impedance relationships

6-3.2. Output Impedance

Output impedance (Z_{OUT}) is defined as the impedance seen by a load at the output of the op-amp. (See Fig. 6-14.) Excessive output impedance can reduce the gain since, in conjunction with the load and feedback resistors, output impedance forms an attenuator network. In general, output impedance of ICs used as op-amps is less than 200 Ω. Generally, input resistances are at least 1000 Ω, with feedback resistance several times higher than 1000 Ω. Thus, the output impedance of a typical IC op-amp has little effect on gain.

If the op-amp is serving primarily as a voltage amplifier (which is usually the case), the effect of output impedance is at a minimum. Output impedance has a more significant effect on design of power amplifiers that must supply large amounts of load current.

Closed-loop output impedance is found by the equation of Fig. 6-14. Thus, it is seen that output impedance increases as frequency increases, since open-gain decreases.

6-3.3. Input Impedance

Input impedance (Z_{IN}) is defined as the impedance seen by a source looking into one input of the op-amp, with the other input grounded. (See Fig. 6-15.) The primary effect of input impedance on design is to reduce

Open-loop

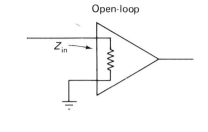

Closed-loop

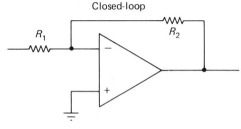

Fig. 6-15 Open-loop and closed-loop input impedance relationships

Closed-loop $Z_{in} \approx R_1 + \dfrac{R_2}{\text{open-loop gain}}$

When open-loop gain = infinity, closed-loop $Z_{in} \approx R_1$

amplifier loop gain. Input impedance will change with temperature and frequency. Generally, input impedance is listed on the datasheet as 25°C and 1 kHz.

If input impedance is quite different from the impedance of the device driving the op-amp, there is a loss of input signal due to the mismatch. However, in practical terms, it is not possible to alter the input impedance of an *IC* op-amp. Thus, if impedance match is critical, either the *IC* or driving source must be changed to effect a match.

6-3.4. Output Voltage Swing

Output voltage swing (V_{OUT} or P-P V_{OUT}) is defined as the peak output voltage swing (referred to zero) that can be obtained without clipping. (See Fig. 6-16.) A symmetrical voltage swing is dependent upon frequency, load current, output impedance, and slew rate. Generally, an increase in frequency, load current, and output impedance will decrease the possible output voltage swing.

An increase in slew rate will increase possible output voltage swing capabilities. Since slew rate is related to compensation techniques (a high-value compensation capacitor reduces slew rate), output voltage swing is also related to compensation (a high-value compensation capacitor reduces output voltage swing at a given frequency).

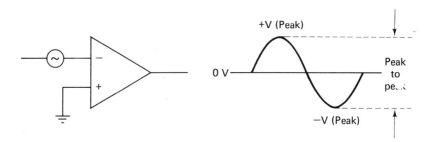

Fig. 6-16   Output voltage swing relationships

6-3.5. Bandwidth and Frequency Range

Bandwidth of an op-amp is usually expressed in terms of open-loop operation. The common term is BW_{OL} at -3 dB. (See Fig. 6-17.) For example, a BW_{OL} of 300 kHz indicates that the open-loop gain of the op-amp drops to a value 3 dB below the flat or low-frequency level at a frequency of 300 kHz.

The frequency range of an op-amp is often listed as "useful frequency range" (such as direct current to 15 MHz). Useful frequency range for an op-amp is similar to the f_T term (Chapters 1 and 2) for discrete transistors. Generally, the high-frequency limit specified for an op-amp is the frequency at which gain drops to unity.

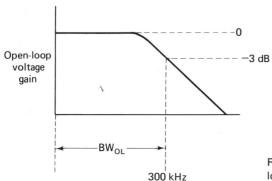

Fig. 6-17 Bandwidth and open-loop gain relationships

6-3.6. Input Common-Mode Voltage Swing

Input common-mode voltage swing (V_{ICM}) is defined as the maximum peak input voltage that can be applied to either input terminal of the op-amp without causing abnormal operation or damage. (See Fig. 6-18.) Some IC op-amp datasheets list a similar term: common-mode input voltage range (V_{CMR}). Usually, V_{ICM} is listed in terms of peak voltage, with positive

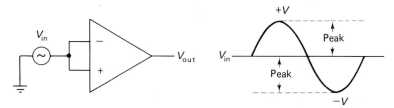

Fig. 6-18   Input common-mode voltage swing relationships

and negative peaks being equal. V_{CMR} is often listed on positive and negative voltages of different value (such as $+1$ V and -3 V).

In practical use, either of these parameters limit the differential signal amplitude that can be applied to the op-amp input. As long as the input signal does not exceed the V_{ICM} or V_{CMR} values (in either the positive or negative direction), there should be no problem.

Note that some IC op-amp datasheets list "single-ended" input voltage signal limits, where the differential input is not to be used.

6-3.7. Common-Mode Rejection Ratio

When common-mode rejection ratio (CMR, CMRR, or CM_{rej}) is listed, the definitions of Sec. 5-3 apply.

6-3.8. Input-Bias Current

Input-bias current (I_i or I_b) is defined as the average value of the two input-bias currents of the op-amp *differential* input stage. (See Fig. 6-19.) Input-bias current is a function of the large signal current gain of the input stage.

In use, the only real significance of input-bias current is that the resultant voltage drop across the input resistors can restrict the input common-mode voltage range at higher impedance levels. The input-bias current produces a voltage drop across the input resistors. This voltage drop must be overcome by the input signal.

Input-bias current decreases as temperature increases.

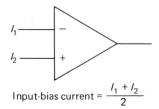

$$\text{Input-bias current} = \frac{I_1 + I_2}{2}$$

Fig. 6-19 Input-bias current

6-3.9. Input-Offset Current

Input-offset current (I_{io}) is defined as the difference in input bias current into the input terminals of a differential op-amp. (See Fig. 6-20.) Input-offset current is an indication of the degree of matching of the input differential stage.

When high impedances are used in design, input-offset current can be of greater importance than input-offset voltage. If the input bias current is different for each input, the voltage drops across the input resistors (or input impedance) will not be equal. If the resistance is large, there is a large unbalance in input voltages. This condition can be minimized by means of a resistance connected between the noninverting (or $+$) input and ground, as shown in Fig. 6-21. The value of this resistor (R_3) should equal the parallel equivalent of the input and feedback resistors (R_1 and R_2), as shown by the equation.

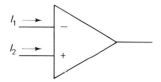

Input-offset current = $I_1 - I_2$ or $I_2 - I_1$ Fig. 6-20 Input-offset current

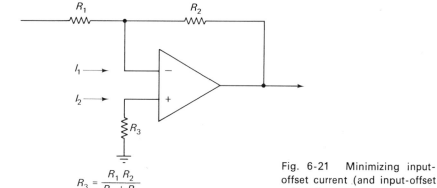

$$R_3 = \frac{R_1 R_2}{R_1 + R_2}$$

Fig. 6-21   Minimizing input-offset current (and input-offset voltage)

In practical design, the trial value for R_3 is based on the equation of Fig. 6-21. Then the value of R_3 is adjusted for *minimum voltage difference at both terminals* (under normal operating conditions) but with no signal.

6-3.10. Input-Offset Voltage

Input-offset voltage (V_{io}) is defined as that voltage which must be applied at the input terminals to obtain zero output voltage. (See Fig. 6-22.)

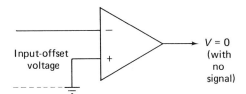

Fig. 6-22 Input-offset voltage

Input-offset voltage indicates the matching tolerance in the differential-amplifier stages. A perfectly matched amplifier requires zero input voltage to produce zero output voltage. Typically, input-offset voltage is on the order of 1 mV for an IC op-amp.

The effect of input-offset voltage on design is that the input signal must overcome the offset voltage before an output is produced. For example, if an op-amp has a 1-mV input-offset voltage and a 1-mV signal is applied, there is no output. If the signal is increased to 2 mV, the op-amp produces only the peaks.

Input-offset voltage is increased by amplifier gain. In the closed-loop condition, the effect of input-offset voltage is increased by the ratio of feedback resistance to input resistance, plus unity (or 1). For example, if the ratio is 100 to 1, the effect of input-offset voltage is increased by 101.

Some IC op-amps include provisions to neutralize any offset. Typically, an external voltage is applied through a potentiometer to a terminal on the IC. The voltage is adjusted until the offset, at the input and output, is zero. However, since this is a special circuit, no general rules can be included concerning voltage or potentiometer values, or external connections. The datasheet must be consulted.

For an IC without offset compensation, the effects of input-offset voltage can be reduced by minimizing input-offset current, as described in Sec. 6-3.9 and Fig. 6-21.

6-3.11. Power-Supply Sensitivity

Power-supply sensitivity ($S+$ and $S-$) is defined as the ratio of change in input-offset voltage to the change in supply voltage producing it, with the remaining supply held constant. (See Fig. 6-23.) Some IC datasheets list a similar parameter: *input-offset voltage sensitivity*. In either case, the parameter is expressed in terms of millivolts per volt or microvolts per volt, representing the change (in millivots or microvolts) of input-offset voltage to a change (in volts) of one power supply. Usually, there is a separate sensitivity parameter for each power supply, with the opposite power supply assumed to be held constant. For example, a typical sensitivity listing is 0.1 mV/V for a positive supply. This implies that with the negative supply held constant, the input-offset voltage changes by 0.1 mV for each 1-V change in positive supply voltage.

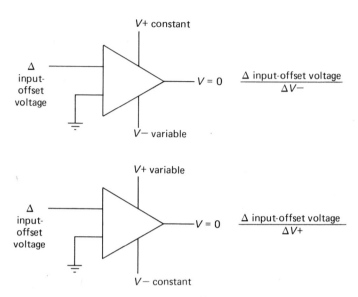

Fig. 6-23   Power-supply sensitivity

The effects of power-supply sensitivity (or input-offset voltage sensitivity) are obvious. If an op-amp has considerable sensitivity to power-supply variations, overall performance is affected by each supply voltage change. The power-supply regulation must be increased to provide correct operation with minimum input signal levels.

6-3.12. Average Temperature Coefficient of Input-Offset Voltage

The parameter (TCV_{io}) is dependent upon the temperature coefficients of various components within the op-amp. Temperature changes affect stage gain, match of differential amplifiers, and so on, and thus change input-offset voltage.

From a practical standpoint, TCV_{io} need be considered only if the parameter is large and the op-amp must be operated under extreme temperatures. For example, if input-offset voltage doubles with an increase to a temperature that is likely to be encountered during normal operation, the higher input-offset voltage should be considered as the "normal" value for design.

6-3.13. Input-Noise Voltage

There are many systems for measuring noise voltage in an op-amp, and equally as many methods used to list the value on datasheets. Some datasheets omit the value entirely. In general, noise is measured with the op-amp in the open-loop condition, with or without compensation, and with the input shorted, or a fixed resistance load at the input terminals.

The input and/or output voltage is measured with a sensitive voltmeter or oscilloscope. Input noise is in the order of a few microvolts, with output noise usually less than 100 mV. Output noise is almost always greater than input noise (because of the amplifier gain).

Except where the noise value is very high, or the input signal is very low, amplifier noise can be ignored. Obviously, 10-μV noise at the input will mask a 10-μV signal. If the signal is raised to 1 mV with the same op-amp, the noise will be unnoticed.

Noise is temperature dependent, as well as dependent upon the method of compensation used.

6-3.14. Power Dissipation

An op-amp datasheet usually lists two power dissipation ratings. One value is the *total device dissipation*, which includes any load current. The other value is *device dissipation* (P_D or P_T), which is defined as the direct-current power dissipated by the op-amp itself (with output at zero and no load).

The device dissipation must be subtracted from the total dissipation to calculate the load dissipation.

For example, if an IC op-amp can dissipate a total of 300 mW (at a given temperature, supply voltage, and with or without a heat sink), and the IC itself dissipates 100 mW, the load cannot exceed 200 mW (300 − 100 = 200 mW).

6-4. OP-AMP POWER SUPPLIES

As a general rule, op-amps require connection to both a positive and negative power supply. This is because most IC op-amps use one or more differential amplifiers in their circuits. When two power suppies are required, the supplies are usually equal or *symmetrical* (such as +9 and −9 V, +20 and −20 V). This is the case with the IC op-amp of Fig. 6-24 which normally operates with +18 and −18 V. A few IC op-amps use unsymmetrical power supplies (+9 and −4.5 V, for example), and there are IC op-amps that require only a single supply. However, such cases are the exception. And, in some cases, it is possible to operate an IC op-amp (that normally requires two supplies) from a single supply by means of special circuits (external to the IC), as discussed in later sections.

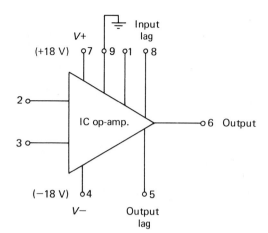

Fig. 6-24   Typical linear IC op-amp operation with symmetrical 18-V power supplies

6-4.1. Labeling of Integrated-Circuit Op-Amp Power Supplies

Unlike most discrete transistor circuits, where it is usual to label one power supply lead positive and the other negative without specifying which (if either) is common to ground, it is necessary that all IC op-amp power-supply voltages be referenced to a common or ground (which may or may not be physical or equipment ground).

Integrated circuit manufacturers do not agree on power-supply labeling. For example, the circuit of Fig. 6-24 uses $V+$ to indicate the positive voltage, and $V-$ to indicate the negative voltage. Another manufacturer might use the symbols V_{CC} and V_{EE} to represent positive and negative, respectively. The IC datasheet must be studied carefully *before applying* any power source.

6-4.2. Typical Integrated-Circuit Op-Amp Power-Supply Connections

Figure 6-25 shows typical power-supply connections for an IC op-amp. The protective diodes shown are recommended for any power supply circuit where the leads could be accidently reversed. The diodes permit current flow only in the appropriate direction. The op-amp requires two power sources (of 18 V each) with the positive lead of one and the negative lead of the other tied to ground or common.

The two capacitors provide for decoupling of the power supply (signal bypass). Usually, disc ceramic capacitors are used. The capacitors should always be connected as close to the IC terminals as practical, not at the power supply terminals. A guideline for IC op-amp power-supply decoupling capacitors is to use values between 0.1 and 0.001 μF.

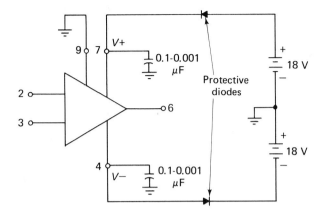

Fig. 6-25 Typical power-supply connections for IC op-amps

6-4.3. Grounding Metal Integrated-Circuit Cases

The metal case of the IC op-amp shown in Fig. 6-25 is connected to terminal 9, *and to no other point in the internal circuit.* Thus, terminal 9 can and should be connected to equipment ground, as well as to the common or ground of the two power supplies.

The metal case of some ICs may be connected to a point in the internal circuit. If so, the case will be at the same voltage as the point of contact. For example, the case might be connected to pin 4 of the IC shown in Figs. 6-24 and 6-25. If so, the case will be below ground (or "hot") by 18 V. If the case is mounted directly on a metal chassis at ground, the IC and power supply can be damaged. Of course, not all ICs have metal cases. Likewise, not all metal cases are connected to the internal circuits. However, this point must be considered *before* connecting a particular IC op-amp.

6-4.4. Calculating Current Required for Integrated-Circuit Op-Amps

The datasheets for IC op-amps usually specify a nominal operating voltage (and possibly a maximum operating voltage), as well as a "total device dissipation." These figures can be used to calculate the current required for a particular IC op-amp. Use simple dc Ohm's law, and divide the power by the voltage to find the current. However, certain points must be considered.

First, use the actual voltage applied to the IC op-amp. The actual voltage should be equal to the nominal operating voltage, but in no event higher than the maximum voltage.

Second, use the total device dissipation. The datasheet may also list other power dissipations, such as "device dissipation," which is defined as the dc power dissipated by the IC itself (with output at zero and no load). The other dissipation figures are always less than the total power dissipation.

6-4.5. Operating Integrated-Circuit Op-Amps from a Single Power Supply

An IC op-amp is generally designed to operate from symmetrical positive and negative power supplies. This results in a high common-mode rejection capability, as well as good low-frequency operation (typically a few hertz down to direct current).

If the loss of very low frequency operation can be tolerated, it is possible to operate IC op-amps from a single power supply, even though designed for dual supplies. Except for the low-frequency loss, the other operating characteristics are unaffected.

The following notes describe a technique that can be used with most IC op-amps to permit operation from a single power supply with a minimum of design compromise. The same maximum device ratings that appear on the datasheet are applicable to the IC when operating from a single polarity power supply and must be observed for normal operation. Likewise, all the considerations discussed thus far in this chapter apply to single-supply operation.

The technique described here is generally referred to as the *split-zener method*. The main concern in setting up for single-supply operation is to maintain the *relative voltage levels*. With an IC op-amp designed for dual-supply operation, there are three reference levels: $+V$, 0, and $-V$. For example, if the datasheet calls for plus and minus 10-V supplies, the three reference levels are $+10$ V, 0 V, and -10 V.

For single-supply operation, these same reference levels can be maintained by using $++V$, $+V$, and ground (that is $+20$ V, $+10$ V and 0 V), where $++V$ represents a voltage level *double* that of $+V$. This is illustrated in Fig. 6-26 where the IC is connected in the split-zener mode. Note that there is no change in the *relative* voltage levels.

Note that the IC of Fig. 6-26 has a ground reference terminal (terminal 3). Not all IC op-amps have such terminals. Some ICs have only $+V$ and $-V$ terminals or leads, even though the two levels are referenced to a common ground. That is, there is no physical ground terminal or lead on the IC, only $+V$ and $-V$ terminals. Figure 6-27 shows the split-zener connections for single-supply operation with such ICs. Here, the input terminals (A and B) are set at one half the total zener supply voltage, the $-V$ terminal is set at ground, and the $+V$ terminal is at the full zener voltage ($+20$ V).

$$R_S = \frac{(\text{maximum supply voltage} - \text{total zener voltage})^2}{\text{safe power dissipation of zeners}}$$

Normal dual-supply
connections

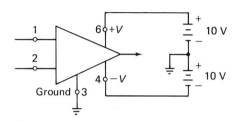

Single-supply
connections

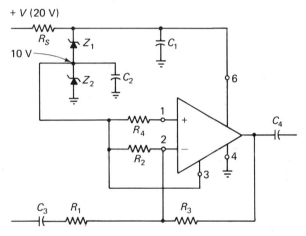

Fig. 6-26 Connections for single-power-supply operation (with ground reference)

Both Figs. 6-26 and 6-27 show a series resistance R_S for the zener diodes. This is standard practice for zener operation. The remaining component values are the same as for dual-supply operation.

The normal IC op-amp frequency-compensation techniques are the same for both types of supplies. The high-frequency limits are essentially the same. However, the low-frequency limit of an IC with a single supply is set by the values of capacitors C_3 and C_4. These capacitors are not required for dual-supply operation. Capacitors C_3 and C_4 are required for single-supply operation since both the input and output of the IC are at a voltage level equal to

$$R_S = \frac{(\text{maximum supply voltage} - \text{total zener voltage})^2}{\text{safe power dissipation of zeners}}$$

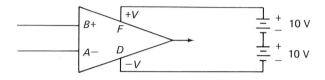

Normal dual-supply
connections

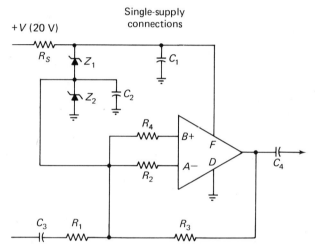

Fig. 6-27   Connections for single-supply operation (without ground reference)

one-half the total zener voltage (or 10 V). *Thus, the IC op-amp cannot be used as a dc amplifier with the single-supply system.*

The closed-loop gain is the same for both types of supplies, and is set by the ratio of R_3/R_1. It may be necessary to use slightly larger values of C_1 and C_2 with the single-supply system, since the impedance of the zeners is probably different than that of the power supply (without zeners).

The value of R_2 is typically 50 to 100 kΩ for an IC op-amp. From a practical standpoint, choose trial values using these guidelines, and then run gain and frequency-response tests. The value of R_4, the input-offset resistance, is chosen to minimize offset error due to impedance unbalance. As an approximate trial value, the resistance of R_4 should be equal to the parallel combination of R_2 and R_3. That is, R_4 is approximately equal to $R_2 R_3/(R_2 + R_3)$.

6-5. TYPICAL INTEGRATED-CIRCUIT OP-AMP CIRCUIT DESIGN

Figure 6-28 is the working schematic of an IC closed-loop op-amp, complete with external circuit parts. The design considerations discussed in all previous sections of this chapter are applicable to the circuit of Fig. 6-28. The following paragraphs provide a specific design example for the circuit.

6-5.1. *Integrated-Circuit Characteristics*

The IC shown in Fig. 6-28 has the following characteristics listed in its datasheet:

Supply voltage: $+10$ and -10 V maximum, $+6$ and -6 V nominal

Total device dissipation: 600 mW, derating 5 mW/°C

Temperature range: -55 to $+125$°C (operating)

Input-offset voltage: 1 mV

Input-offset current: 0.5 μA

Input-offset voltage sensitivity: 0.25 mV/V

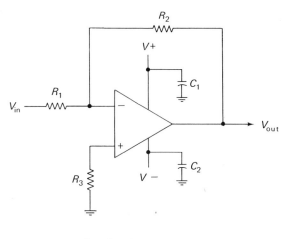

$$C_1 = C_2 = 0.1 - 0.001 \ \mu\text{F}$$

$$V_{out} \approx V_{in} \times \frac{R_2}{R_1}$$

$$R_3 \approx \frac{R_1 R_2}{R_1 + R_2}$$

Fig. 6-28 Typical IC closed-loop op-amp

Device dissipation: 100 mW

Open-loop voltage gain: 60 dB

Open-loop bandwidth: 300 kHz

Common-mode rejection: 100 dB

Maximum output voltage swing: 7 V (peak-to-peak)

Input impedance: 15 kΩ

Output impedance: 200 Ω

Useful frequency range: dc to 15 MHz

Maximum input signal: $+1, -4$ V

6-5.2. Design Example

Assume that the circuit of Fig. 6-28 is to provide a voltage gain of 100 (40 dB), the input signal is 10 mV (rms), the input source impedance is not specified, the output load impedance is 100 Ω, the ambient temperature is 25°C, the frequency range is direct current to 200 kHz, and the power supply is subject to 10 per cent variation.

Note that the frequency and phase-compensation components and values are omitted from Fig. 6-28. As discussed in Sec. 6-2, the compensation method and values found on the IC datasheet must be used. Even these datasheet values should be confirmed by frequency-response test of the final circuit. If no compensation values are available, use the guideline values found in Sec. 6-2, and test the results over the frequency range of interest.

Supply voltage. The positive and negative supply voltages shoud both be 6 V, since this is the nominal value listed. Most IC datasheets will list certain characteristics as maximum (temperature range, total dissipation, maximum supply voltage, maximum input signal), and then list the remaining characteristics as "typical" with a given "nominal" supply voltage.

In no event can the supply voltage exceed the 10-V maximum. Since the available supply voltage is subject to 10 per cent variation, or 6.6-V maximum, it is well within the 10-V limit.

Decoupling or bypass capacitors. The values of C_1 and C_2 should be found on the datasheet. In the absence of a value, use 0.1 μF for any frequency up to 10 MHz. If this value produces a response problem at any frequency (high or low), try any value between 0.001 and 0.1 μF.

Closed-loop resistance. The value of R_2 should be 100 times the value of R_1 to obtain the desired gain of 100. The value of R_1 should be selected so that the voltage drop across R_1 (with the nominal input bias current) is comparable with the input signal (never larger than the input signal). A 100-Ω value for R_1 produces a 0.5 mV drop with the nominal 5-μA bias current. Such a

0.5-mV drop is less than 10 per cent of the 10-mV input signal. Thus, the fixed drop across R_1 should have no appreciable effect on the input signal. With a 100-Ω value for R_1, the value of R_2 must be 10 kΩ (100 × 100 gain = 10 kΩ).

Offset minimizing resistance. The value of R_3 is found using the equation of Fig. 6-28, once the values of R_1 and R_2 are established. Note that the value of R_3 works out to about 99 Ω, using the Fig. 6-28 equation:

$$\frac{R_1 R_2}{R_1 + R_2} = \frac{100 \times 10,000}{100 + 10,000} \approx 99 \ \Omega$$

Thus, a simple *trial value* for R_3 is always *slightly less* than the R_1 value. The final value of R_3 should be such that the no-signal voltages at both inputs are equal.

Comparison of circuit characteristics. Once the values of the external circuit components have been selected, the characteristics of the IC and the external closed-loop circuit should be checked against the requirements of the design example. The following is a summary of the comparison.

Gain. The closed-loop gain should always be less than the open-loop gain. The required closed-loop gain is 100 (40 dB) at a frequency up to 200 kHz. That is, the closed-loop circuit should have a flat frequency response of 40 dB to at least 200 kHz. The open-loop gain is 60 dB, dropping 3 dB down at 300 kHz. Thus, the closed-loop gain is well within tolerance.

Input voltage. The peak input voltage must not exceed the rated maximum input signal. In this case, the rated maximum is $+1$ and -4 V, whereas the input signal is 10 mV rms or approximately 14 mV (peak). This is well below the $+1$-, -4-V maximum limit.

When the rated maximum input signal is an uneven positive and negative value, always use the lowest value for total swing of the input signal. In this case, the input swings from $+14$ to -14 V. An input signal that started from zero could swing as much as and ± 1 V without damaging the IC. An input signal that started from -2 V could swing as much as ± 2 V (from zero to -4 V).

Output voltage. The peak-to-peak output voltage must not exceed the rated maximum output voltage swing (with the required input signal and selected amount of gain).

In this case, the rated maximum output is 7 V P-P, whereas the actual output is approximately 2800 mV (10 mV rms input = 28 mV P-P × gain of 100 = 2800 mV P-P output). Thus, the anticipated output should be well within the rated maximum.

With a gain of 100, the input could go as high as 70 mV P-P (25 mV rms).

Output impedance. Ideally, the closed-loop output impedance should be as low as possible, and always less than the load impedance. The closed-loop output impedance can be found using the equation of Sec. 6-3.2. In this case, the output impedance is

$$\frac{200}{1 + 1000 \times \dfrac{100}{100 + 10,000}}, \quad \text{or approximately } 20 \, \Omega$$

Output power. The output power of an IC op-amp is usually computed on the basis of rms output voltage and output load.

In this case, the output voltage is 1 V rms (10 mV × 100 gain = 1000 mV or 1 V). The load resistance or impedance is 100 Ω, as stated in the design assumption. Thus, the output power is

$$\frac{E^2}{R} = \frac{1^2}{100} = 0.01 \, \text{W} = 10 \, \text{mW}$$

Since the IC dissipation is 100 mW and the total device dissipation is 600 mW, a 10-mW output is well within tolerance. If the case temperature is maintained at 25°C, it is possible to have a 500-mW power output.

6-6. INTEGRATED-CIRCUIT OP-AMP APPLICATIONS

In this section we discuss a few of the many applications for the basic IC op-amp. The reader will note that the power supply and phase/frequency compensation connections are omitted from the schematics in this section. In all the applications it is assumed that the IC is connected to a power source as described in Sec. 6-4. Likewise, it is assumed that a suitable phase/frequency compensation scheme has been selected for the IC, as described in Sec. 6-2. Unless otherwise stated, all the design considerations for the basic IC op-amp described thus far in this chapter apply to each application.

Before going into specific applications, we shall discuss some basic IC op-amp characteristics that are related to all applications. These characteristics include the virtual-ground concept, overload protection, and dc stabilization.

6-6.1. Virtual-Ground Concept

When the positive input of an op-amp is grounded, a concept of *virtual ground* is applied to the negative input. (See Fig. 6-29.) Actually, the dc level at the negative input of an op-amp is very close to ground. When an input signal is applied, the signal tends to move the base away from ground. However, the negative feedback from the output of the amplifier resists this tendency.

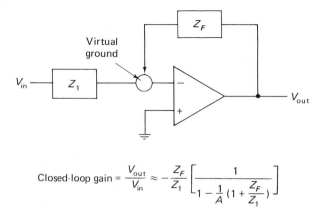

$$\text{Closed-loop gain} = \frac{V_{out}}{V_{in}} \approx -\frac{Z_F}{Z_1}\left[\cfrac{1}{1-\frac{1}{A}(1+\frac{Z_F}{Z_1})}\right]$$

A = open-loop gain

Fig. 6-29 Theoretical feedback circuit and gain equation for op-amp showing virtual ground concept

The amount that the negative input voltage varies with a signal depends on the open-loop gain of the amplifier; the higher the gain, the less the negative input voltage varies. This is the same for any feedback amplifier, as described in previous chapters. (An increase in feedback decreases distortion.)

With the high open-loop gain normally found in an op-amp, the negative input voltage varies only slightly under closed-loop conditions. It is convenient to assume that for all practical purposes the negative input voltage does not change with signal. Thus, it appears as though the negative input is grounded. The term *virtual ground* is used to indicate that although the input of the amplifier appears grounded it is actually not. (Many equations for the functions performed by an op-amp can be derived most easily by the use of the virtual-ground concept.)

It should be noted that since a virtual ground exists at the negative input, the input impedance of the amplifier is essentially determined by the value of the Z_i component. For example, when an op-amp is used in laboratory test equipment, the input component is typically a 50-Ω resistor. Thus, the source sees 50 Ω, no matter what signal level is applied to the input.

6-6.2. Direct-Current Stability and Zero Correction

Since most op-amps are direct coupled, they are subject to various form of drift. For example, a dc amplifier cannot tell the difference between a change in the supply voltage or a change in the signal voltage. Because of the instability problems, many op-amps have a zero correction circuit (in addition to the chopper stabilization circuit discussed in Chapter 4).

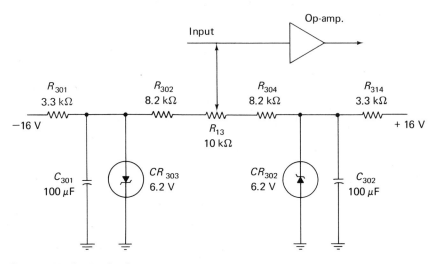

Courtesy Hewlett-Packard

Fig. 6-30   Zero-correction circuit for an op-amp

Generally, the zero correction circuit is external to the op-amp, but can be made part of the internal circuit if desired.

In theory, the output voltage of an op-amp depends on the ratio of input and feedback resistors, as well as the input voltage. If the input voltage is zero, the output voltage will be zero. In practice, however, it will be found that there is a small, unwanted drift voltage, which appears at the amplifier output terminals, even when the input terminals are short circuits (no signal). With solid-stage op-amps, the offset voltage may amount to tens of millivolts, and often comes from transistor leakage currents or static charges.

The undesired voltage (often referred to as zero offset) has both a truly random component (very low frequency noise) and a component that is temperature sensitive. The temperature-dependent component can be taken care of easily by injecting an extremely small offset voltage or current into the input circuit, which cancels the internally generated offset (or offset due to poor design). The zero-correction circuit of Fig. 6-30 is used with an op-amp that is part of a laboratory instrument.

6-6.3. Overload Protection for Op-Amps

Op-amps are often provided with overload-protection circuits. This prevents damage to the transistors, which can occur on overloads due to saturation. The protection also prevents the various ac coupling capacitors (if any) from charging up. Such capacitors are usually large enough in value

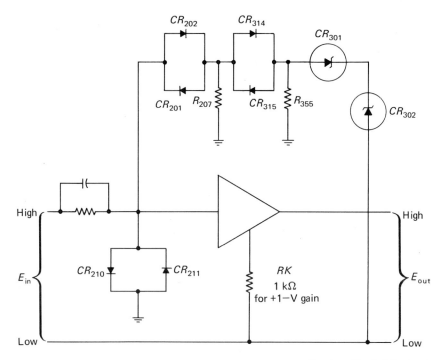

Courtesy Hewlett-Packard

Fig. 6-31   Overload protection circuit for an op-amp

to ensure good low-frequency response. If these capacitors become fully charged, the transient recovery time of the amplifier is long. With proper overload protection, recovery time can be kept to about 20 μs for a typical solid-state op-amp.

Figure 6-31 is the overload protection circuit for a typical op-amp. Input diodes CR_{210} and CR_{211} prevent the input voltage from ever exceeding ± 0.6 V, thus protecting the input transistors from being destroyed. Zener diodes CR_{301} and CR_{302} have nominal breakdown voltages of 10 V, so that whenever the input signal exceeds about ± 11.8 V either CR_{301} or CR_{302} will exhibit zener breakdown at about 10 V. Feedback current from the output terminal cannot flow back to the input unless one of CR_{314} or CR_{315} and one of CR_{201} or CR_{202} are also turned on. Thus, when the output voltage reaches 11.8 V ($10 + 0.6 + 0.6 + 0.6$), large amounts of current from the output terminal are available to oppose the incoming overload current, and keep the entire op-amp from going into saturation.

CR_{301} and CR_{302} cannot be used alone since the reverse leakage of these zener diodes will feed input currents on the order of 1 μA directly into the input. The remaining four diodes (CR_{201}, CR_{202}, CR_{314}, and CR_{315}) and resistors R_{207} and R_{355} shunt the normal zener-diode reverse leakage current to ground.

6-6.4. Op-Amp Summing Amplifier (Adder)

Figure 6-32 is the working schematic of an IC op-amp used as a summing amplifier (or analog adder). Summation of a number of voltages can be accomplished using this circuit. (Voltage summation was one of the original uses for operational amplifiers in computer work.) The output of the circuit is the sum of the various input voltages (a total of four in this case), multiplied by any circuit gain. Generally, gain is set so that the output is at some given voltage value, when all inputs are at their maximum value. In other cases, the values are selected for unity gain.

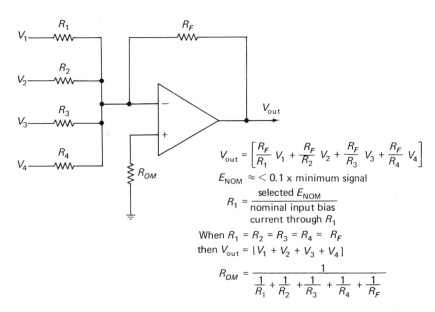

$$V_{out} = \left[\frac{R_F}{R_1} V_1 + \frac{R_F}{R_2} V_2 + \frac{R_F}{R_3} V_3 + \frac{R_F}{R_4} V_4 \right]$$

$E_{NOM} \approx\ < 0.1 \times \text{minimum signal}$

$$R_1 = \frac{\text{selected } E_{NOM}}{\substack{\text{nominal input bias} \\ \text{current through } R_1}}$$

When $R_1 = R_2 = R_3 = R_4 = R_F$

then $V_{out} = [V_1 + V_2 + V_3 + V_4]$

$$R_{OM} = \frac{1}{\frac{1}{R_1} + \frac{1}{R_2} + \frac{1}{R_3} + \frac{1}{R_4} + \frac{1}{R_F}}$$

Fig. 6-32 Summing amplifier using op-amp

6-6.5. Op-Amp Scaling Amplifier (Weighted Adder)

Figure 6-33 is the working schematic of an IC op-amp used as a scaling amplifier or weighted adder. A scaling amplifier is essentially the same as a summing amplifier, except that the inputs to be summed are *weighted* or compensated to produce a given output range, or a given relationship between inputs.

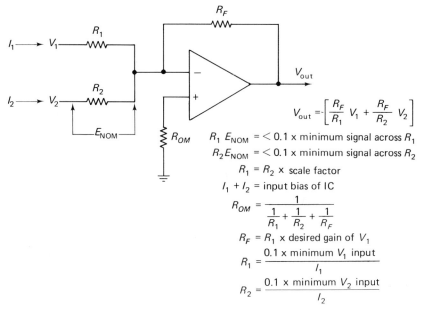

$$V_{out} = -\left[\frac{R_F}{R_1}V_1 + \frac{R_F}{R_2}V_2\right]$$

$R_1 E_{NOM} = <0.1 \times$ minimum signal across R_1

$R_2 E_{NOM} = <0.1 \times$ minimum signal across R_2

$R_1 = R_2 \times$ scale factor

$I_1 + I_2 =$ input bias of IC

$$R_{OM} = \frac{1}{\frac{1}{R_1} + \frac{1}{R_2} + \frac{1}{R_F}}$$

$R_F = R_1 \times$ desired gain of V_1

$$R_1 = \frac{0.1 \times \text{minimum } V_1 \text{ input}}{I_1}$$

$$R_2 = \frac{0.1 \times \text{minimum } V_2 \text{ input}}{I_2}$$

Fig. 6-33 Scaling amplifier (weighted adder) using op-amp

For example, assume that there are two inputs to be summed, and that one input has a nominal voltage range *five times* that of the other. Now assume that it is desired to have the *output voltage range be the same* for both inputs. This can be done by making the input resistance for the low voltage input one fifth the value of the high input resistance. Since the feedback resistance is the same for both inputs, and gain (or output) is set by the ratio of feedback-to-input resistance, the low input is multiplied five times as much as the high input. Thus, the output range is the same for both inputs.

In other cases, a scaling amplifier is used to make two equal inputs produce two outputs of different voltage ranges. Likewise, the scaling amplifier can be weighted so that two unequal inputs are made more (or less) unequal by a given *scale factor*. For example, if the inputs are normally 7 to 1, they can be made 3 to 1, 8 to 1, or any practical value within the limits of the IC op-amp.

6-6.6. Op-Amp Difference Amplifier

Figure 6-34 is the working schematic of an IC op-amp used as a difference amplifier and/or subtractor. One signal voltage is subtracted from another through simultaneous applications of signals to both inputs.

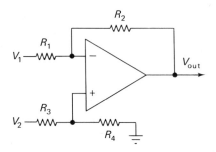

$$V_{out} = -(\frac{R_2}{R_1} \times V_1) + \left[(\frac{R_4}{R_3 + R_4})(\frac{R_1 + R_2}{R_1}) V_2\right]$$

when $R_1 = R_2 = R_3 = R_4$

then $V_{out} = V_2 - V_1$

$R_4 = R_2 = R_1 \times$ gain

$R_3 = \dfrac{R_1}{\text{gain}}$

Fig. 6-34   Difference amplifier and/or subtractor using op-amp

If the values of all resistors are the same, the output is equal to the voltage at the positive (noninverting) input, *less* the voltage at the negative (inverting) input. The output also represents the difference between the two input voltages. Thus, the same circuit can be used as a subtractor or difference amplifier.

Generally, all resistors (R_1 through R_4) are made the same value when the circuit is to be used as a subtractor. Using all four resistors of the same value provides the greatest accuracy.

When some gain is required, the circuit is usually considered as a difference amplifier (rather than a subtractor). The gain is directly proportional to the ratio of R_2/R_1. Under these conditions, the output is approximately equal to the algebraic sum (or difference) of the gains for the two input voltages, as shown by the equation of Fig. 6-34.

6-6.7. Op-Amp Unity-Gain Amplifier (Voltage Follower or Source Follower)

Figure 6-35 is the working schematic of an IC op-amp used as a unity-gain amplifier (also known as a voltage follower or a source follower). There are no feedback or input resistances in the circuit. Instead, the output is *fed back directly to the inverting input.* Signal input is applied directly to the noninverting input. With this arrangement, the output voltage equals the input voltage (or may be slightly less than the input voltage). However, the

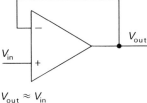

$$V_{out} \approx V_{in}$$

Closed-loop Z_{in} ≈ open-loop gain x open-loop Z_{in} op. amp.

Fig. 6-35   Unity-gain amplifier using op-amp

$$\text{Closed-loop } Z_{out} \approx \frac{\text{open-loop } Z_{out} \text{ of op. amp.}}{\text{open-loop gain}}$$

input impedance is very high, with the output impedance very low (as shown by the equations of Fig. 6-35).

6-6.8. Op-Amp High-Input-Impedance Amplifier

Figure 6-36 is the working schematic of an IC op-amp used as a high-input-impedance amplifier. The high-input-impedance and low-output-impedance features of the unity-gain amplifier (Sec. 6-6.7) are combined with modest gain.

Note that the circuit of Fig. 6-36 is similar to the basic op-amp, except that there is no input offset compensating resistance (in series with the noninverting input). This results in a tradeoff of higher input impedance, with some increase in the output-offset voltage. In the basic op-amp, an offset compensating resistance is used to nullify the input-offset voltage of the IC. This (theoretically) results in no offset at the output. The output of the basic op-

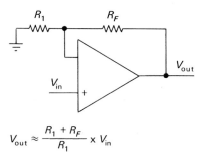

$$V_{out} \approx \frac{R_1 + R_F}{R_1} \times V_{in}$$

$$\text{Closed-loop } Z_{out} \approx \frac{\text{open-loop } Z_{out} \text{ of op. amp.}}{1 + \text{open-loop gain } (\frac{R_1}{R_1 + R_F})}$$

$$R_F \approx (\text{gain} - 1) \times R_1$$

Closed-loop Z_{in} ≈ open-loop Z_{in} of op. amp. x closed-loop gain

Fig. 6-36   High input impedance amplifier using op-amp

amp is at 0 V, in spite of the tremendous gain. In the unity-gain amplifier there is no offset compensating resistance, but, since there is no gain, the output is at the same offset as the input. Typically, input offset is on the order of 1 mV. This figure should not be critical for the output of a typical unity-gain-amplifier application.

With the circuit of Fig. 6-36, the offset compensation resistance is omitted. The output is therefore offset by an amount equal to the input-offset voltage of the IC multiplied by the closed-loop gain. However, since the circuit of Fig. 6-36 is to be used for modest gains, a modest output offset results.

6-6.9. Op-Amp Unity-Gain Amplifier with Fast Response (Good Slew Rate)

One of the problems with a unity-gain amplifier is that the slew rate is very poor. That is, the amplifier response time is very slow, and the power bandwidth is decreased, as discussed in Sec. 6-3.1 The reason for poor slew rate with unity gain is that most IC datasheets recommend a large value compensating capacitor for unity gain. For example, a typical datasheet recommendation for compensating capacitance is 0.01 μF (with a gain of 100). All other factors being equal (but with unity gain), the same datasheet recommends a compensating capacitance of 1 μF.

There are several methods used to provide fast response time (high slew rate) and a good power bandwidth with unity gain. Two such methods are summarized here.

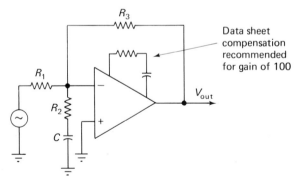

$R_1 = R_3 = R_2 \times 100 \text{ (gain)}$

$R_2 = \dfrac{R_1}{100}$

$C = \dfrac{1}{6.28 \times R_2 \times F}$

Slew rate $\approx$ slew rate for gain of 100

Fig. 6-37 Unity-gain IC op-amp with fast response (good slew rate) using IC datasheet phase compensation

Using integrated-circuit datasheet phase compensation. The circuit of Fig. 6-37 shows a method of connecting an IC op-amp for unity gain, but with high slew rate (fast response time and good power bandwidth). With this circuit, the phase compensation recommended on the IC op-amp datasheet is used, but with modification. Instead of using the unity-gain compensation (usually a high-value compensating capacitor), use the datasheet phase compensation recommended for a gain of 100 (generally a much lower value capacitor). Then select values of R_1 and R_3 to provide unity gain. That is, R_1 and R_3 must be the same value. As shown by the equations, the value of R_1 and R_3 must be approximately 100 times the value of R_2 (when the datasheet phase compensation for a gain of 100 is chosen). Thus, R_1 and R_3 must be fairly high values for practical purposes.

Using input phase compensation. The circuit of Fig. 6-38 shows a method of connecting an IC op-amp for unity gain, but with high slew rate, using input compensation. With this circuit, the phase compensation recommended on the IC datasheet is not used. Instead, the input phase compensation system of Sec. 6-2.3 is used.

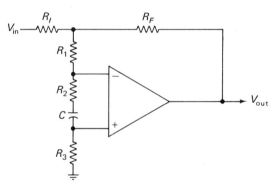

Fig. 6-38 Unity-gain IC op-amp with fast response (good slew rate) using input phase compensation

R_1, R_2, R_3, C = see text
$R_I = R_F = 0.25 \times R_1$

The first step is to compensate the op-amp by modifying the open-loop input impedance, as described in Sec. 6-2.3.

Next, select values of R_I and R_F to provide unity gain (both R_I and R_F must be the same value). Using the equations of Fig. 6-38, the values of R_I and R_F should be approximately one-fourth the value of R_1 (and R_3). Assuming that the values of R_1 and R_3 are 43 kΩ (using the example of Sec. 6-2.3), 10 kΩ is a good trial value for R_I and R_F.

6-6.10. Op-Amp Narrow Bandpass Amplifier

Figure 6-39 is the working schematic of an IC op-amp used as a narrow bandpass amplifier (or tuned peaking amplifier).

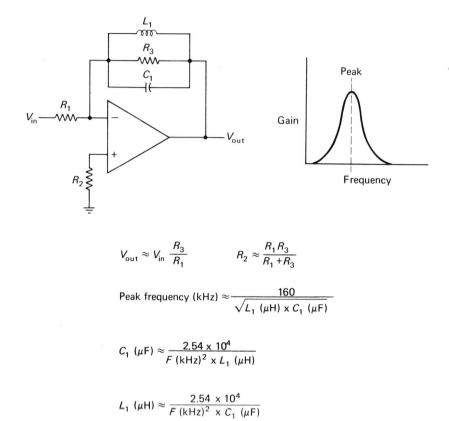

$$V_{out} \approx V_{in} \frac{R_3}{R_1} \qquad R_2 \approx \frac{R_1 R_3}{R_1 + R_3}$$

$$\text{Peak frequency (kHz)} \approx \frac{160}{\sqrt{L_1 \ (\mu H) \times C_1 \ (\mu F)}}$$

$$C_1 \ (\mu F) \approx \frac{2.54 \times 10^4}{F \ (kHz)^2 \times L_1 \ (\mu H)}$$

$$L_1 \ (\mu H) \approx \frac{2.54 \times 10^4}{F \ (kHz)^2 \times C_1 \ (\mu F)}$$

Fig. 6-39   Narrow bandpass amplifier (tuned peaking) using op-amp

Circuit gain is determined by the ratio of R_1 and R_F in the usual manner. However, the frequency at which maximum gain occurs (or the narrowband peak) is the resonant frequency of the $L_1 C_1$ circuit. Capacitor C_1 and inductance L_1 form a parallel-resonant circuit that rejects the resonant frequency. Therefore, there is minimum feedback (and maximum gain) at the resonant frequency.

6-6.11. Op-Amp Wide Bandpass Amplifier

Figure 6-40 is the working schematic of an IC op-amp used as a wide bandpass amplifier.

Maximum circuit gain is determined by the ratio of R_R and R_F. That is, the gain of the passband or *flat portion* of the response curve is set by R_F/R_R.

Minimum circuit gain is determined by the ratio of R_1 and R_F in the usual manner.

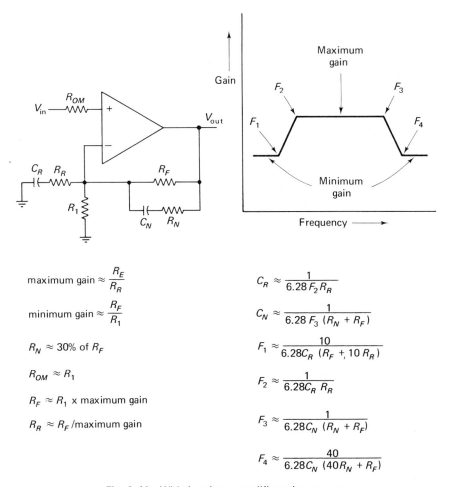

$$\text{maximum gain} \approx \frac{R_F}{R_R}$$

$$\text{minimum gain} \approx \frac{R_F}{R_1}$$

$$R_N \approx 30\% \text{ of } R_F$$

$$R_{OM} \approx R_1$$

$$R_F \approx R_1 \times \text{maximum gain}$$

$$R_R \approx R_F / \text{maximum gain}$$

$$C_R \approx \frac{1}{6.28 \, F_2 \, R_R}$$

$$C_N \approx \frac{1}{6.28 \, F_3 \, (R_N + R_F)}$$

$$F_1 \approx \frac{10}{6.28 C_R \, (R_F + 10 R_R)}$$

$$F_2 \approx \frac{1}{6.28 C_R \, R_R}$$

$$F_3 \approx \frac{1}{6.28 C_N \, (R_N + R_F)}$$

$$F_4 \approx \frac{40}{6.28 C_N \, (40 R_N + R_F)}$$

Fig. 6-40 Wide bandpass amplifier using op-amp

The frequencies at which rolloff starts and ends (at both high- and low-frequency limits) are determined by impedances of the various circuit combinations, as shown by the equations of Fig. 6-40.

Note that if phase compensation is required for the basic IC op-amp the compensation values must be based on the minimum gain, and not on the passband gain.

6-6.12. Op-Amp Integration Amplifier

Figure 6-41 is the working schematic of an IC op-amp used as an integration amplifier (or integrator). Integration of various signals (usual-

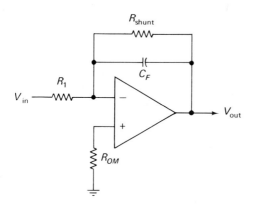

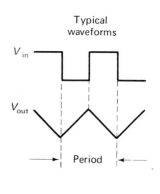

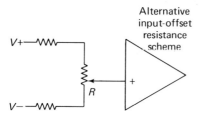

$$V_{out} \approx \frac{1}{R_1 C_F} \int V_{in} \, \Delta time$$

$R_1 \times C_F \approx$ period of signal to be integrated

Alternative
input-offset
resistance
scheme

$$C_F \approx \frac{period}{R_1}$$

$$R_{shunt} \approx 10 \times R_1, \qquad R_{OM} \approx \frac{R_1 R_{shunt}}{R_1 + R_{shunt}}$$

V_{out} shifted by $+90°$ from V_{in}

Fig. 6-41   Integration amplifier (integrator) using op-amp

ly squarewaves) can be accomplished using this circuit. The output voltage from the amplifier is inversely proportional to the time constant of the feedback network, and directly proportional to the integral of the input voltage.

The value of R_1 is chosen on the basis of input bias current and voltage drop. The value of the $R_1 C_F$ time constant must be *approximately equal to the period* of the signal to be integrated. The value of the $R_{shunt} C_F$ time constant must be *substantially larger than the period* of the signal to be integrated (approximately 10 times longer). Thus, R_{shunt} is 10 times R_1.

Keep in mind that R_{shunt} and C_F form an impedance which is frequency sensitive (that is, most noticed at low frequencies).

6-6.13. Op-Amp Differentiation Amplifier

Figure 6-42 is the working schematic of an IC op-amp used as a differentiation amplifier (or differentiator). Differentiation of various signals (usually squarewaves, or sawtooth and sloping waves) can be accom-

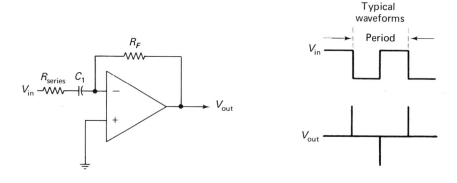

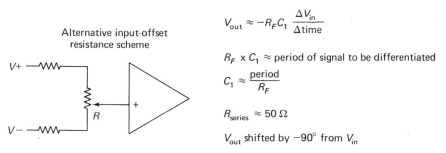

$$V_{out} \approx -R_F C_1 \frac{\Delta V_{in}}{\Delta \text{time}}$$

$R_F \times C_1 \approx$ period of signal to be differentiated

$$C_1 \approx \frac{\text{period}}{R_F}$$

$R_{series} \approx 50\,\Omega$

V_{out} shifted by $-90°$ from V_{in}

Fig. 6-42   Differentiation amplifier (differentiator) using op-amp

plished using this circuit. The output voltage from the amplifier is inversely proportional to the *time rate of change* of the input voltage.

The value of the $R_F C_1$ time constant should be *approximately equal* to the period of the signal to be differentiated. In practical applications, the time constant must be chosen on a trial-and-error basis to obtain a reasonable output level.

The main problem in design of differentiating amplifiers is that the *gain increases with frequency*. Thus, differentiators are very susceptible to high-frequency noise. The classic remedy for this effect is to connect a small resistor (on the order of 50 Ω) in series with the input capacitor so that the high-frequency gain is decreased. The addition of the resistor results in a more realistic model of the differentiator, because a resistance is always added in series with the input capacitance by the signal source impedance.

Conversely, in some applications a differentiator may be advantageously used to detect the presence of distortion or high-frequency noise in the signal.

A differentiator can often detect hidden information that is not detected in the original signal.

Differentiation permits slight changes in input slope to produce very significant changes in output. An example of this feature is in determining the linearity of a sweep sawtooth waveform. Nonlinearity results from changes in slope of waveform. Therefore, if nonlinearity is present, the differentiated waveform (amplifier output) indicates the points of nonlinearity quite clearly. (However, it should be noted that repetitive waveforms with a rise and fall of differing slopes can show erroneous waveforms.)

6-6.14. Op-Amp Voltage-to-Current Converter

Figure 6-43 is the working schematic of an IC op-amp used as a voltage-to-current converter (or transadmittance amplifier). This circuit is used to supply a current (to a variable load) that is proportional to the voltage applied at the input of the amplifier (rather than proportional to the

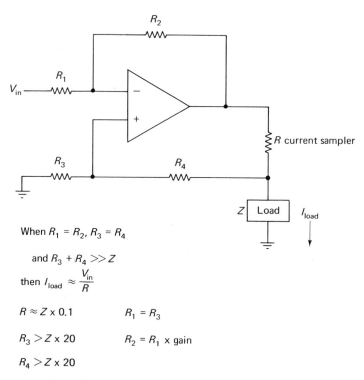

When $R_1 = R_2$, $R_3 = R_4$

and $R_3 + R_4 \gg Z$

then $I_{load} \approx \dfrac{V_{in}}{R}$

$R \approx Z \times 0.1$ $R_1 = R_3$

$R_3 > Z \times 20$ $R_2 = R_1 \times \text{gain}$

$R_4 > Z \times 20$

Fig. 6-43   Voltage-to-current converter using op-amp

load). The current supplied to the load is *relatively independent* of the load characteristics. This circuit is essentially a *current feedback* amplifier.

Current sampling resistor R is used to provide the feedback to the positive input. When R_1, R_2, R_3, and R_4 are all the same value, the feedback maintains the voltage across R at the same value as the input voltage. When R_2 is made larger than R_1, the voltage across R remains constant at a value equal to the ratio R_2/R_1.

If a constant input voltage is applied to the amplifier, the voltage across R also remains constant, regardless of the load (with close tolerances). If the voltage across R remains constant, the current through R must also remain constant. With R_3 and R_4 normally much larger than the load impedance, the current through the load remains nearly constant, regardless of a change in impedance.

The most satisfactory configuration for the circuit of Fig. 6-43 is where the op-amp is operated as a unity-gain amplifier with the values of R_1 through R_4 all the same. This requires that the input voltage be sufficient to produce the desired current (or power) for the load. If the input voltage or signal is not sufficient, the values of R_1 and R_2 must be selected to provide the necessary gain.

The value of R must be selected to limit the output power to a value within the capability of the op-amp. Output power is found by $I^2 \times (R + \text{load})$. For example, if the IC op-amp is rated at 600-mW total dissipation, with 100-mW dissipation for the basic IC, the total output power must be limited to 500 mW. As a guideline, make the value of R approximately one tenth the load.

6-6.15. Op-Amp Voltage-to-Voltage Amplifier

Figure 6-44 is the working schematic of an IC op-amp used as a voltage-to-voltage converter (voltage-gain amplifier). This circuit is similar to the voltage-to-current converter (Sec. 6-6.14), except that the load and the current sensing resistor are transposed. The voltage across the load is relatively independent of the load characteristics.

The most satisfactory configuration for the circuit of Fig. 6-44 is where the IC is operated as a unity-gain amplifier with the values of R_1 through R_4 all the same. This requires that the input voltage be sufficient to produce the desired current (or power) for the load. If the input voltage or signal is not sufficient, the values of R_1 and R_2 must be selected to provide the necessary gain.

The value of R must be selected to limit the output power to a value within the capability of the IC, as is the case for the voltage-to-current converter.

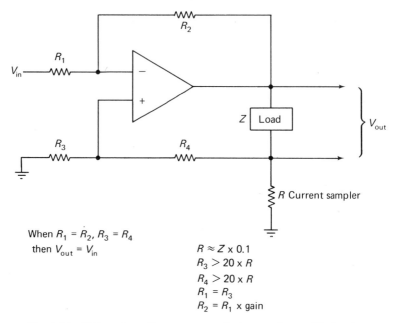

When $R_1 = R_2$, $R_3 = R_4$
then $V_{out} = V_{in}$

$R \approx Z \times 0.1$
$R_3 > 20 \times R$
$R_4 > 20 \times R$
$R_1 = R_3$
$R_2 = R_1 \times$ gain

Fig. 6-44   Voltage-to-voltage converter (voltage-gain amplifier) using op-amp

6-6.16. Op-Amp Parallel-T Filter

Figure 6-45 is the working schematic of an IC op-amp used as a low-frequency filter. The operating principle involved is similar to that of a parallel-T oscillator. However, the function is that of a *narrowband filter* (tuned peaking amplifier) described in Sec. 6-6.10. The filter circuit described in Sec. 6-6.10 uses an inductance as part of the resonant circuit. At very low frequencies, the high values of inductance (and capacitance) required make a circuit similar to that of Sec. 6-6.10 impractical. Thus, for low frequencies, the parallel-T (or twin-T as it is sometimes called) filter is generally a better choice.

In the circuit of Fig. 6-45, gain is set by open-loop gain of the op-amp. For this reason, the parallel-T filter is somewhat less stable than the tuned filter.

The frequency at which maximum gain occurs (or narrowband peak) is set by values of R and C. Any combination of R and C can be used, provided that they work out to the desired frequency. However, the value of R should be selected on the basis of the load resistance (or load impedance). The value of R and the load are, in effect, in parallel. If the value of R is many times (at

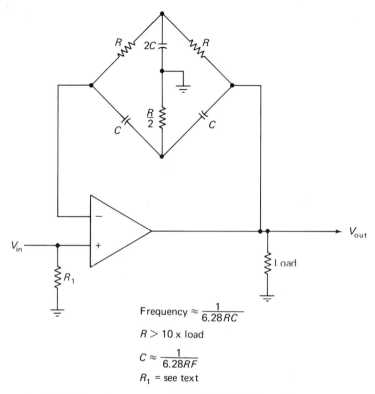

$$\text{Frequency} \approx \frac{1}{6.28RC}$$

$R > 10 \times \text{load}$

$$C \approx \frac{1}{6.28RF}$$

$R_1 = \text{see text}$

Fig. 6-45　Low-frequency parallel-T (twin-T) filter using op-amp

least 10) that of the load, the net parallel resistance is just slightly less than the load. Thus, the output current requirements for the op-amp are increased only slightly (for a given voltage).

6-6.17. Op-Amp Log and Antilog Amplifiers

Figure 6-46 is the working schematic of an IC op-amp used as a log (logarithmic) amplifier. A log amplifier is nonlinear so that a large input variation produces only a small output variation. This is shown by the curve of Fig. 6-47, where the input varies from 1 mV to 100 V, but produces an output variation from about 350 to 640 mV (an approximate 300-mV output swing). Note that the output appears as a straight line on Fig. 6-47, since the horizontal lines represent logarithmic variations (5 decades in this case).

The amplifier of Fig. 6-46 compresses the 5 decades of information into a small output swing. One circuit requiring such a log amplifier is a display which reads out data that span many orders of magnitude in a single range.

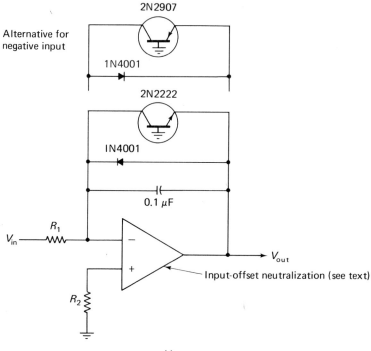

$$R_1 = R_2 \geqq \frac{\max V_{in}}{\text{max input current of IC and transistor}}$$

$$R_1 \leqq \frac{\min V_{in}}{\text{Input bias current of IC}}$$

Fig. 6-46 Logarithmic amplifier using op-amp

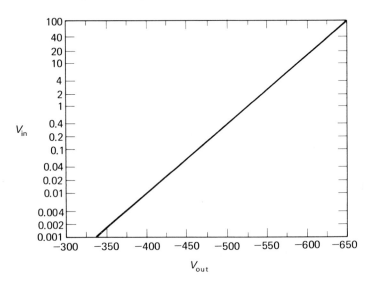

Fig. 6-47 Input versus response of logarithmic amplifier

An *NPN* transistor is used when the input is positive. A negative input requires a *PNP*, as shown in Fig. 6-46.

The circuit uses the base–emitter junction of a transistor to provide the logarithmic response. The principle of operation relies on the exponential relationship of the base–emitter voltage and collector current of the transistor. All transistors do not exhibit good logarithmic characteristics, so care must be taken in selecting the proper transistor. Likewise, since the transistor junction is temperature sensitive, the amplifier response is also subject to temperature variations.

Capacitor C_1 across the feedback transistor is necessary to reduce the ac gain (and thus reduce noise pickup). Use a value of 0.1 μF for C_1 at frequencies up to about 10 MHz. The diode CR_1 protects the transistor against excessive reverse base–emitter voltages, should the polarity of the input voltage be reversed accidentally.

A log amplifier generally requires an IC that has provisions for neutralizing input offset voltage (Sec. 6-3.10). The effects of offset voltage are noticed at the low limit of the input range. This is shown in Fig. 6-48.

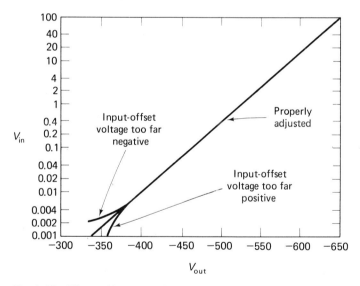

Fig. 6-48   Effects of improper offset adjustment on logarithmic amplifier

If an IC op-amp without input-offset provisions must be used, it is possible to provide offset neutralization with the circuit of Fig. 6-49. This circuit uses the available values of V_{CC} and V_{EE}, and provides both coarse and fine neutralization of the input-offset voltage.

The circuit of Fig. 6-49 also provides for neutralization or compensation of the input bias current. As a rule, an IC op-amp with the lowest possible

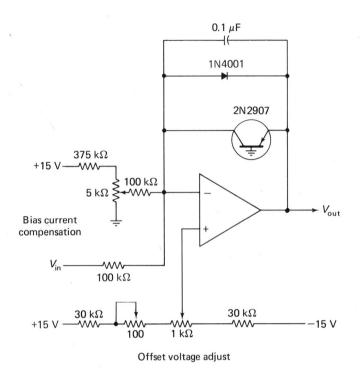

Fig. 6-49 Alternative method to provide offset neutralization (for ICs without input-offset provisions)

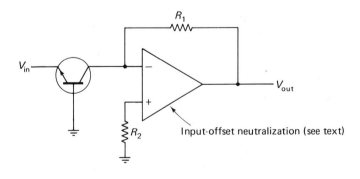

$R_1 = R_2$ = see text

Fig. 6-50 Antilogarithmic amplifier using op-amp

input bias current is best for log amplifiers. This permits operation with low input signal voltages.

Figure 6-50 is the working schematic of an IC op-amp used as an *antilog amplifier*. The antilog amplifier is the complement of a log amplifier (the transistor is in the input circuit rather than as the feedback element). In an antilog circuit, larger input voltages produce progressively higher amplifier gains. Because the input is applied across the transistor base–emitter junction, input voltages are generally 1 V or less. The maximum circuit output voltage is limited by the maximum voltage swing of the IC op-amp.

6-6.18. Op-Amp High-Impedance Bridge Amplifier

Figure 6-51 is the working schematic of an IC op-amp used as a high-impedance bridge amplifier. The circuit is not limited to use with a bridge. However, a bridge of any type provides greater accuracy when its output is fed to a high-impedance amplifier. Under these conditions, little or no current is drawn from the bridge, and the bridge voltage output is amplified as necessary to provide a given reading.

Amplification for the entire circuit is provided by IC_3, and is dependent upon the ratio of R_2/R_1. IC_1 and IC_2 act as voltage followers, and provide no amplification. However, IC_1 and IC_2 do provide a high impedance to the bridge (easily 100 kΩ, and above).

The value of R_1 is chosen on the basis of input bias current and voltage drop, as described previously in this chapter. The value of R_2 is selected to

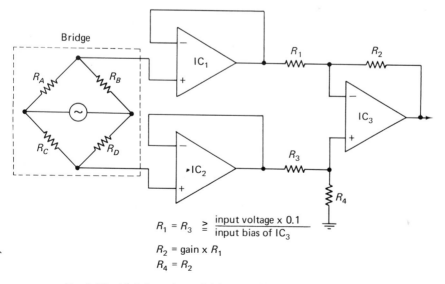

$$R_1 = R_3 \geq \frac{\text{input voltage} \times 0.1}{\text{input bias of } IC_3}$$

$$R_2 = \text{gain} \times R_1$$

$$R_4 = R_2$$

Fig. 6-51 High-impedance bridge amplifier using three op-amps

provide the desired gain. Resistors R_3 and R_4 should be equal to the values of R_1 and R_2, respectively.

6-6.19. Op-Amp for Differential Input and Differential Output

Figure 6-52 is the working schematic of two IC op-amps connected to provide differential-in/differential-out amplification. Generally, IC op-amps have a single-ended output. There are cases where a differential output is required. For example, the circuit of Fig. 6-52 can be used at the input of a system where there is considerable noise pickup at the input leads, but the input differential signal is very small. Since the noise signals are common-mode, they are not amplified. However, the small differential input is amplified.

The circuit of Fig. 6-52 can be formed with two identical IC op-amps or a dual-channel IC op-amp. The dual-channel IC is preferable since both channels have identical characteristics (gain, bias, etc.). This is because both channels are fabricated on the same semiconductor chip. However, the circuit

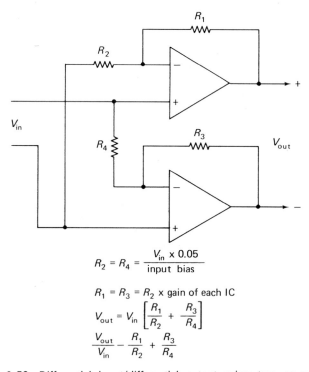

$$R_2 = R_4 = \frac{V_{in} \times 0.05}{input\ bias}$$

$$R_1 = R_3 = R_2 \times gain\ of\ each\ IC$$

$$V_{out} = V_{in} \left[\frac{R_1}{R_2} + \frac{R_3}{R_4} \right]$$

$$\frac{V_{out}}{V_{in}} - \frac{R_1}{R_2} + \frac{R_3}{R_4}$$

Fig. 6-52 Differential input/differential output using two op-amps

will work satisfactorily if the two ICs are carefully matched as to characteristics (particularly in regards to input-bias and input-offset voltage).

As shown by the equations, each IC (or each channel) provides one half the total differential output gain. Thus, if each IC provides a gain of 10, the differential output is 20 times the differential input.

The output voltage swing is double that of the individual ICs. Generally, the maximum output swing of a single-ended IC op-amp is slightly less than the $V_{CC} - V_{EE}$ voltages. Using the circuit of Fig. 6-52, the maximum differential output swing is twice that of the $V_{CC} - V_{EE}$ voltages. For example, if the $V_{CC} - V_{EE}$ voltages are 6 V, the maximum differential output is slightly less than 12 V.

The values of R_2 and R_4 are chosen on the basis of input bias current and voltage drop, as normal. The values of R_1 and R_3 are chosen to provide the desired gain. Resistors R_3 and R_4 should be equal in value to R_1 and R_2, respectively.

6-6.20. Op-Amp with Zero Offset Suppression

Figure 6-53 is the working schematic of an IC op-amp used to provide amplification of a small signal riding on a large, fixed, dc level. For

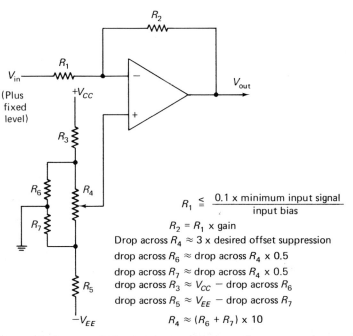

$$R_1 \leq \frac{0.1 \times \text{minimum input signal}}{\text{input bias}}$$

$R_2 = R_1 \times$ gain

Drop across $R_4 \approx 3 \times$ desired offset suppression

drop across $R_6 \approx$ drop across $R_4 \times 0.5$

drop across $R_7 \approx$ drop across $R_4 \times 0.5$

drop across $R_3 \approx V_{CC} -$ drop across R_6

drop across $R_5 \approx V_{EE} -$ drop across R_7

$R_4 \approx (R_6 + R_7) \times 10$

Fig. 6-53   Integrated-circuit op-amp with zero offset suppression to provide amplification of small signal riding on large, fixed level

example, assume that the input signal varies between 2 and 10 mV, and that the signal source never drops below $+5$ V. That is, the source is $+5$ V with no signal and $+5.002$ to $+5.010$ V with signal. Now assume that the output is to vary between 200 mV and 1 V.

The obvious solution is to apply a fixed $+5$ V to the noninverting input. This will offset the $+5$ V at the inverting input and result in 0-V output (under no-signal conditions). This solution ignores the fact that the IC op-amp probably has some characteristic input offset V_{io}, or assumes that the IC has some provision for neutralizing V_{io}. The solution also assumes that the signal is riding on exactly $+5$ V.

If V_{io} cannot be ignored (say because it is large in relation to the signal), or if the fixed dc voltage is subject to possible change, the alternative offset circuit of Fig. 6-53 should be used. The circuit is a simple resistance network that makes use of existing V_{CC} and V_{EE} voltages. In use, potentiometer R_4 is adjusted to provide zero output from the IC under no-signal conditions. That is, the 2- to 10-mV signal is removed, but the $+5$ V remains at the inverting input, while R_4 is adjusted for zero at the IC output.

The values for offset networks are not critical. However, the values should be selected so that a minimum of current is drawn from the $V_{CC} - V_{EE}$ supplies, and a minimum of current should flow through R_4.

Although it may not be obvious, the principles shown in Fig. 6-53 can be applied to most of the IC op-amp applications described in this chapter. That is, any zero offset (at the IC input and output) can be suppressed by applying a fixed (or adjustable) voltage of correct polarity and amplitude to the opposite input.

6-6.21. Op-Amp Track-and-Hold Amplifier

Figure 6-54 is the working schematic of two IC op-amps used as a track-and-hold amplifier (also known as a *sample-and-hold* amplifier).

With this circuit, when voltages are applied to the gate inputs, the diode bridge conducts, and the output voltage tracks the input voltage. However, the polarity of the output signal is reversed from that of the input. If it is necessary to have an output that tracks the input directly, an IC connected as a unity-gain inverter can be used at the output.

With conventional silicon diodes, which have a normal voltage drop of about 0.7 V, the gate voltage must be about 1.5 V, plus the maximum input voltage. When the gate voltage polarity is reversed from that shown in Fig. 6-54 by an external switching system, the diode bridge stops conducting. Due to the charge on capacitor C_1, the value of the output voltage is equal to that of the input voltage just prior to switching the bridge off. The output voltage remains at this value for a time period, the length of which is dependent on the value of C_1, the diode leakage current, and the input bias current of the

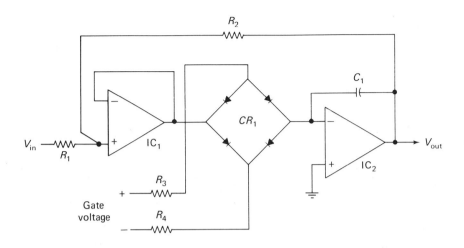

$$R_1 \lesssim \frac{V_{in} \times 0.1}{\text{input bias of IC}}$$

$R_2 = R_1 \times$ gain

$R_3 = R_4 = 1 \text{ k}\Omega$

gate voltage $\approx 1.5 + \text{max } V_{in}$

$$C_1 \approx \frac{t (I_R + I_B)}{\Delta_{out}}$$

t = time in seconds for no change in V_{out}

$I_R = CR_1$ diode leakage

$I_B = IC_2$ input bias

Δ_{out} = difference in output voltage

Fig. 6-54   Track-and-hold (sample-and-hold) amplifier using op-amp

IC, as shown by the equation. Neither of the currents can be altered. However, the time period can be set to a given value by proper selection of capacitor C_1 value.

The value of R_1 is chosen on the basis of input bias current and voltage drop. Since the usual purpose of the circuit is to track the input voltage, no gain is required. Thus, R_2 should be the same value as R_1, at least as the first trial value. However, if the circuit shows some loss during test (output voltage is less than input voltage), increase the value of R_2 as necessary to produce the desired output.

The values of R_3 and R_4 are not critical. Use the 1-kΩ values for both resistors, assuming that standard silicon diodes are used and that the gate voltage does not exceed about 7 V.

7. AMPLIFIER TEST
AND
TROUBLESHOOTING

This chapter is devoted to test and troubleshooting techniques for all types of amplifiers. It is assumed that the reader is already familiar with the use of common electronic test equipment and with basic solid-state troubleshooting techniques. These subjects are covered thoroughly in the author's *Handbook of Electronic Test Equipment* (Prentice-Hall, Inc., Englewood Cliffs, N.J., 1971); and *Handbook of Practical Solid-State Troubleshooting* (Prentice-Hall, Inc., Englewood Cliffs, N.J., 1971), respectively. The following sections describe test and troubleshooting procedures that apply specifically to amplifiers.

7-1. BASIC AUDIO-AMPLIFIER TESTS

The following paragraphs do not describe every possible test to which audio circuits can be subjected. However, they do include all basic tests necessary for "typical" audio-amplifier circuit operation.

7-1.1. Frequency Response

The frequency response of an audio amplifier, or filter, can be measured with an audio signal generator and a meter or oscilloscope. When a meter is used, the signal generator is tuned to various fequencies, and the resultant circuit output response is measured at each frequency. The results are then plotted in the form of a graph or *response curve*, such as shown in Fig. 7-1. The procedure is essentially the same when an oscilloscope is used to measure audio-circuit frequency response. However, an oscilloscope gives the added benefit of visual distortion analysis, as discussed in Sec. 7-1.10.

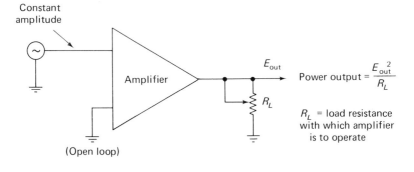

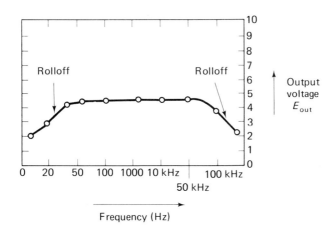

Fig. 7-1　Linear IC frequency-response test connections and typical response curve

The basic frequency-response measurement procedure (with either meter or oscilloscope) is to apply a *constant-amplitude* signal while monitoring the circuit output. The input signal is varied in frequency (but not amplitude) across the entire operating range of the circuit. Any well-designed audio circuit should have a constant response from about 20 Hz to 20 kHz. With direct-coupled amplifiers, the response can be extended from a few hertz (or possibly from direct current) on up to 100 kHz (and higher). The voltage output at various frequencies across the range is plotted on a graph, as follows:

1. Connect the equipment as shown in Fig. 7-1.
2. Initially, set the generator output frequency to the low end of the range. Then set the generator output amplitude to the desired input level.
3. In the absence of a realistic test input voltage, set the generator output to an arbitrary value. A simple method of finding a satisfactory input level is to

monitor the circuit output (with the meter or oscilloscope) and increase the generator output at the circuit center frequency (or at 1 kHz) until the circuit is overdriven. This point is indicated when further increases in generator output do not cause further increases in meter reading (or the output waveform peaks begin to flatten on the oscilloscope display). Set the generator output *just below* this point. Then return the meter or oscilloscope to monitor the generator voltage (at circuit input) and measure the voltage. Keep the generator at this voltage throughout the test.

4. If the circuit is provided with any operating or adjustment controls (volume, loudness, gain, treble, balance, etc.), set these controls to some arbitrary point when making the initial frequency-response measurement. The response measurements can then be repeated at different control settings if desired.

5. Record the circuit output voltage on the graph. Without changing the generator output amplitude, increase the generator frequency by some fixed amount, and record the new circuit output voltage. The amount of frequency increase between each measurement is an arbitrary matter. Use an increase of 10 Hz where rolloff occurs and 100 Hz at the middle frequencies.

6. Repeat the process, checking and recording the amplifier output voltage at each of the check points in order to obtain a frequency-response curve. With a typical audio amplifier, the curve will resemble that of Fig. 7-1, with a flat portion across the middle frequencies and a rolloff at each end. A bandpass filter has a similar response curve. High-pass and low-pass filters produce curves with rolloff at one end only. (High pass has a rolloff at the low end, and vice versa.)

7. After the initial frequency-response check, the effect of operating or adjustment controls should be checked. Volume, loudness, and gain controls should have the same effect all across the frequency range. Treble and bass controls may also have some effect at all frequencies. However, a treble control should have the greatest effect at the high end, while a bass control affects the low end most.

8. Note that generator output may vary with changes in frequency, a fact often overlooked in making a frequency-response test of any circuit. Even precision laboratory generators can vary in output with changes in frequency, thus resulting in considerable error. It is recommended that the generator output be monitored after each change in frequency (some audio generators have a built-in output meter). Then, if necessary, the generator output amplitude can be reset to the correct value. Within extremes, it is more important that the generator output *amplitude remain constant* rather than at some specific value when making a frequency-response check.

7-1.2. Voltage-Gain Measurement

Voltage-gain measurement in an audio amplifier is made in the same way as frequency response. The ratio of output voltage divided by input voltage (at any given frequency, or across the entire frequency range) is the voltage gain. Since the input voltage (generator output) is held constant for a frequency response test, a voltage gain curve should be identical to a frequency-response curve.

7-1.3. Power Output and Gain Measurement

The power output of an audio amplifier is found by noting the output voltage E_{OUT} across the load resistance R_L (Fig. 7-1), at any frequency, or across the entire frequency range. Power output is E_{OUT}^2/R_L.

To find power gain of an amplifier, it is necessary to find both the input and output power. Input power is found in the same way as output power, except that the impedance at the input must be known (or calculated). Calculating input impedance is not always practical in some amplifiers, especially in designs where input impedance is dependent upon transistor gain. (The procedure for finding input impedance of an amplifier is described in Sec. 7-1.7.) With input power known (or estimated), the power gain is the ratio of output power to input power.

In some applications the specification *input sensitivity* is used. Input-sensitivity specifications require a minimum power output with a given *voltage input* (such as 100-W output with 1-V rms input).

7-1.4. Power-Bandwidth Measurement

Many audio-amplifier design specifications include a power-bandwidth factor. Such specifications require that the audio amplifier deliver a given power output across a given frequency range. For example, a circuit may produce full power output up to 20 kHz, even though the frequency response is flat up to 100 kHz. That is, voltage (without load) remains constant up to 100 kHz, whereas power output (across a normal load) remains constant up to 20 kHz.

7-1.5. Load-Sensitivity Measurement

An audio-amplifier circuit of any design is sensitive to changes in load, especially power amplifiers. An amplifier produces maximum power when the output impedance of the amplifier is the same as the load impedance.

The circuit for load-sensitivity measurement is the same as for frequency response (Fig. 7-1), except that load resistance R_L is variable. (Never use a wirewound load resistance. The reactance can result in considerable error.)

Measure the power output at various load impedance/output impedance ratios. That is, set R_L to various resistance values, including a value equal to the amplifier output impedance, and note the voltage and/or power gain at each setting. Then repeat the test at various frequencies. Figure 7-2 is a typical

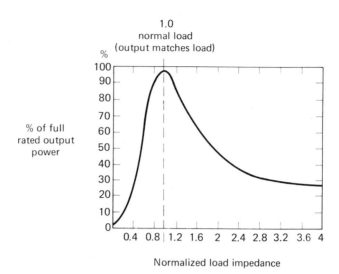

Fig. 7-2   Output power versus load impedance (showing effect of match and mismatch between output and load)

load-sensitivity response curve. Note that if the load is twice the output impedance (as indicated by a 2.0 ratio in Fig. 7-2) the output power is reduced to approximately 50 per cent.

7-1.6. Dynamic-Output-Impedance Measurement

The load-sensitivity test can be reversed to find the dynamic output impedance of an amplifier circuit. The connections (Fig. 7-1) and procedures are the same, except that R_L is varied until *maximum output power is found*. Power is removed and R_L is disconnected from the circuit. The dc resistance of R_L is equal to the dynamic output impedance. Of course, the value applies only at the frequency of measurement. The test can be repeated across the entire frequency range if desired.

7-1.7. Dynamic-Input-Impedance Measurement

To find the dynamic input impedance of an amplifier, use the circuit of Fig. 7-3. The test conditions are identical to those for frequency response, power output, and so on. That is, the same audio generator operating load, meter or oscilloscope, and frequencies are used.

Adjust the signal source to the frequency (or frequencies) at which the circuit is to be operated. Move switch S between positions A and B, while adjusting resistance R until the voltage reading is the *same in both positions* of S. Disconnect R, and measure the dc resistance of R, which is then equal to the dynamic impedance of the amplifier input.

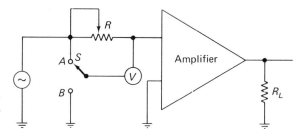

Fig. 7-3 Linear IC dynamic-input-impedance test connections

Accuracy of the impedance measurement is dependent upon the accuracy with which the dc resistance is measured. A noninductive resistance *must be used.* The impedance found by this method applies only to the frequency used during test.

7-1.8. Signal Tracing

An oscilloscope is the most logical instrument for checking amplifier circuits, whether they are complete amplifier systems, or a single stage. The oscilloscope will duplicate every function of an electronic voltmeter in troubleshooting, signal tracing, and performance testing amplifier circuits. In addition, the oscilloscope offers the advantage of a visual display for common audio-amplifier conditions as distortion, hum, noise, ripple, and oscillation.

An oscilloscope is used in a manner similar to that of a voltmeter when signal tracing amplifier circuits. A signal is introduced into the input by the signal generator. The amplitude and waveform of the input signal are measured on the oscilloscope. The oscilloscope probe is then moved to the input and output of each stage, in turn, until the final output is reached. The gain of each stage is measured as a voltage on the oscilloscope. In addition, it is possible to observe any change in waveform from that applied to the input. Thus, stage gain and distortion (if any) are established quickly with an oscilloscope.

7-1.9. Checking Distortion by Sinewave Analysis

The connections for amplifier circuit signal tracing with sinewaves are shown in Fig. 7-4.

The procedure for checking amplifier distortion by means of sinewaves is essentially the same as that described in Sec. 7-1.8. The primary concern, however, is deviation of the amplifier (or stage) output waveform from the input waveform. If there is no change (except in amplitude), there is no distortion. If there is a change in the waveform, the nature of the change often reveals the cause of distortion. For example, the presence of second or third harmonics distorts the fundamental.

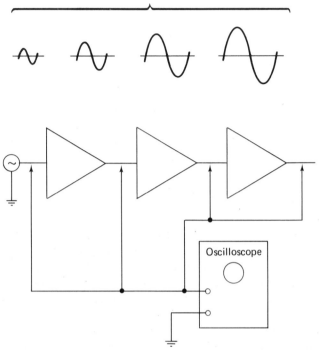

Fig. 7-4   Basic signal tracing through amplifier stages using sinewaves
and an oscilloscope

In practice, analyzing sinewaves to pinpoint amplifier problems that produce distortion is a difficult job requiring considerable experience. Unless the
distortion is severe, it may pass unnoticed. Sinewaves are best used where
harmonic distortion or *intermodulation distortion* meters are combined with
the oscilloscope for distortion analysis. If an oscilloscope is to be used alone,
squarewaves provide the best basis for distortion analysis. (The reverse is
true for frequency-response and power measurements.)

7-1.10. Checking Distortion by Squarewave Analysis

The procedure for checking distortion by means of squarewaves is essentially the same as for sinewaves. Distortion analysis is more
effective with squarewaves because of their high odd-hamonic content, and
because it is easier to see a deviation from a straight line with sharp corners
than from a curving line.

As in the case of sinewave distortion testing, squarewaves are introduced

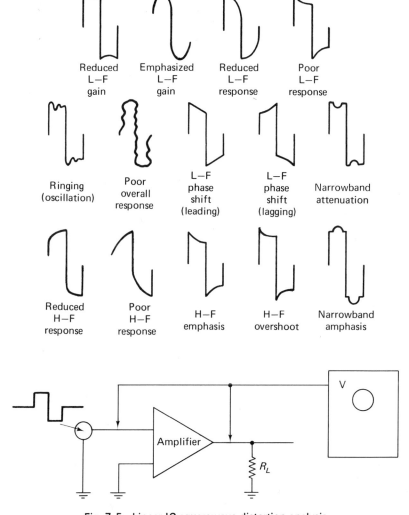

Fig. 7-5   Linear IC squarewave distortion analysis

into the circuit input, while the output is monitored on an oscilloscope. (See Fig. 7-5.) The primary concern is deviation of the amplifier (or stage) output waveform from the input waveform (which is also monitored on the oscilloscope). If the oscilloscope has the dual-trace feature, the input and output can be monitored simultaneously. If there is a change in waveform, the nature of the change often reveals the cause of distortion.

The third, fifth, seventh, and ninth harmonics of a clean squarewave are emphasized. If an amplifier passes a given frequency and produces a clean squarewave output, it is safe to assume that the frequency response is good up to at least *nine times* the squarewave frequency.

7-1.11. Harmonic-Distortion Measurement

No matter what amplifier circuit is used, or how well the circuit is designed, there is always the possibility of odd or even harmonics being present with the fundamental. These harmonics combine with the fundamental and produce distortion, as is the case when any two signals are combined. The effects of second- and third-harmonic distortion are shown in Fig. 7-6.

Commercial harmonic-distrotion meters operate on the *fundamental suppression* principle. As shown in Fig. 7-6, a sinewave is applied to the amplifier input, and the output is measured on the oscilloscope. The output is then applied through a filter that suppresses the fundamental frequency. Any output from the filter is then the result of harmonics.

The output is also displayed on the oscilloscope. (Some commercial harmonic-distortion meters use a built-in meter instead of, or in addition to, an external oscilloscope.) When the oscilloscope is used, the frequency of the filter output signal is checked to determine harmonic content. For example, if the input is 1 kHz and the output (after filtering) is 3 kHz, it would indicate third-harmonic distortion.

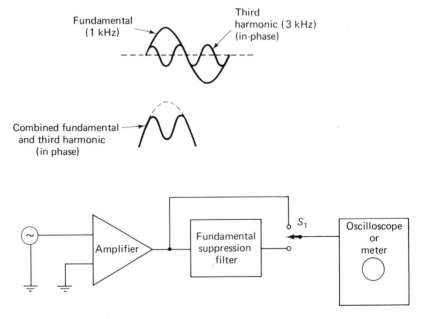

Fig. 7-6 Harmonic-distortion analysis

The *percentage* of harmonic distortion is also determined by this method. For example, if the output (without filter) is 100 mV and with filter it is 3 mV, a 3 per cent harmonic distortion is indicated.

In some commercial harmonic-distortion meters the filter is tunable so that the amplifier can be tested over a wide range of fundamental frequencies. In other harmonic-distortion meters the filter is fixed in frequency, but can be detuned slightly to produce a sharp null.

7-1.12. Intermodulation-Distortion Measurement

When two signals of different frequency are mixed in an amplifier, there is a possibility that the lower-frequency signal will *amplitude modulate* the higher-frequency signal. This produces a form of distortion known as *intermodulation distortion*.

Commercial intermodulation-distortion meters consist of a signal generator and high-pass filter, as shown in Fig. 7-7. The signal generator portion of the meter produces a high-frequency signal (usually about 7 kHz) that is modulated by a low-frequency signal (usually 60 Hz). The mixed signals are applied to the circuit input. The amplifier output is connected through a high-pass filter to the oscilloscope vertical channel. The high-pass filter removes the low-frequency (60-Hz) signal. Thus, the only signal appearing on the oscilloscope vertical channel should be the high-frequency (7-kHz) signal. If any 60-Hz signal is present on the display, it is being passed through as modulation on the 7-kHz signal.

Figure 7-7 also shows an intermodualtion test circuit that can be fabricated in the shop or laboratory. Note that the high-pass filter is designed to pass

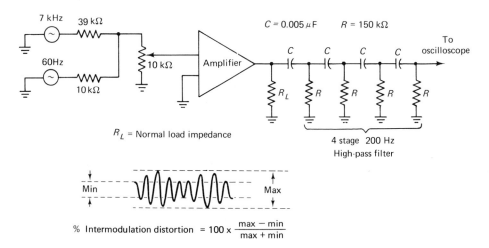

$$\% \text{ Intermodulation distortion } = 100 \times \frac{\text{max} - \text{min}}{\text{max} + \text{min}}$$

Fig. 7-7   Intermodulation-distortion analysis

signals above about 200 Hz. The purpose of the 39- and 10-kΩ resistors is to set the 60-Hz signal at four times that of the 7-kHz signal. Most audio generators provide for a line-frequency output (60 Hz) that can be used as the low-frequency modulation source.

If the laboratory circuit of Fig. 7-7 is used instead of a commercial meter, set the generator line-frequency output to 2 V (if adjustable). Then set the generator audio output (7 kHz) to 2 V. If the line-frequency output is not adjustable, measure the value and then set the generator audio output to the same value.

The percentage of intermodulation distortion can be calculated using the equation of Fig. 7-7.

7-1.13. Background-Noise Measurement

If the vertical channel of an oscilloscope is sufficiently sensitive, an oscilloscope can be used to check and measure the background noise level of an amplifier, as well as to check for the presence of hum, oscillation, and the like. The oscilloscope vertical channel should be capable of a measurable deflection with about 1 mV (or less), since this is the background noise level of many amplifiers.

The basic procedure consists of measuring amplifier output with the volume or gain control (if any) at maximum, but without an input signal. The oscil-

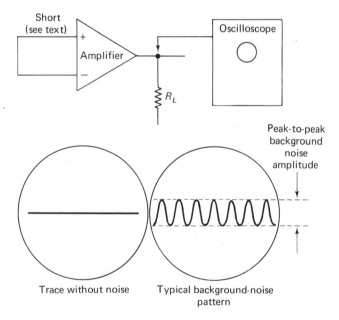

Fig. 7-8   Amplifier background-noise test connections

loscope is superior to a voltmeter for noise-level measurement since the frequency and nature of the noise (or other signal) are displayed visually.

The basic connections for background-noise-level measurement are shown in Fig. 7-8. The oscilloscope gain or sensitivity control is increased until there is a noise or "hash" indication.

It is possible that a noise indication could be caused by pickup in the leads. If in doubt, disconnect the leads from the amplifier, but not from the oscilloscope.

If it is suspected that there is a 60 Hz line hum present in the amplifier output (picked up from the power supply or any other source), set the oscilloscope "Sync" control to line (or whatever other control is required to synchronize the oscilloscope trace at the line frequency). If a stationary signal pattern appears, it is due to the line hum.

If a signal appears that is not at the line frequency, it can be due to oscillation in the amplifier, or stray pickup. Short the amplifier input terminals. If the signal remains, it is probably oscillation in the amplifier.

7-1.14. Phase-Shift and Feedback Measurements

In any amplifier there is some phase shift between input and output signals. This is usually not cirtical for audio-amplifier circuits. One exception is in operational amplifiers where feedback from output to input is used to control gain, as discussed in Chapter 6. For that reason, the procedures for measurement of phase-shift and feedback levels in amplifier circuits are described in Sec. 7-4.

7-1.15. Transformer Characteristics

If an audio-amplifier design does not prove satisfactory, the fault may be with the transformer. The obvious test is to measure the transformer windings for opens, shorts, and the proper resistance value with an ohmmeter. In addition to basic resistance checks, it is possible to test a transformer's proper polarity markings, regulation, impedance ratio, and center-tap balance with a voltmeter.

Phase relationships. When two supposedly identical transformers must be operated in parallel, as they are in some audio loudspeaker systems, and the transformers are not marked as to phase or polarity, the phase relationship of the transformers can be checked using a voltmeter and power source.

The test circuit for phase relationships is shown in Fig. 7-9. Audio-amplifier transformers can be tested using an audio generator as the source voltage. As shown, the transformers are connected in *proper phase* if the meter reading is zero. The transformers are out of phase if the secondary output voltage is *double* that of the normal secondary output. This condition can be

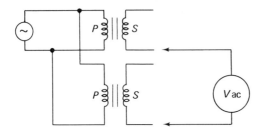

Fig. 7-9 Measuring transformer phase relationships

corrected by reversing either the primary or secondary leads (but not both) of *one transformer* (but not both transformers).

If the meter indicates some secondary voltage, it is possible that one transformer has greater output than the other. This condition will result in considerable local current (current between transformers) in the secondary winding and will produce a power loss (and possible damage to the transformers if the local current is large).

Polarity markings. Many transformers are marked as to polarity or phase. These markings may consist of dots, color-coded wires, or some similar system. Unfortunately, transformer polarity markings are not always standard. This can prove very confusing, particularly in experimental work.

Generally, transformer polarities are indicated on schematics as dots next to the terminals. When standard markings are used, the dots mean that if electrons are flowing into the terminal with the dot, the electrons will flow out of the secondary terminal with the dot. The dots have the same polarity so far as the external circuits are concerned. No matter what system is used, the dots or other markings show *relative phase*, since instantaneous polarities are changing across the transformer windings.

From a practical standpoint, there are only two problems of concern: the relationship of the primary to the secondary, and the relationship of markings on one transformer to those on another.

The phase relationship of primary to secondary can be found using the test circuit of Fig. 7-10. First check the voltage across terminals 1 and 3, then across 1 and 4 (or 1 and 2). Assume that there is 3 V across the primary with 7 V across the secondary. If the windings are as shown in Fig. 7-10a, the 3 V is added to the 7 V and appears as 10 V across terminals 1 and 3. If the windings are as shown in Fig. 7-10b, the voltages oppose each other, and appear as 4 V (7 − 3) across terminals 1 and 3.

The phase relationship of one transformer marking to another can be found using the test circuit of Fig. 7-11. Assume that there is a 3-V output from the secondary of transformer *A*, and a 7-V output from transformer *B*. If the markings are consistent on both transformers, the two voltages oppose, and 4 V is indicated. If the markings are not consistent, the two voltages add, resulting in a 10-V reading.

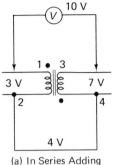

(a) In Series Adding

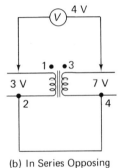

Fig. 7-10 Checking transformer polarity markings

(b) In Series Opposing

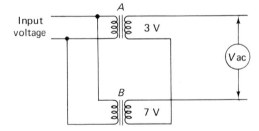

Fig. 7-11 Testing consistency of transformer polarities or phase markings

Impedance ratio. The impedance ratio of a transformer is the square of the winding ratio. For example, if the winding ratio of a transformer is 15 to 1, the impedance ratio is 225 to 1. Any impedance value placed across one winding is reflected back onto the other winding by a value equal to the impedance ratio. Assuming an impedance ratio of 225 to 1 and a 1800-Ω impedance placed on the primary, the secondary has a reflected impedance of 8 Ω (1800/225 = 8). Likewise, if a 10-Ω impedance is placed on the secondary, the primary has a reflected impedance of 2250 (225 × 10 = 2250).

Impedance ratio is related to turns ratio (primary to secondary). However, turns-ratio information is not always available, so the ratio must be calculated using a test circuit as shown in Fig. 7-12.

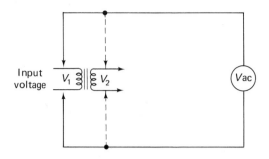

$$\text{Turns ratio} = \frac{V_1}{V_2} \text{ or } \frac{V_2}{V_1}$$

$$\text{Impedance ratio} = (\frac{V_1}{V_2})^2 \text{ or } (\frac{V_2}{V_1})^2$$

Fig. 7-12 Measuring transformer turns and impedance ratio

Measure both the primary and secondary voltage. Divide the larger voltage by the smaller, noting which is primary and which is secondary. For convenience, set either the primary or secondary to some exact voltage.

The *turns ratio* is equal to one voltage divided by the other.

The *impedance ratio* is the square of the turns ratio.

For example, assume that the primary shows 11.5 V with 2.3 V at the secondary. This indicates a 5-to-1 turns ratio (11.5/2.3 = 5), and a 25-to-1 impedance ratio ($5^2 = 25$).

Winding balance. There is always some imbalance in center-tapped transformers (such as those used in class *B* audio power amplifiers). That is, the turns ratio and impedance ratio are not exactly the same on both sides of the center tap. A large imbalance can impair operation of any circuit, but especially a push–pull class B amplifier where an imbalance can result in distortion.

It is possible to find a large imbalance by measuring the dc resistance on either side of the center tap. However, a small imbalance might not show up with dc resistance measurements.

It is usually more practical to measure the *voltage* on both sides of a center tap, as shown in Fig. 7-13. If the voltages are equal, the transformer winding is balanced. If a large imbalance is indicated by a large voltage difference, the winding should then be checked with an ohmmeter for shorted turns, poor design, and so on.

Fig. 7-13 Measuring transformer winding balance

7-2. ANALYZING AMPLIFIER OPERATION WITH TEST RESULTS

The following notes apply primarily to solving *design problems* (poor frequency response, lack of gain, etc.) in amplifiers, but can also be applied to analyzing circuits of existing commercial amplifiers.

When design circuits of any kind fail to perform properly (or as hoped they would perform), a planned procedure for isolating the problem is very helpful. Keep in mind that transistor circuit troubleshooting is difficult at best. This is especially true when the circuit involves more than one stage, since the stages are interdependent.

Failure in design circuits. A special problem arises in analyzing failure of design circuits. The first requirement in logical troubleshooting is a thorough knowledge of the circuit's performance when *operating normally*. However, a failure in a trial circuit, just designed, can be the result of component failure or improper trial values for components. For example, an existing amplifier may show low gain based on past performance. A newly designed amplifier circuit may show the same results simply because it is the best gain possible with the selected trial components.

To minimize the problem in newly designed amplifiers, try to isolate problems on a stage-by-stage basis. For example, if the circuit has two or more stages and gain is low for the overall circuit, measure the gain for each stage. With trouble isolated to a particular stage, try to determine which half of the stage is at fault.

Stage input–output relationship. Any transistor stage has two halves, input and output. Generally, the input is base–emitter (or gate–source for a FET), with the emitter–collector (source–drain) acting as the output. Keep in mind that a defect in one half will affect the other half. An obvious example of this is where low input current (base) produces low output current (collector). Sometimes less obvious is the case where output affects input. For example, in a stage with an emitter resistor (for feedback stabilization), an open collector appears to reduce the input impedance.

Loop feedback problems. Circuits with loop feedback (overall feedback) present a particular problem. A closed feedback loop causes all stages to respond as a unit, making it difficult to know which stage is at fault. This problem can be solved by opening the feedback loop. To do so, however, creates another problem, since the operating Q point of one (or possibly all) stages is disturbed.

Saturated and/or cut-off transistors. Look for any transistor that is *full on* (collector or drain voltage very low) or *full off* (collector or drain voltage very high, probably near the supply voltage). Either of these conditions in a linear amplifier of any type is the result of component failure or improper design. In effect, the transistor is operating as a switch, rather than an amplifier.

Bias problems. Design faults that appear under no-signal conditions (such as inproper Q point) are generally the result of improper biss relationships. Faults that appear only when a signal is applied can be caused by poor bias, but can also result from wrong component values.

A high-gain circuit is generally more difficult to bias than a low-gain circuit. Use the following procedure when it is difficult to find a good bias point (or Q point) for a high-gain amplifier.

Increase the signal input until the output waveform appears as a squarewave (Fig. 7-14). That is, overdrive the amplifier circuit.

Keep reducing the input signal, while adjusting the bias, until both positive and negative peaks are clipped by the same amount.

If it is impossible to find any bias point where the signal peaks can be clipped symmetrically, a defect in design or components can be suspected.

Unsymmetrical clipping can be caused by operating the transistor on a nonlinear portion of its transfer curve of load line (improper bias), or by the fact that the transistor does not have a linear curve (or a very short linear curve). Try a different transistor. If the results are the same, change the circuit trial values (particularly collector, emitter, and base resistances).

Oscillation and feedback problems. If it is impossible to obtain any Q point except very near full on or full off, this can be caused by excess *positive feed-*

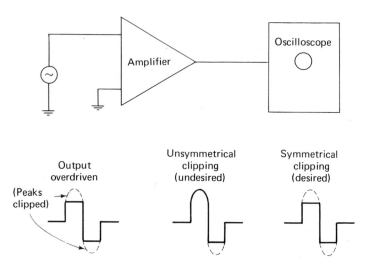

Fig. 7-14   Determining proper bias (Q point) by means of oscilloscope displays

back. While this condition is desirable in a multivibrator or oscillator, it must be avoided in linear amplifiers.

High-gain amplifiers should be carefully checked to find any tendency to oscillate or to exhibit abnormal noise. High-gain amplifiers may also be very sensitive to supply-voltage changes. When an amplifier design circuit has been completed, it is often helpful to repeat all the basic test procedures with various supply voltages. Unless othwerwise specified by design requirements, any amplifier should peform equally well with a ± 10 per cent supply-voltage variation.

If oscillation occurs in any amplifier circuit (high or low gain), try moving the input and output leads. Even a well-designed amplifier may oscillate simply because input and output leads are close together. It may be necessary to physically relocate parts or to shield parts of an amplifier to prevent feedback that results in oscillation.

Low-frequency oscillation is often the result of poor supply-voltage filtering or too many stages connected to the same supply-voltage point. Try isolating the stages with separate supply-voltage filter capacitors.

Poor low-frequency response. The most common cause of poor low-frequency response is *low capacitor values*. The design procedures of Chapters 1 and 2 provide for an approximate 1-dB loss at the low-frequency limit. If a greater loss can be tolerated, a lower capacitor value can be used. (This applies to coupling capacitors, emitter-bypass capacitors, or source-bypass capacitors, but not to power supply or decoupling capacitors.) If better low-frequency response is desired, all other factors being equal, *increase* capacitor values.

Poor high-frequency response. The most common cause of poor high-frequency response is the *input capacitance of transistors*. As frequency increases, transistor input capacitance decreases, changing the input impedance. This change in input impedance usually results in decreased gain, all other factors remaining equal. Generally, poor high-frequency response is not a problem over the audio range (up to about 20 kHz), but can be a problem beyond about 200 kHz. The only practical solutions are to accept reduced stage gain or change transistors.

7-3. BASIC RADIO-FREQUENCY AMPLIFIER TESTS

The following paragraphs describe test procedures for RF amplifiers. The first paragraphs are devoted to test and measurement procedures for the resonant circuits used at radio frequencies (resonant-frequency measurements, Q measurements, etc.). The remaining sections cover test procedures for RF amplifiers in such typical applications as transmitters and receivers.

For best results, RF amplifiers are tested in final form (with all components soldered in place). This will show if there is any change in circuit characteristics due to the physical relocation of components (as a result of design changes or troubleshooting and repair procedures).

Often there is capacitance or inductance between components, from components to wiring, and between wires. These stray "components" can add to the reactance and impedance of circuit components. When the physical locations of parts and wiring are changed, the stray reactances change, and alter circuit performance.

7-3.1. Basic Radio-Frequency Voltage Measurement

When the voltages to be measured are at radio frequencies and are beyond the frequency capabilities of a meter or oscilloscope, an RF *probe* is required. Such probes rectify the RF signals into a dc output that is almost equal to the peak RF voltage. The dc output of the probe is then applied to the meter or oscilloscope and is displayed as a voltage readout in the normal manner.

If a probe is available as an accessory for a particular meter or oscilloscope, use that probe in favor of any homemade probe. The manufacturer's probe is matched to the meter or oscilloscope in calibration, frequency compensation, or the like. If a probe is not available for a particular meter, the following notes can be used to make a probe for RF voltage measurement.

The half-wave probe (Fig. 7-15) provides an output to the meter (or oscilloscope) that is *approximately* equal to the peak value of the voltage being measured. Since most meters are calibrated to read in rms values, the probe output must be reduced to 0.707 of the peak value by means of R_1. A variable (noninductive) resistor can be substituted for R_1 during calibration, and then replaced by a fixed resistor of the correct value.

To calibrate the probe, apply an RF voltage of precise, known amplitude

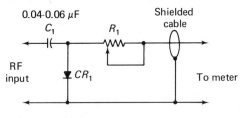

CR$_1$ = IN34 or equivalent

$R_1 \approx$ 10-20 kΩ for V_{OM}

$R_1 \approx$ 1 MΩ for electronic voltmeter

Fig. 7-15   Half-wave probe for RF measurements

at the RF input terminals. Adjust R_1 until the meter reads 0.707 times the known input voltage. For example, with 10-V radio frequency at the input, adjust R_1 for a 7.07-V reading on the meter. Replace variable R_1 with a fixed resistance of corresponding value (or leave R_1 set at the correct value). Repeat the test over the anticipated frequency range. The probe of Fig. 7-15 should provide satisfactory results up to about 250 MHz. Beyond that frequency, always use the probe supplied with the meter or oscilloscpe. Keep in mind that the meter must be set to read *direct current*, since the probe output is direct current.

7-3.2. *Measuring LC Circuit Resonant Frequency*

The circuit for measuring resonant frequency of an *LC* circuit is shown in Fig. 7-16.

To use the circuit, adjust the unmodulated RF generator output amplitude for a convenient indication on the meter. Then, starting at a frequency well below the lowest possible frequency of the *LC* circuit, slowly increase the generator output frequency.

For a parallel-resonant *LC* circuit, watch the meter for a maximum or peak indication.

For a series-resonant *LC* circuit, watch the meter for a minimum or dip indication.

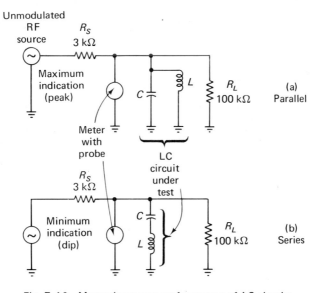

Fig. 7-16   Measuring resonant frequency of *LC* circuits

The resonant frequency of the *LC* circuit is the one at which there is a *maximum (for parallel)* or *minimum (for series)* indication on the meter.

Note that there may be peak or dip indications at harmonics of the resonant frequency. The test is most efficient when the approximate resonant frequency is known.

To broaden the response (so that the peak or dip can be approached more slowly), increase the value of R_L from 100 kΩ. (An increase in R_L lowers the *LC* circuit *Q*). To sharpen the response, lower the value of R_L.

7-3.3. *Measuring Inductance of a Coil*

The circuit for measuring inductance of a coil is shown in Fig. 7-17.

To use the circuit, adjust the unmodulated RF generator output amplitude for a convenient indication on the meter. Then, starting at a frequency well below the lowest possible resonant frequency of the *LC* combination under test, slowly increase the generator output frequency.

The resonant frequency of the *LC* circuit is the one at which there is a maximum indication on the meter. Using this resonant frequency and a known capacitance value, calculate the unknown inductance using the equation of Fig. 7-17. To simplify calculations, use a convenient capacitance value, such as 100 pF, 1000 pF, or the like.

Note that the procedure can be reversed to find an unknown capacitance value when a known inductance value is available.

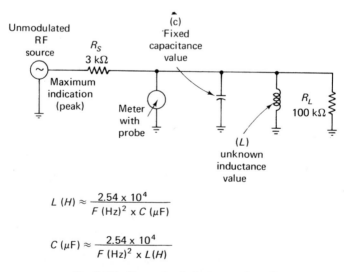

$$L \ (H) \approx \frac{2.54 \times 10^4}{F \ (Hz)^2 \times C \ (\mu F)}$$

$$C \ (\mu F) \approx \frac{2.54 \times 10^4}{F \ (Hz)^2 \times L \ (H)}$$

Fig. 7-17   Measuring inductance of a coil

Increase the value of R_L to broaden the peak indication, if desired (or vice versa).

7-3.4. Measuring Self-Resonance and Distributed Capacitance of a Coil

There is distributed capacitance in any coil, which can combine with the coil's inductance to form a resonant circuit. Although the self-resonant frequency may be quite high in relation to the operating frequency at which the coil is used, the self-resonant frequency may be near a harmonic of the operating frequency. This limits the usefulness of the coil in an *LC* circuit. Some coils, particularly RF chokes, may have more than one self-resonant frequency.

The circuit for measuring self-resonance and distributed capacitance of a coil is shown in Fig. 7-18.

To use the circuit, adjust the unmodulated RF generator output amplitude for a convenient indication on the meter. Tune the generator over its entire frequency range, starting at the lowest frequency. Watch for either *peak or dip* indications on the meter. Either a peak or dip indicates that the inductance is at a self-resonant point. The generator output frequency at this point is the self-resonant frequency (or a harmonic).

Make certain that peak or dip indications are not the result of changes in generator output level. Cover the entire frequency range of the generator, or at least from the lowest frequency up to the third harmonic of the highest frequency involved in circuit design or operation.

Once the resonant frequency (or frequencies) has been found, calculate the distributed capacitance using the equation of Fig. 7-19.

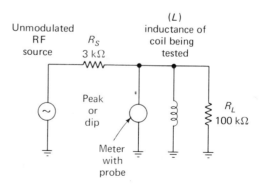

Fig. 7-18   Measuring self-resonance and distributed capacitance of a coil

$$C\ (\mu F) \approx \frac{2.54 \times 10^4}{F\ (Hz)^2 \times L\ (H)}$$

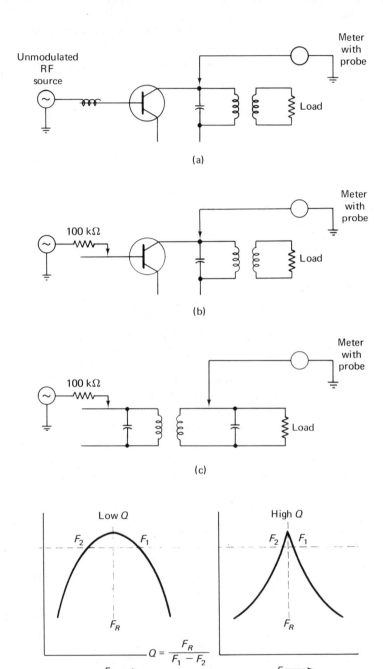

F_R = peak resonant frequency

Fig. 7-19   Measuring Q of resonant circuits

7-3.5. Measuring Q of Resonant Circuits

The Q of a resonant circuit sets the circuit bandwidth. That is, a high Q circuit has a narrow bandwidth (sharp tuning), whereas a low Q circuit has a wide bandwidth (broad tuning). The most practical measurement of resonant circuit Q is to measure the bandwidth at the resonant frequency. The circuits for measuring bandwidth (or Q) of resonant circuits are shown in Fig. 7-19.

Figure 7-19a shows the test circuit in which the signal generator is connected directly to the input of a complete stage, and Fig. 7-19b shows the indirect method of connecting the signal generator to the input.

When the stage or circuit has sufficient gain to provide a good reading on the meter with a nominal output from the generator, the indirect method (with isolating resistor) is preferred. Any signal generator has some output impedance (typically 50 Ω). When this resistance is connected directly to the tuned circuit, the Q is lowered and the response becomes broader. (In some cases, the generator output impedance can seriously detune the circuit.)

Figure 7-19c shows the test circuit for a single component (such as an IF transformer).

When the resonant circuit is normally used with a load, the most realistic Q measurement is made with the circuit terminated in that load value. A fixed resistance can be used to simulate the load. The Q of a resonant circuit is often dependent upon the load value.

To use the circuit, adjust the unmodulated RF generator output amplitude for a convenient indication on the meter. Tune the signal generator to the circuit resonance frequency; then tune the generator for *maximum reading* on the meter. Note the generator frequency.

Tune the generator below resonance until the meter reading is 0.707 of the maximum reading. Note the generator frequency (this is frequency F_2). To make the calculation more convenient, adjust the generator output level so that the meter reading is some even value, such as 1 V or 10 V, after the generator is tuned for maximum. This will make it easy to find the 0.707 mark.

Tune the generator above resonance until the meter reading is 0.707 of the maximum reading. Note the generator frequency (this is frequency F_1).

Calculate the circuit Q using the equation of Fig. 7-19.

7-3.6. Measuring Impedance of Resonant Circuits

Any resonant circuit has some impedance at the resonant frequency. The impedance changes with frequency. This includes transformers (tuned and untuned), tank circuits, and so on. In theory, a series-resonant circuit has zero impedance, and a parallel-resonant circuit has infinite impedance. In practical circuits, this is impossible, since there is always some resistance in the circuit.

In practical amplifier work it is often convenient to find the actual impedance of a completed circuit, at a given frequency. Also, it may be necessary to find the impedance of a component so that circuit values can be designed around the impedance. For example, an IF transformer presents an impedance at both its primary and secondary. These values may not be specified.

The impedance of a resonant circuit or component can be measured using a signal generator and a meter with an RF probe. An electronic voltmeter provides the least loading effect on the circuit, thus providing the most accurate indication.

The procedure for impedance measurement at radio frequencies is the same as for audio frequencies, as discussed in Sec. 7-1.7, except as follows:

An RF generator must be used as the signal source. The meter must be provided with an RF probe. If the circuit or component under measurement has both an input and output (such as a transformer), the opposite side or winding must be terminated in its normal load. A fixed resistance can be used to simulate the resistance. If the impedance of a tuned circuit is to be measured, tune the circuit to peak or dip; then measure the impedance at resonance. Once the resonant impedance is found, the generator can be tuned to other frequencies to find the corresponding impedance.

7-3.7. Testing Transmitter Radio-Frequency Amplifier Circuits

It is possible to test and adjust transmitter RF amplifiers using a meter with an RF probe. If an RF probe is not available, or as an alternative, it is possible to use a test circuit such as shown in Fig. 7-20. This circuit is essentially a pick-up coil, which is placed near the RF amplifier inductance, and a rectifier that converts the radio frequency into a dc voltage for measurement on a meter.

Figure 7-21 shows the basic circuit for test and measurement of RF amplifier circuits. If the amplifier being measured is without an oscillator, a drive signal must be supplied by means of a signal generator. Use an unmodulated signal at the correct operating frequency.

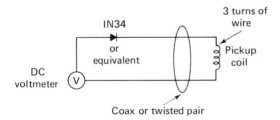

Fig. 7-20 Circuit for pickup and measurement of RF signals

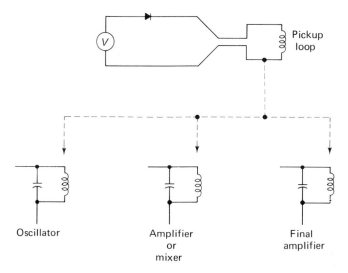

Fig. 7-21   Testing transmitter RF amplifier circuits

In turn, connect the meter to each amplifier stage. Start with the first stage (this will be the oscillator if the circuit under test is a complete transmitter), and work toward the final or output amplifier.

A voltage indication should be obtained at each stage. Usually, the voltage indication increases with each amplifier stage. Some stages may be frequency multipliers and provide no voltage amplification.

If a particular amplifier stage is to be tuned, adjust the tuning control for a maximum reading on the meter. If the stage is to be operated with a load (such as the final amplifier into an antenna), the load should be connected, or a simulated load should be used. A fixed resistance provides a good simulated load at frequencies up to about 250 MHz.

It should be noted that this tuning method or measurement technique does not guarantee each stage is at the desired operating frequency. It is possible to get maximum readings on harmonics. However, it is conventional to design RF transmitter amplifier circuits so that they will not tune to both the desired operating frequency and a harmonic. Generally, RF amplifier tank circuits tune on either side of the desired frequency, but not to a harmonic (unless the circuit is seriously detuned, or the design calculations are hopelessly inaccurate).

7-3.8. Testing Receiver Radio-Frequency Circuits

It is possible to test and adjust receiver RF circuits using a meter and signal generator. Both AM and FM receivers require alignment of the IF and RF amplifiers. An FM receiver also requires alignment of the detector stage (discriminator or ratio detector). If a complete receiver is being tested, and the receiver includes an AVC–AGC circuit, the AGC must be disabled (by means of a fixed bias of opposite polarity to the signal produced by the detector). The bias is typically on the order of a few volts and is placed on the AGC line during alignment.

FM detector alignment. The circuit for FM detector alignment is shown in Fig. 7-22.

Adjust the signal generator frequency to the intermediate frequency (usually 10.7 MHz). Use an unmodulated output from the signal generator.

Adjust the secondary winding (either capacitor or tuning slug) of the discriminator transformer (which is the output of the final IF amplifier or IF amplifier limiter) for *zero reading* on the meter. Adjust the transformer

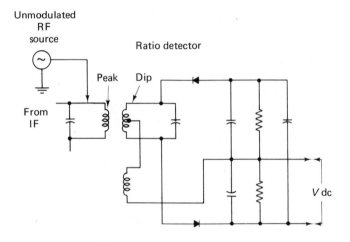

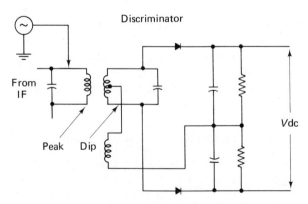

Fig. 7-22 Frequency-modulated detector alignment circuit

slightly each way and make sure the meter moves smoothly above and below the exact zero mark. (A meter with a zero-center scale is most helpful when adjusting FM detectors.)

Adjust the signal generator to some point below the intermediate frequency (to 10.625 MHz for an FM detector with a 10.7-MHz intermediate frequency). Note the meter reading.

Adjust the signal generator to some point above the intermediate frequency *exactly* equal to the amount set below the intermediate frequency. For example, if the generator is set to 0.075 MHz below the intermediate frequency set the generator to 0.075 MHz above it (or 10.625 and 10.775 MHz)

The meter should read approximately the same on both sides of the intermediate frequency, except the polarity is reversed. For example, if the meter reads 7 scale divisions below zero and 7 scale divisions above zero, the FM detector is balanced. If an FM detector cannot be balanced, the fault is usually a serious mismatch in diodes or other components.

Return the generator to the intermediate frequency (10.7 MHz) and adjust the primary winding of the discriminator transformer for *maximum reading* on the meter. This sets the primary winding at the correct resonant frequency of the intermediate frequency.

AM and FM alignment. The alignment procedures for the IF amplifier stages of an AM receiver are essentially the same as those of an FM receiver. However, the meter must be connected at different points in the corresponding detector, as shown in Fig. 7-23. In either case, the meter is set to measure direct current, and the RF probe is not used. In those cases where the IF stages are being tested without a detector (such as during design), an RF probe is required. As shown in Fig. 7-23, the RF probe is connected to the secondary of the final IF output transformer.

Set the meter to measure direct current and connect it to the appropriate test point (with or without an RF probe, as applicable). Adjust the generator frequency to the receiver intermediate frequency (typically 10.7 MHz for FM and 455 kHz for AM). Use an unmodulated RF signal.

Adjust the windings of the IF transformers (capacitor or tuning slug) in turn, starting with the *last stage and working toward the first stage*. Adjust each winding *for maximum reading*. Repeat the procedure to make sure that there is no interaction between adjustments (usually there is some interaction).

AM and FM radio-frequency amplifier alignment. The alignment procedures for the RF stages (RF amplifier, local oscillator, mixer/converter) of an AM receiver are essentially the same as for an FM receiver. Again, it is a matter of connecting the meter to the appropriate test point. The same test points used for IF alignment can be used for aligning the RF stages, as shown in Fig. 7-24. However, if an individual RF stage is to be aligned, the meter must be connected to the secondary winding of the RF stage output transformer through an RF proble.

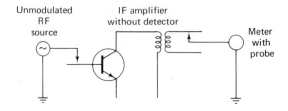

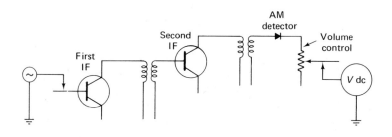

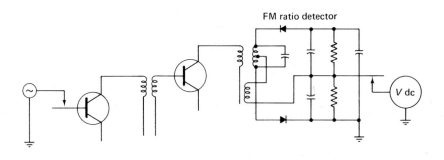

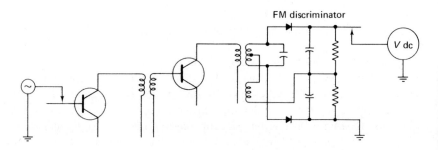

Fig. 7-23   Intermediate-frequency alignment for AM and FM receivers

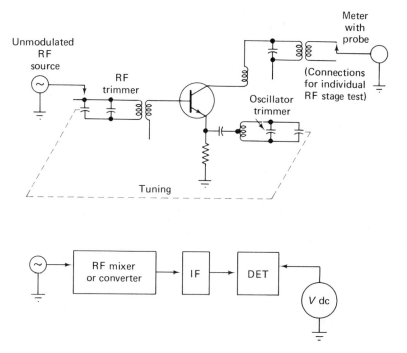

Fig. 7-24 Alignment of RF amplifier and local oscillator (converter–mixer)

Set the meter to measure direct current and connect it to the appropriate test point (with or without an RF probe, as applicable).

Adjust the generator frequency to some point near the high end of the receiver operating frequency (typically 107 MHz for a broadcast FM receiver, and 1400 kHz for an AM broadcast receiver). Use an unmodulated output from the signal generator.

Adjust the RF stage trimmer for maximum reading on the meter.

Adjust the generator frequency to the low end of the receiver operating frequency (typically 90 MHz for FM and 600 kHz for AM).

Adjust the oscillator stage trimmer for maximum reading on the meter.

Repeat the procedure to make sure the resonant circuits "track" across the entire tuning range.

7-4. BASIC OPERATIONAL-AMPLIFIER TESTS

The following paragraphs describe test procedures for op-amps. The tests are primarily for use with IC op-amps, but also apply to discrete-component op-amps.

The basic procedures are essentially the same as for audio amplifiers (Sec. 7-1), except for the differences noted in the following paragraphs.

7-4.1. Frequency Response

Frequency response for an op-amp is tested in essentially the same way as for an audio amplifier, except that the frequency range is extended beyond the audio range. Both open-loop and closed-loop frequency response should be measured with the same load.

7-4.2. Voltage Gain

Voltage-gain measurement for an op-amp is the same as for an audio amplifier, except that the basic IC has a maximum input and output voltage limit, neither of which can be exceeded without possible damage to the IC and/or clipping of the waveform.

Note the frequency at which the open-loop voltage gain drops 3 dB down from the low-frequency value. This is the open-loop bandwidth.

Keep in mind that the open-loop voltage gain and bandwidth are characteristics of the basic IC. Closed-loop gain is (or should be) dependent upon the ratio of feedback and input resistances, while closed-loop bandwidth is essentially dependent upon phase compensation values.

7-4.3. Power Output, Gain, and Bandwidth

Most IC op-amps are not designed as power amplifiers. However, their power output, gain, and bandwidth can be measured in the same way as audio amplifiers (Secs. 7-1.3, 7-1.4).

Keep in mind that an IC has a power dissipation of its own, which must be subtracted from the total device dissipation to find the available power output.

7-4.4. Load Sensitivity

Since an IC op-amp is generally not used as a power amplifier, load sensitivity is not critical. However, if it should be necessary to measure the load sensitivity, use the procedure of Sec. 7-1.5.

7-4.5. Input and Output Impedance

Dynamic input and output impedance of an IC op-amp can be found using the procedures of Sec. 7-1.6 and 7-1.7. Keep in mind that closed-loop impedances will differ from open-loop impedances.

7-4.6. Distortion

Distortion measurements for IC op-amps are the same as for audio amplifiers (Secs. 7-1.9 through 7-1.12). However, distortion requirements for op-amps are usually not critical.

7-4.7. Background Noise

Background noise measurements for IC op-amps are the same as for audio amplifiers (Sec. 7-1.13). Generally, background noise should be measured under open-loop conditions. Some datasheets specify that both input and output voltages be measured. When input voltage is to be measured, a fixed resistance (usually 50 Ω) is connected between the input terminals.

7-4.8. Feedback Measurement

Since op-amp characteristics are based on the use of feedback signals, it is often convenient to measure feedback voltage at a given frequency with given operating conditions.

The basic feedback measurement connections are shown in Fig. 7-25. The most accurate measurement is made when the feedback lead is terminated in the *normal operating impedance.*

If an input resistance is used in the normal circuit, and this resistance is considerably lower than the IC input impedance, use the resistance value.

If in doubt, measure the input impedance of the IC (Sec. 7-4.5); then terminate the feedback lead in that value to measure feedback voltage.

7-4.9. Input Bias Current

Input bias current can be measured using the circuit of Fig. 7-26. Any resistance value for R_1 and R_2 can be used, provided the value

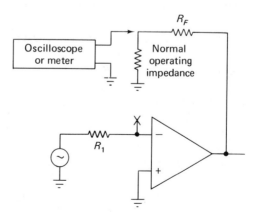

Fig. 7-25 Feedback measurement of op-amp

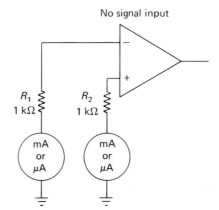

No signal input

Fig. 7-26   Input bias current measurement of op-amp

produces a measurable voltage drop. A value of 1 kΩ is realistic for R_1 and R_2.

Once the voltage drop is found, the input bias current can be calculated. For example, if the voltage drop is 3 mV across 1 kΩ, the input bias current is 3 μA.

In theory, the input bias current should be the same for both inputs. In practice, the bias currents should be almost equal. Any great difference in input bias is the result of unbalance in the *input differential amplifier* of the IC, and can seriously affect design.

7-4.10. Input-Offset Voltage and Current

There are a number of ways in which input-offset voltage can be measured. The simplest is to short both inputs to ground and measure the output voltage (if any). However, this does not establish input-offset current.

Input-offset voltage and current can be measured using the circuit of Fig. 7-27.

As shown, the output is alternately measured with R_3 shorted and with R_3 in the circuit. The two output voltages are recorded as E_1 (S_1 closed, R_3 shorted), and E_2 (S_1 open, R_3 in the circuit).

With the two output voltages recorded, the input-offset voltage and input-offset current can be calculated using the equations of Fig. 7-27. For example, assume that $R_1 = 51\ \Omega$, $R_2 = 5.1\ \text{k}\Omega$ $R_3 = 100\ \text{k}\Omega$, $E_1 = 83\ \text{mV}$, and $E_2 = 363\ \text{mV}$.

$$\text{input-offset voltage} = \frac{83\ \text{mV}}{100} = 0.83\ \text{mV}$$

$$\text{input-offset current} = \frac{280\ \text{mV}}{100\ \text{k}\Omega\ (1 + 100)} = 0.0277\ \mu\text{A}$$

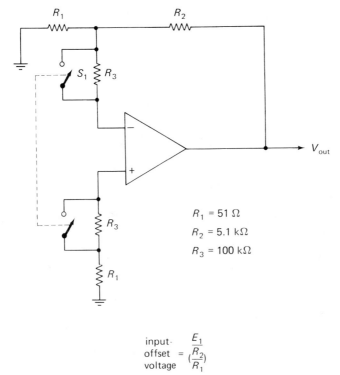

$$R_1 = 51 \ \Omega$$
$$R_2 = 5.1 \ k\Omega$$
$$R_3 = 100 \ k\Omega$$

$$\text{input-offset voltage} = \frac{E_1}{\left(\dfrac{R_2}{R_1}\right)}$$

$$\text{input-offset current} = \frac{E_2 - E_1}{R_3 \left(1 + \dfrac{R_2}{R_1}\right)}$$

$E_1 = V_{\text{out}}$ with S_1 closed (R_3 shorted)

$E_2 = V_{\text{out}}$ with S_1 open (R_3 in circuit)

Fig. 7-27 Input-offset voltage and current measurement

7-4.11. Common-Mode Rejection

Common-mode rejection can be measured using the circuit of Fig. 7-28.

Find the open-loop gain under identical conditions of frequency, input, and so forth, as described in Secs. 7-4.1 and 7-4.2.

Then connect the IC in the common-mode circuit of Fig. 7-28. Increase the common-mode voltage (V_{IN}) until a measurable output V_{OUT} is obtained. Be careful not to exceed the maximum specified input common-mode voltage swing. If no value is specified, do not exceed the normal input voltage of the IC.

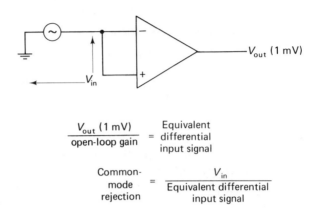

$$\frac{V_{out}\ (1\ mV)}{\text{open-loop gain}} = \begin{array}{l}\text{Equivalent}\\\text{differential}\\\text{input signal}\end{array}$$

$$\begin{array}{c}\text{Common-}\\\text{mode}\\\text{rejection}\end{array} = \frac{V_{in}}{\begin{array}{c}\text{Equivalent differential}\\\text{input signal}\end{array}}$$

Fig. 7-28   Common-mode rejection measurement

To simplify calculation, increase the input voltage until the output is 1 mV. With an open-loop gain of 100, this will provide a differential input signal of 0.00001 V. Then measure the input voltage. Move the input voltage decimal point over five places to find the CMR.

7-4.12. Slew Rate

An easy way to observe and measure the slew rate of an op-amp is to measure the *slope* of the output waveform produced by a square-

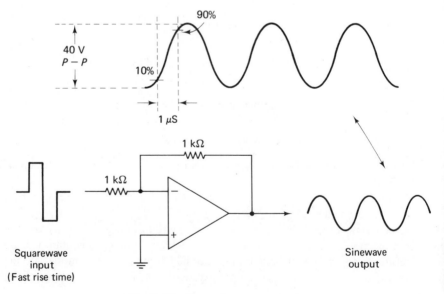

Fig. 7-29   Slew-rate measurement

wave input signal, as shown in Fig. 7-29. The input squarewave must have a rise time that *exceeds* the slew rate capability of the amplifier. Thus, the output does not appear as a squarewave, but as an integrated wave.

In the example shown in Fig. 7-29, the output voltage rises (and falls) about 40 V in 1 μs.

Note that slew rate increases with higher gain.

7-4.13. Power-Supply Sensitivity

Power-supply sensitivity can be measured using the circuit of Fig. 7-27 (the same test circuit as for input-offset voltage, Sec. 7-4.10).

The procedure is the same as for measurement of input-offset voltage, except that one supply voltage is changed (in 1-V steps) while the other supply voltage is held constant. The amount of change in input-offset voltage for a 1-V change in one power supply is the power-supply sensitivity (or input-offset voltage sensitivity).

For example, assume that the normal positive and negative supplies are 10 V, and the input offset voltage is 7 mV. With the positive supply held constant, the negative supply is reduced to 9 V. Under these conditions, assure that the input-offset voltage if 5 mV. This means that the negative power-supply sensitivity is 2 mV/V. The test should be repeated over a wide range (in 1-V steps) if the op-amp is to be operated under conditions where the power supply may vary by a large amount.

7-4.14. Phase Shift

Because of the feedback principle involved in op-amps, phase shift between input and output of an op-amp is far more critical than with an audio amplifier. In the ideal open-loop op-amp, the output is 180° out of phase with the negative input and in phase with the positive input.

Phase shift is measured with an oscilloscope. The most accurate method is with a dual-trace oscilloscope.

Dual-trace phase measurement. The dual-trace procedure is essentially one of displaying both input and output signals on the oscilloscope screen simultaneously, measuring the distance (in screen scale divisions) between related points on the two traces, and then converting the distance into phase.

The test connections for dual-trace phase measurement are shown in Fig. 7-30. For most accurate results, the cables connecting input and output signals should be of the same length and characteristics. At higher frequencies a difference in cable length or characteristics could introduce a phase shift.

The oscilloscope controls are adjusted until one cycle of the input signal occupies exactly nine divisions (9 cm horizontally) of the screen. Then the phase factor of the input signal is found. For example, if 9 cm represents one complete cycle or 360°, 1 cm represents 40° (360/9 = 40).

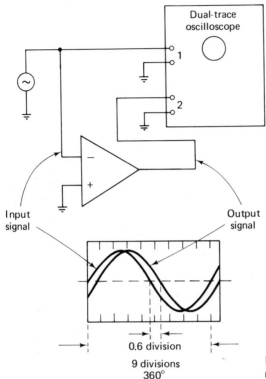

Fig. 7-30   Dual-trace phase measurement

With the phase factor established, the norizontal distance between corresponding points on the two waveforms (input and output signals) is measured. This measured distance is multiplied by the phase factor of 40°/cm to find the exact amount of phase difference. For example, assume a horizontal difference of 0.6 cm with a phase factor of 40° as shown in Fig. 7-30. Multiply the horizontal difference by the phase factor to find the phase difference (0.6 × 40 = 24° phase shift between input and output signals).

7-5. BASIC FEEDBACK AMPLIFIER TROUBLESHOOTING

Troubleshooting amplifiers without feedback is a relatively simple procedure. The input and output waveforms of each stage can be monitored on an oscilloscope (Sec. 7-1.8). Any stage showing an abnormal waveform (in amplitude, waveshape, etc.) or the absence of an output waveform (in amplitude, waveshape, etc.) or the absence of an output waveform, with a known, good input signal, points to a defect in that stage. Voltage measurements on all transistor elements will then pinpoint the problem.

Troubleshooting an amplifier with feedback is a more difficult task. Such problems as measurement of gain can be a particular concern. For example, if you try opening the loop to make gain measurements, you usually find so much gain that the amplifier saturates, and the measurements are meaningless. On the other hand, if you start making waveform measurements on a working closed-loop system, you often find the input and output signals are normal (or near normal), while inside the loop many of the waveforms are distorted. For this reason feedback loops, especially internal-stage feedback loops, require special attention.

Typical feedback-amplifier circuit. Figure 7-31 is the schematic of a basic feedback amplifier. Note the various waveforms around the circuit. These waveforms are similar to those that appear if the amplifier is used with sine-waves. Note that there is an approximate 15 per cent distortion *inside* the feedback loop (between Q_1 and Q_2), but only a 0.5 per cent distortion at the output. This is only slightly greater distortion than at the input (0.3 per cent). Open-loop gain for this circuit is approximately 4300; closed-loop gain is approximately 1000. The gain ratio (open loop to closed loop) of 4 to 1 is typical for feedback amplifiers used in laboratory work.

Amplification of signals. Transistors in feedback amplifiers behave just like transistors in any other circuit. That is, the transistors respond to all the same rules for gain and input–output impedance. Specifically, each transistor amplifies the signal appearing *between its emitter and base*. It is here that the

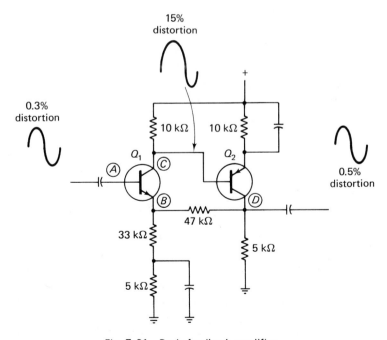

Fig. 7-31 Basic feedback amplifier

greatest difference occurs between gain stages in feedback amplifiers and gain stages in nonfeedback (open-loop) amplifiers.

Difference in open-loop and closed-loop gain. Transistor Q_1 in Fig. 7-31 has a varying signal on *both the emitter and base* rather than on one element. In a nonfeedback amplifier, the signal usually only varies at one element, *either* the emitter or base. Since most systems use negative feedback, the signals at both the base and emitter are in phase. The resultant gain is much less than when one of these elements is fixed (no feedback, open loop).

This accounts for the great amplifier gain increase when the loop is opened. Either the base or the emitter of the transistor stops moving, and the base-emitter control elements see a much larger effective input signal. Assume that a perfect input signal is applied to the input (point A of Fig. 7-1). If the amplifier is perfect (produces no distortion), the signal returning to B will also be undistorted. Since the system uses negative feedback, the signal that travels around the loop a second time is undistorted as well. If the amplifier is not perfect (assume an extreme case of clipping distortion), the returning signal will show that effect of distortion, as in Fig. 7-32.

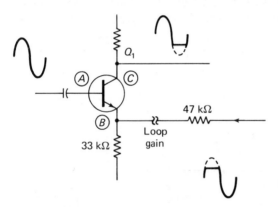

Fig. 7-32   Amplifier-induced distortion in signal returning to point B

To simplify the explanation, assume that the clipping is introduced in Q_1 and that Q_2 is perfect. Now the signals applied to the base and emitter of Q_1 are not identical. The resultant applied signal at the control point of Q_1 will be quite distorted. In effect, the distortion will be a mirror image of the distortion introduced by Q_1. Transistor Q_1 then amplifies this distortion and adds in its own counterdistortion. The result then, after many trips around the loop, is that there can be distortion inside the loop, but it is counterbalanced by the feedback system. The final output from Q_2 is undistorted, or relatively free of amplifier-induced distortion. The higher the amplification and the greater the feedback, the more effective this cancellation becomes, and the lower the output distortion becomes.

This last fact marks the basic difference in troubleshooting a feedback amplifier. In any amplifier there are three basic causes of distortion: *overdriving, operating the transistor at the wrong bias point*, and the inherent nonlinearity of any solid-state device.

Overdriving can be the result of many causes (too much input signal, too much gain in the previous stage, etc.). However, the net result is that the output signal is clipped on one peak due to the transistor being driven into saturation, and on the other peak by driving the transistor below cutoff.

Operating at the wrong bias point can also produce clipping, but of only one peak. For example, if the input signal is 1 V and the transistor is biased at 1 V, the input will swing from 0.5 to 1.5 V. Assume that the transistor saturates at any point above 1.6 V and is cut off at any point below 0.4 V. No problem will occur with the correct bias (1 V).

But now assume that the bias point is shifted (due to component aging, transistor leakage, etc.) to 1.3 V. The input now swings from 0.8 to 1.8 V, and the transistor saturates when one peak goes from 1.6 to 1.8 V. If, on the other hand, the bias point is shifted down to 0.7 V, the input swings from 0.2 to 1.2 V, and the opposite peak is clipped as the transistor goes into cutoff.

Even if the transistor is not overdriven, it is still possible to *operate a transistor on a nonlinear portion* of its curve due to wrong bias. All transistors have some portion of their input–output curve that is more linear than other portions. That is, the output increases (or decreases) directly in proportion to input. An increase of 10 per cent at the input produces an increase of 10 per cent at the output. Ideally, transistors are operated at the center of this linear curve. If the bias point is changed, the transistor can operate on a portion of the curve that is less linear than the desired point.

The inherent nonlinearity of any solid-state device (diode, transistor, etc.) can produce distortion even if a stage is not overdriven and is properly biased. That is, the output never increases (or decreases) directly in proportion to the input. For example, an increase of 10 per cent at the input can produce an increase of 13 per cent (or 7 per cent) at the output. This is one of the main reasons for feedback in amplifiers where low distortion is required.

In summary, a negative feedback loop operates to minimize distortion, in addition to stabilizing gain. The feedback takeoff point has the minimum distortion of any point within the loop. From a practical troubleshooting standpoint, if the *final output* distortion and the overall gain are within limits, all the stages within the loop can be considered as operating properly. Even if there is some abnormal gain in one or more of the stages, the overall feedback system has compensated for the problem. Of course, if the overall gain and/or distortion are not within limits, the individual stages must be checked.

7-5.1. Feedback-Amplifier Troubleshooting Notes

Most feedback-amplifier problems can be pinpointed by waveform measurements and voltage measurements. The following notes should be given special attention when troubleshooting any feedback amplifier circuit.

Opening the loop. Some troubleshooting literature recommends that the loop be opened and the circuits checked under no-feedback conditions. In some cases, this can cause circuit damage. Even if there is no damage, the technique is rarely effective. Open-loop gain is usually so high that some stage will block or distort badly. If the technique is used, as it must be for some circuits (typically an op-amp), keep in mind that distortion will be increased. That is, a normally closed-loop amplifier can show considerable distortion when operated as an open loop, even though the amplifier is good.

Measuring stage gain. Care should be taken when measuring the gain of amplifier stages in a feedback amplifier. For example, in Fig. 7-31 if you measured the signal at the base of Q_1, the base-to-ground voltage would not be the same as the input voltage. To get the correct value, connect the low side of the measuring device (ac voltmeter or oscilloscope) to the emitter and the other lead to the base (see Fig. 7-33). In effect, measure the signal across the base–emitter junction. This will include the effect of the feedback signal.

As a general safety precaution, never connect the ground lead of a voltmeter or oscilloscope to the base of a transistor, unless that lead connects back to an *isolated inner chassis*. The reason is because large ac ground loop currents can flow through the base–emitter junction (and then to ground) and easily blow out the transistor.

Low-gain problems. As discussed, low gain in a feedback amplifier can also result is distortion. That is, if gain is normal in a feedback amplifier, some distortion can be overcome. With low gain the feedback may not be able to bring the distortion within limits. Of course, low gain by itself is sufficient cause to troubleshoot amplifier (feedback or not).

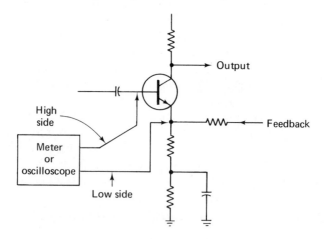

Fig. 7-33   Measuring input signal voltage or waveforms

Take the classic failure pattern of a solid-state feedback amplifier that was working properly, but now the output is low by about 10 per cent. This indicates a general deterioration of performance, rather than a major breakdown.

Keep in mind that most feedback amplifiers have a very high open-loop gain that is set to some specific value by the ratio of resistors (feedback resistor to load resistor). If the closed-loop gain is low, it usually means that the open-loop gain has fallen far enough so that the resistors no longer set the gain. For example, if the ac beta of Q_2 in Fig. 7-31 is lowered, the open-loop gain is lowered. Also the lower beta lowers the input impedance of Q_2, which, in turn, reduces the effective value of the load resistor for Q_1. This also has the effect of lowering overall gain.

In troubleshooting such a situation, if waveforms indicate low gain and element voltages are normal, try replacing the transistors. Of course, never overlook the possibility of open or badly leaking emitter-bypass capacitors. If the capacitors are open or leaking (acting as a resistance in parallel with the emitter resistor), there will be considerable negative feedback and little ac gain. Of course, a completely shorted emitter-bypass capacitor will produce an abnormal dc voltage indication at the transistor emitter.

Distortion problems. As discussed, distortion can be caused by improper bias, overdriving (too much gain), or underdriving (too little gain, preventing the feedback signal from countering the distortion). One problem often overlooked in a feedback amplifier with a distortion failure pattern is overdriving due to transistor leakage. (The problem of transistor leakage is discussed further in Sec. 7-6.)

Generally, it is assumed that the collector–base leakage will reduce gain, since the leakage is in opposition to the signal current flow. While this is true in the case of a single stage, it may not be true where more than one feedback stage is involved.

Whenever there is collector–base leakage, the base assumes a voltage nearer to that of the collector (nearer, than is the case without leakage). This increases transistor forward bias and increases transistor current flow. An increase in the transistor current causes a lower h_{ib} (ac input resistance, grounded-base configuration), which causes the stage gain to go up. At the same time, a reduction in h_{ib} causes a reduction in common-emitter input resistance, which may or may not cause a gain reduction (depending on where the transistor is located in the amplifier).

If the feedback amplifier is direct coupled, the effects of feedback are increased. This is because the operating point (base bias) of the following stage is changed, possibly resulting in distortion. For example, the collector of Q_1 is connected directly to the base of Q_2. If Q_1 starts to leak (or the collector–base leakage increases with age), the base of Q_2 (as well as the collector of Q_1) will shift its Q point (no-signal voltage level).

7-6. EFFECTS OF LEAKAGE ON AMPLIFIER GAIN

When there is considerable leakage in a solid-state amplifier, the gain is reduced to zero and/or the signal waveform is drastically distorted. Such a condition also produces abnormal waveforms and transistor voltages. These indications make the problem easy, or relatively easy, to locate. The real difficulty occurs when there is just enough leakage to reduce amplifier gain, but not enough leakage to seriously distort the waveform or produce transistor voltages that are way off.

Collector–base leakage is the most common form of transistor leakage and produces a classic condition of low gain (in a single stage). When there is any collector–base leakage, the transistor is forward biased, or the forward bias is increased. This condition is shown in Fig. 7-34.

Collector–base leakage has the same effect as a resistance between the collector and base. The base assumes the same polarity as the collector (although at a lower value), and the transistor is forward biased. If leakage is sufficient, the forward bias can be enough to drive the transistor into or near saturation. When a transistor is operated at or near the saturation point, the gain is reduced *for a single stage*. This is shown in the curve of Fig. 7-35.

If the normal transistor element voltages are known, excessive transistor leakage can be spotted easily, since all the transistor voltages will be off. For

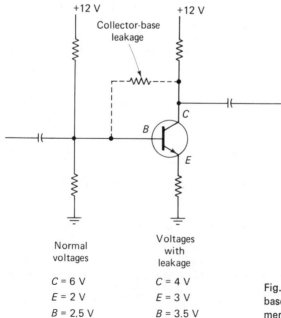

Normal
voltages

Voltages
with
leakage

$C = 6$ V
$E = 2$ V
$B = 2.5$ V

$C = 4$ V
$E = 3$ V
$B = 3.5$ V

Fig. 7-34 Effect of collector–base leakage on transistor element voltages

Gain reduction (dB)
0
10
20
30
40
50
60
70
80
90
100
7
8
9
10
11
12
13
14
Collector current (mA)

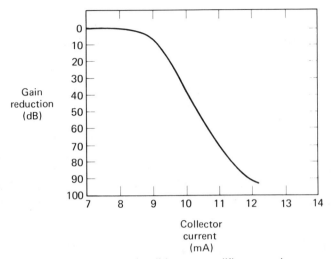

Fig. 7-35 Relative gain of solid-state amplifier at various average collector-current levels

example, in Fig. 7-34 the base and emitter will be high and the collector will be low when measured in reference to ground. However, if the normal operating voltages are not known, the transistor can appear to be good, since all the *voltage relationships are normal.* That is, the collector–base junction is reverse biased (collector more positive than base for an *NPN*) and the emitter–base junction is forward biased (emitter less positive than base for *NPN*).

A simple way to check transistor leakage is shown in Fig. 7-36. Measure

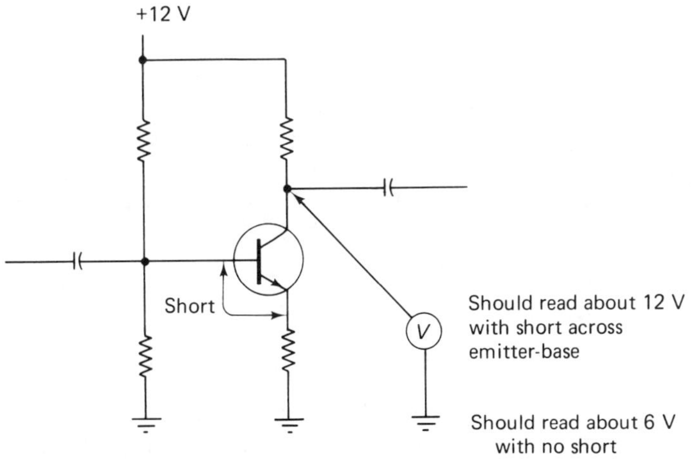

+12 V
Short
V
Should read about 12 V with short across emitter-base
Should read about 6 V with no short

Fig. 7-36 Checking for transistor leakage in amplifier circuit

the collector voltage to ground. Then short the base to the emitter and re-measure the collector voltage. If the transistor is not leaking, the base–emitter short will turn the transistor off, and the collector voltage will rise to the same value as the supply. If there is any leakage, a current path will remain (through the emitter resistor, emitter–base short, collector–base leakage path, and collector resistor). There will be some voltage drop across the collector resistor, and the collector will have a voltage at some value lower than the supply.

Note that most meters draw current, and this current passes through the collector resistor. This can lead to some confusion, particularly if the meter draws heavy current (has a low ohms-per-volt rating). To eliminate any doubt, connect the meter to the supply through a resistor with the same value as the collector resistor. The drop, if any, should be the same as when the transistor collector is measured to ground. If the drop is much lower when the collector is measured, the transistor is leaking.

As an example, assume that in the circuit of Fig. 7-36 the supply is 12 V, the collector resistance is 2 kΩ, and the collector measures 4 V with respect to ground. This means that there is an 8-V drop across the collector resistor and a collector current of 4 mA (8/2000 = 4 mA). Normally, the collector is operated at about one-half the supply voltage or 6 V. However, simply because the collector is at 4 instead of 6 V does not make the circuit faulty. Some circuits are designed that way. Therefore, the transistor must be checked for leakage. Now assume that the collector voltage rises to 10 V when the base and emitter are shorted. This indicates that the transistor is cutting off, but there is still some current flow through the collector resistor, about 1 mA (2/2000 = 1 mA).

A 1-mA current flow is high for a meter. However, to confirm a leaking transistor, connect the meter through a 2-kΩ resistor to the 12-V supply, preferably at the same point where the collector resistor connects to the supply. Now assume that the indication is 11.7 V through the external resistor. This indicates that there is some transistor leakage. The amount can be estimated as follows: 11.7 − 10.5 = 1.2-V drop; 1.2/2000 = 0.6 mA. However, from a practical troubleshooting standpoint, the presence of any current flow with the transistor supposedly cut off is sufficient cause to replace the transistor.

7-7. EFFECT OF CAPACITORS IN SOLID-STATE AMPLIFIERS

Although the functions of capacitors in solid-state amplifiers are similar to those invacuum-tube equipment, the results produced by capacitor failure are not necessarily the same. An emitter-bypass capacitor is a good example.

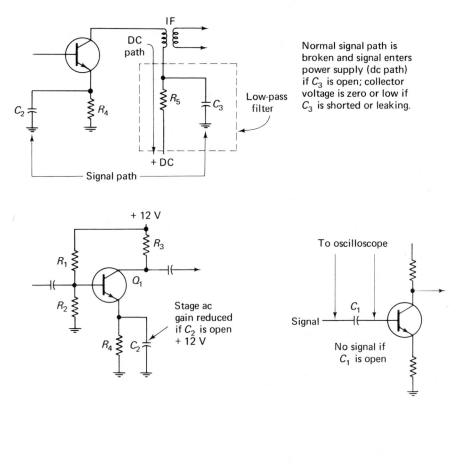

Normal signal path is broken and signal enters power supply (dc path) if C_3 is open; collector voltage is zero or low if C_3 is shorted or leaking.

Stage ac gain reduced if C_2 is open

No signal if C_1 is open

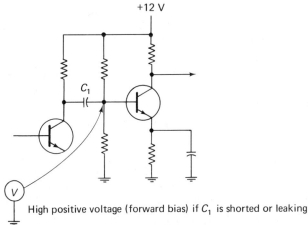

High positive voltage (forward bias) if C_1 is shorted or leaking

Fig. 7-37 Effects of capacitor failure in solid-state amplifier circuits

The emitter resistor in a solid-state amplifier (such as R_4 in Fig. 7-37) is used to stabilize the transistor dc gain and prevent thermal runaway. With an emitter resistor in the circuit, any increase in collector current produces a greater drop in voltage across the resistor. When all other factors remain the same, this change in emitter voltage reduces the base–emitter forward-bias differential, thus tending to reduce collector current flow.

When circuit stability is more important than gain, the emitter resistor is not bypassed. When ac or signal gain must be high, the emitter resistance is bypassed to permit passage of the signal. If the emitter-bypass capacitor is open, stage gain is reduced drastically, although the transistor dc voltages remain substantially the same. Thus, if there is a *low-gain symptom* in any solid-state amplifier with an emitter bypass, and the voltages appear normal, *check the bypass capacitor*. This can be done by shunting the bypass with a known good capacitor of the same value. As a precaution, shut off the power before connecting the shunt capacitor; then reapply power. This will prevent damage to the transistor (due to large current surges).

The functions of coupling and decoupling capacitors in solid-state amplifiers are essentially the same as for vacuum-tube epuipment. However, the capacitance values are much larger, particularly when the amplifier must pass very low frequencies. Electrolytics are usually required to get the large capacitance values. From a practical standpoint, electrolytics tend to have more leakage than mica or ceramic capacitors. However, good-quality electrolytics (typically the bantam type found in solid-state circuits) have leakage of less than 10 μA at normal operating voltages.

The function of C_1 in Fig. 7-37 is to pass signals from the previous stage to the base of Q_1. If C_1 is shorted or leaking badly, the voltage from the previous stage is applied to the base of Q_1. This forward biases Q_1, causing heavy current flow and possible burnout of the transistor. In any event, Q_1 is driven into saturation, and stage gain is reduced. If C_1 is open, there will be little or no change in the voltages at Q_1, but the signal from the previous stage will not appear at the base of Q_1.

From a troubleshooting standpoint, a shorted or leaking C_1 will show up as abnormal voltages (and probably as distortion of the signal waveform). If C_1 is suspected of being shorted or leaky, replace C_1. An open C_1 will show up as a lack of signal at the base of Q_1, with a normal signal at the previous stage. If an open C_1 is suspected, replace C_1 or try shunting C_1 with a known good capacitor, whichever is convenient.

The function of C_3 in Fig. 7-37 is to pass operating signal frequencies to ground (to provide a return path) and to prevent signals from enteiing the power-supply line or other circuits connected to the line. In effect, C_3 and R_5 form a low-pass filter that passes dc and very low frequency signals (well below the operating frequency of the circuit) through the power-supply line. Higher-frequency signals are passed to ground and do not enter the power-supply line.

If C_3 is shorted or leaking badly, the power-supply voltage will be shorted to ground or greatly reduced. This reduction of collector voltage makes the stage totally inoperative or will reduce the output, depending on the amount of leakage in C_3.

If C_3 is open, there will be little or no change in the voltages at Q_1. However, the signals will appear in the power-supply line. Also, signal gain will be reduced and the signal waveform will be distorted. In some cases, at higher signal frequencies, the signal simply cannot pass through the power-supply circuits. Since there is no path through an open C_3, the signal will not appear on the collector circuit in any form. From a practical standpoint, the results of an open C_3 will depend on the values of R_5 (and other power-supply components) as well as on the signal frequency involved.

7-8. BASIC OP-AMP TROUBLESHOOTING

From a practical troubleshooting standpoint, it is often necessary to service op-amps by working with *external feedback* components. In the case of IC packages, the external components are the only ones that can be tested or replaced. Even with printed circuit card op-amps, troubleshooting starts by isolating the problem to the external components or the amplifier components. That is, the amplifier is tested as a separate function first. If the amplifier performs properly, the trouble is isolated to the external components, and vice versa.

7-8.1. Failure Patterns for Op-Amps

Major disasters are relatively rare in well-protected op-amps, since input overloads never drive the circuit into saturation. Likewise, when such major failures occur, they are relatively easy to troubleshoot. That is, the problems are easy to spot by normal signal tracing with waveforms or voltage measurements at the transistor elements. For example, a major failure will usually show up as a normal input, but with no output, at a particular amplifier stage (or at the input and final output of an IC op-amp).

However, op-amps are often plagued with such problems as hum, drift, and noise. The following paragraphs describe the most likely causes for such problems, with practical approaches for locating the faults.

7-8.2. Hum and Ripple Problems

In solid-state op-amps any hum or ripple almost always comes from the dc power supplies feeding the amplifier. This is unlike vacuum-tube op-amps, where hum and ripple can come from heater-to-cathode leakage. A possible exception is when hum is picked up due to poor shielding or badly grounded leads.

The first step in locating a hum or ripple problem is to short the input terminals and monitor the output with an oscilloscope. If the hum or ripple is removed when the input terminals are shorted, the hum is probably being picked up by the leads or at the terminal. Look for loose shields, loose ground terminals, and cold solder joints where lead shielding is attached to chassis or feed-through terminals.

If the hum or ripple is not removed when the input terminals are shorted, the hum is probably coming from the power supply. Monitor the power-supply voltages at the point where they enter the amplifier. If the power supply is showing an abnormal amount of ripple, the problem is in the power supply. However, since the amplifier has considerable gain, the ripple as monitored at the amplifier output may be much greater than at the power supply.

7-8.3. Drift and Noise Problems

Drift and noise problems in op-amps are perhaps the most common complaint. There are several places to look in trying to track the causes of noise and drift.

Unstable power supplies. Op-amps are *extremely sensitive* to power-supply stability. For example, with solid-state op-amps the typical dual-power-supply voltages (required for differential amplifiers) are on the order of ± 12 V or ± 15 V. For satisfactory amplifier operation, the drift should be less than 1 mV/min (or less in some special op-amps used with data-processing equipment).

Because of the low voltages involved, power-supply-stability measurements are best made with a five- or six-place *digital* voltmeter. Such a meter can be connected to the monitoring point and checked at least once every minute, or over at least a 5-min interval. If the drift is less than 1 mV/min over this time interval, the power supply is probably satisfactory for typical op-amp use.

Noisy zero-correction circuits. Another possible source of output noise in op-amps is the zero-correction circuit, which takes voltages from both the positive and negative power supplies, and provides a small dc current to oppose the internally generated offset. (See Fig. 6-30.)

If zener diodes are used to regulate some portion of the zero-correction supply voltage (as is done by CR_{303} and CR_{304} in Fig. 6-30), the zeners should checked carefully for drift and noise. Keep in mind that any noise (or other signal) at the zero-correction circuit is injected into the amplifier at the point of *highest gain* (usually at the first-stage input).

Contaminated printed circuit boards. Another frequent source of output drift is contamination of the printed circuit (PC) boards on which non-IC op-amps are often mounted. Contamination problems even apply to IC op-amps. Although the IC package containing the complete op-amp is sealed, the *summing junction* is not sealed. (Typically, the summing junctions are placed on Teflon terminals near the IC package on the PC board.)

When you remember that the input current at the summing junction is typically 10^{-11} A, it is easy to see why any contamination from fingerprints (providing leakage paths into the junction) can cause annoying output instability. Op-amp circuit boards should be handled as little as possible, and then only while wearing cotton gloves.

Great care should be taken never to touch the summing junction terminals with bare hands. Boards suspected of being contaminated should be washed carefully with a clean degreasing solvent, and dried with warm, dry air. (Never blow them dry with an air hose, as air lines invariably contain oil and water.)

Leakage in the overload-protection circuitry. Another place to look for causes of output noise and drift is the overload-protection circuitry. If an amplifier has been subjected to repeated serious overloads, there is a possibility of finding high and unstable reverse leakage currents in the overload-protection circuits.

For example, in the circuit of Fig. 6-31 in Chapter 6, the reverse currents can be monitored by measuring the voltages across R_{207} and R_{355}. Substantial voltage across these resistors indicates considerable reverse leakage. If such voltages are measured in an op-amp with a noise symptom, try replacing the diodes, one at a time.

Unstable choppers. Perhaps the most common source of output instability in chopper-stabilized op-amps is instability of the chopper itself. In the case of electromechanical (vibrator-type) choppers, the problem is essentially one of variations in dwell time (the time during which the contacts are closed on each half of the cycle).

With photoelectric choppers, such as described in Sec. 4-5.1 and illustrated in Fig. 4-14, instabilities in the firing voltage of the neon lamps will produce the same symptoms of erratic output voltages. However, since choppers modulate low-level signals, it is virtually impossible to ckeck choppers except by substitution. Any variation in musical note of an electromechanical chopper is a sure sign it should be replaced. However, since photoelectric choppers are a sealed unit, they must be checked by substitution.

7-8.4. General Troubleshooting Hints for Op-Amps

Due to their extremely high open-loop gain, troubleshooting op-amps can be quite difficult. The basic test connections are shown in Fig. 7-38. The input is shorted, the drift output (if any) is monitored on a digital voltmeter, and the hum and noise (if any) are monitored on an oscilloscope.

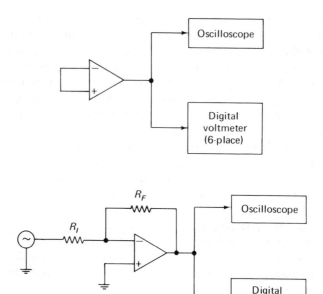

$$R_F = R_I \times 10$$

Fig. 7-38   Basic troubleshooting connections for op-amp

If the op-amp is used in a circuit where feedback is not by means of a resistor (such as an integrator where the feedback is through a capacitor), a resistor should be inserted in the feedback path. The feedback resistor should be 10 times the value of the input resistor. Thus, the op-amp will operate with a gain of 10.

The same test connections can be used to check the range of zero-adjust circuitry and the clipping level of the diode-protection circuitry.

To check the zero-correction range, vary the zero control from one end of its range to the other while observing the op-amp output on the digital voltmeter. (The digital voltmeter should have a sensitivity of at least 1 mV.) If the zero control is provided with steps (instead of, or in addition to, a variable control), check to see that each step produces the same sized step in output voltage. Also check the output stability at each step. If the output voltage appears to be unstable at any particular step, look for poor contacts on the switch, poor solder connections, or a defective resistor connected to the switch contact.

Sometimes it will be found that the range of the zero-adjust control is not sufficient to cause limiting of the amplifier (by means of protective overload circuitry). If this is the case, apply a small dc voltage of known stability (preferably from a battery) to the input terminals of the amplifier. Check to see that limiting occurs at the amplifier output at the correct values for both polarities of the input voltage.

Unbalance between the positive and negative overload breakdown voltages will suggest either an open or shorted diode in the overload-protection loop, or saturation taking place internally within the op-amp. In either case, leave the output driven lightly into saturation, and measure dc voltage appearing across *all elements* in the overload-protection circuit, plus the operating points of the various stages within the amplifier proper (unless the amplifier is an IC where internal stages cannot be measured). This should pinpoint the unbalance problem.

INDEX

409